Zoophysiology and Ecology
Volume 10

Hans-Ulrich Thiele

Carabid Beetles in Their Environments

A Study on Habitat Selection by Adaptations in Physiology and Behaviour

With 152 Figures

Springer-Verlag
Berlin Heidelberg New York 1977

Professor Dr. HANS-ULRICH THIELE
Zoologisches Institut der Universität
Weyertal 119, 5000 Köln 41/FRG

Translated by:
JOY WIESER
Madleinweg 19, A-6064 Rum/Innsbruck

For explanation of the cover motive see legend to Figs. 2a, 2c, 3b (pages 4–6).

ISBN-13:978-3-642-81156-2 e-ISBN-13:978-3-642-81154-8
DOI: 10.1007/978-3-642-81154-8

Library of Congress Cataloging in Publication Data. Thiele, Hans-Ulrich. Carabid beetles in their environments. (Zoophysiology and ecology; v. 10) 1. Carabidae—Ecology. 2. Habitat selection. 3. Adaptation (Physiology). 4. Insects—Ecology. I. Title. QL596.C2T47.595.7'62.77–9924.

Softcover reprint of the hardcover 1st edition 1977

This book is dedicated to
Carl H. Lindroth

Preface

With the increasing numbers of research workers and groups of investigators devoting themselves to the ecology of carabids I felt that the time had come to take stock of the existing knowledge in this field and to endeavour to weld my personal results and those of other workers into a comprehensive picture. It was with these aims in mind that the following study was conceived. A further goal was to attempt to show to what extent research on carabids can contribute to the larger fields of research encompassing ecology, ethology and evolution. In my opinion the investigations on carabids permit us to draw conclusions of general applicability and, as such, comparable with those made in recent years upon other groups of animals.

I am well aware of the risk involved nowadays in attempting, on one's own, to integrate results from a wide variety of scientific disciplines into a meaningful whole, and for this reason I am always grateful for corrections and for additional information.

It is impossible for me to mention by name all of the colleagues who have given me their support in the preparation of the book. Reprints of their publications have been placed at my disposal by almost all of the authors cited, as well as by others whose names and works have been omitted merely in order to prevent the book from taking on encyclopedic proportions. I am nevertheless indebted to them all for their cooperation.

It is neither possible, nor is it my intention to attempt an exhaustive survey of the literature relating to the ecology of carabids. Emphasis will be placed upon studies that have appeared since 1950, mainly those of the 60 s and 70 s. In this way it is hoped to cover ecological investigations that have been carried out subsequent to the appearance of Lindroth's book "Die fennoskandischen Carabiden" (1945–49). In his third volume (1949) Lindroth covers nearly every aspect of the ecology of carabids on the basis of his own large number of experiments. The results are discussed critically and are combined with an almost exhaustive presentation and discussion of the older literature. The book covers in all nearly 1000 pages. In a series of studies on "The Ground Beetles of Canada and Alaska" (1961–1969) Lindroth has also provided a substantial basis for the systematics of carabids of a considerable part of North America. This extensive study also contains a wealth of information concerning the ecology of individual species.

I have endeavoured to cover the literature that appeared before 1975, but later publications could only partly be considered. All literal quotations in the text are given in English; some of them are translated from papers written in other languages. In the references original titles are cited from Dutch, English, French, German and Italian; papers in other languages are quoted with the titles of their summaries in the languages mentioned above.

Special thanks are due to the undergraduate and graduate students who were my collaborators in the carabid studies: Dr. Jürgen Becker, Horst Bittner, Dr. Hans-Jörg Ferenz, Dipl. Biol. Werner Hölters, Günter Jarmer, Petra Kasischke, Dr. Helmut Kirchner, Doris Koch, Dipl. Biol. Johannes Könen, Hanns Kreckwitz, Dr. Ingomar Krehan, Dr. Karl-Heinz Lampe, the late Dr. August Wilhelm Lauterbach, Dr. Horst Lehmann, Gisela Leyk, Dr. Siegfried Löser, Dr. Christian Neudecker, Dr. Ulrich Neumann, Dr. Wilfried Paarmann, Klaus Schlinger, Ludwig Seegers and Hans-Eberhard Weiss. My grateful thanks are also due to the assistants who have helped me in the laboratory and in the delicate task of rearing the carabids: Mrs. Nora Päschel (Dipl. Biol.), Mrs. Mary-Lou Kastoun and Mrs. Ellen P. Fiedler, B. S. Mrs. Ursula Jung deserves thanks for the careful drawing of some carabids.

I am very much indebted to the Deutsche Forschungsgemeinschaft, which since 1968 supported our work by several grants.

Prof. Dr. Carl H. Lindroth, Lund (Sweden), whose work has provided me with a wealth of stimulation, was so kind as to undertake a critical study of my manuscript. I would like to offer him my sincere thanks for his efforts. I am also much indebted to the Editor, Professor Dr. Jürgen Jacobs, for a profitable discussion of the manuscript.

It is a pleasure to express my appreciation to Mrs. Joy Wieser, B. Sc., of Innsbruck, who has translated the German manuscript into English. Her painstaking engagement with the subject matter has produced a translation that conveys in every way my original intentions. Inestimable help in the production of the manuscript was afforded by the patient cooperation of Mrs. Ellen P. Fiedler. My thanks also go to my daughters Astrid and Brita for their help with the preparation of the index and the correction of the proofs. I should like to offer my warmest thanks to Dr. Konrad F. Springer and his staff for their meticulous care and helpfulness in getting the book into print.

Finally it is a real pleasure for me to express my thanks to my wife Elke who has encouraged me in my work and who gave innumerable suggestions for it and for the preparation of the manuscript of this book. Moreover I thank her for all her patience with carabids and with carabidologists.

Cologne, July 1977 HANS-ULRICH THIELE

Contents

Chapter 3

The Connections Between Carabids and Biotic Factors in the Ecosystem 49

Introduction

Today, most biological disciplines are concerned with one particular function or basic structure of the living organism. The choice of any one species for the study of such processes or structures is governed mainly by the technical question of suitability for maintenance and investigation under laboratory conditions.

This book has a different goal. In it, attention will be centered upon one group of insects, the ground beetles (Carabidae). Comprising a single family, they exhibit a relatively high degree of uniformity in morphological structure. It is our intention to assemble the knowledge that has so far been won from widely differing branches of science concerning the relationships of these beetles to their environment.

For several reasons, the family of carabid beetles, in the opinion of the present author, is a well-suited group from which to draw up an interim balance of the autecological findings so far obtained. Carabids have long been a favourite object of investigation, both to professional and amateur scientists, so that the systematics of the group has been studied fairly exhaustively, especially in Europe and to some extent in other faunal regions. In addition, the geography of species belonging to this family has been soundly investigated in many regions. The use of pitfall traps (Barber traps) has considerably enhanced our knowledge of the ecological distribution of carabids as an epigaeic group, and we are now in possession of an appreciable body of reliable quantitative data.

On this foundation it has been possible to develop experimental ecological research on carabids in both field and laboratory. Experimentation with ground beetles, which are in the main predaceous, but often also carrion-consuming species, is not associated with many of the complications that result from the variable physiological state of the host plant when employing phytophages. The relatively meagre inventory of social behaviour in carabids makes them much simpler objects of research than, for example, ants and termites.

Many carabids run along the ground, climb seldom and are either totally incapable of flight or fly spontaneously on rare occasions only. Such characteristics predestine them for use in preference experiments in factor gradients. Increasing success is being reported in breeding various species under controlled laboratory conditions. In overcoming what were long held to be insuperable difficulties in breeding carabids, one of the main obstacles to the experimental study of their ecology has been removed.

The question which we have to answer is: why do certain species occur in a narrowly limited habitat, but not in an apparently similar environment perhaps only a few meters away? Put concisely the question is: why do organisms live where they live? In other words, in what manner is a species adapted to its environment and to its way of life, and how have such adaptations come about in the course of phylogenesis? In many cases there is no morphological explanation of the differing ecological distribution of related species. Perhaps it is to be sought in individual physiological requirements or in behavioural differences.

The problem was already formulated by Darwin in 1860: "...we feel so little surprise at one of two species closely allied in habits being rare and the other abundant in the same district; or, again, that one should be abundant in one district, and another, filling the same place in the economy of nature, should be abundant in a neighbouring district, differing very little in its conditions. If asked how this is, one immediately replies that it is determined by some slight difference in climate, food, or the number of enemies: yet how rarely, if ever, we can point out the precise cause and manner of action of the check!"

Since, unlike Darwin's contemporaries, we are in fact surprised by the above statements, we propose to consider them more closely. There seems to be justification for the use of ground beetles in investigating these questions: only a few insect families can equal or exceed in numbers the 25,000 species of carabids.[1] Preferably, questions concerning the nature and causes of adaptations should be studied in a group that can be regarded as a "phylogenetic success." Not only do carabids represent about 3% of all insects and 2.5% of all animal species[2], they also play an important role in the nutritional chain and have aroused hopes, in some cases with a certain degree of justification, in biological pest control. They are to a large extent homogeneous with respect to body structure and few special macro- or micromorphological adaptations to their innumerable habitats have been observed. Carabids are thus eminently suitable for studying the ecophysiological adaptations by means of which they cope with the demands of the environment to which they are exposed, and which are responsible for their having become so successful as a group, penetrating, despite a simple structure, environments all over the world.

---

[1] This figure was given by Klots and Klots (1959). Basilewsky (1972 and personal communication) considers this figure to be obsolete and estimated the number of carabid species in 1973 at 40,000, which ranks them as the insect family comprising the most species of all.

[2] The percentages are based on the assumption that 25,000 carabid species, about 750,000 insect species and about 1 million animal species in all have been described. Sabrovsky (1952) calculated a figure of about 686,000 insect species described up to 1948. In 1972 he arrived at an estimate of 926,000 species so far described (personal communication to Prof. O. Kraus, Hamburg, to whom I am indebted for this information). Taking the more recent figures as a basis, the carabids account for 4.3% of all insect species.

Chapter 1

Variations in the Body Structure of Carabids in Adaptation to Environment and Mode of Life

A. Macromorphological Variations

Carabids belong to the suborder Adephaga of the order Coleoptera. As compared with the 200 families comprising the other suborder, the Polyphaga, the Adephaga as a group are poor with respect to number of families. To it belong several aquatic families, the most important of which are the Dytiscidae, Gyrinidae, Haliplidae and Hygrobiidae.

Terrestrial Adephaga (or "Geadephaga"), on the other hand, are represented almost exclusively by the Carabidae (including the Cicindelinae which we regard as a subfamily of the Carabidae, see p. 353). The Cicindelinae may be a sister group of the other carabids, as recently pointed out by Regenfuss (1975), but this assumption is hypothetical. The form of the antenna cleaners places the carabids with the Cicindelinae in the group of the so-called "Anisochaeta", a sister group of the "Isochaeta". The latter comprise a few relic groups of the Geadephaga, i.e., the Metriidae, Ozaenidae, and as the most interesting the Paussidae, which will be considered in connection with the evolution of the carabids (see Chap. 9.A.). The remaining family of the Geadephaga, the Rhysodidae, are regarded as an aberrant surviving group of the primitive Adephagan stock (Regenfuss, 1975).

In comparison with the Polyphaga, the Adephaga as an entire group are fairly uniform in morphology and are regarded as monophyletic in origin. The carabid family exhibits especially little variation in basic body structure throughout its wealth of species. As in all Adephaga the thighs of the hindlegs are fused with the first sternites. The number of sternites is almost always six, only increasing to eight in the Brachinini. The antennae are filiform and invariably 11-jointed. The legs follow a particularly consistent pattern, and the fact that they have, without exception, evolved as slender running legs has led to the German name "Laufkäfer" (running beetles). The legs always have five tarsal segments.

Carabids represent, with a considerable degree of uniformity, the life form type of the ground predator. They additionally consume carrion and plant material to extents varying from species to species. With a few exceptions, only slight variations on this type are found among the large number of species comprising the family.

The most pronounced morphological variations are connected with *specialized modes of nutrition* (Figs. 1–4). For example, species of the tribe Cychrini are extreme snail predators. Thorax and head are much narrowed and elongated, thus enabling the beetle to crawl into the snail shell and pull out the soft

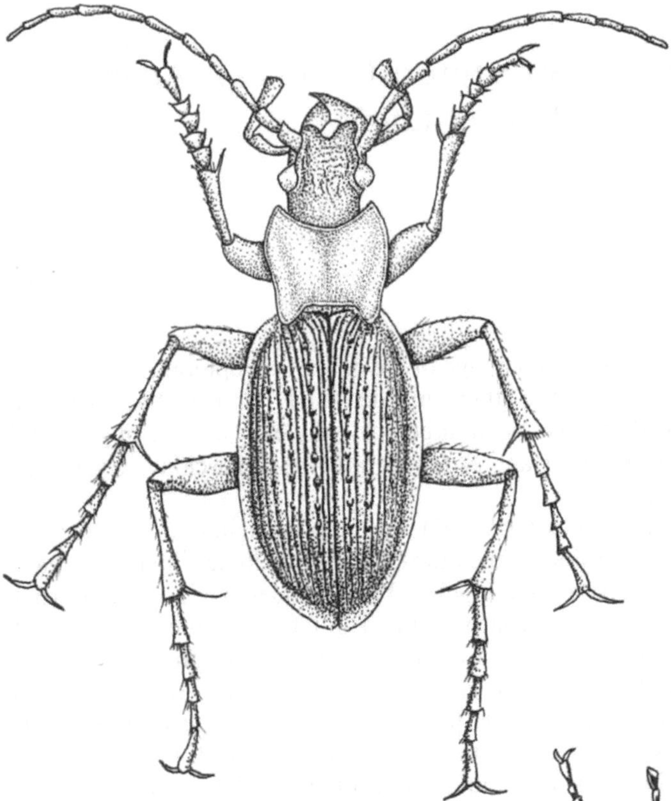

Fig. 1. *Carabus problematicus*, "normal type" of a carabid. Original size ca. 25 mm. Figs. 1–4: Drawings by U. Jung

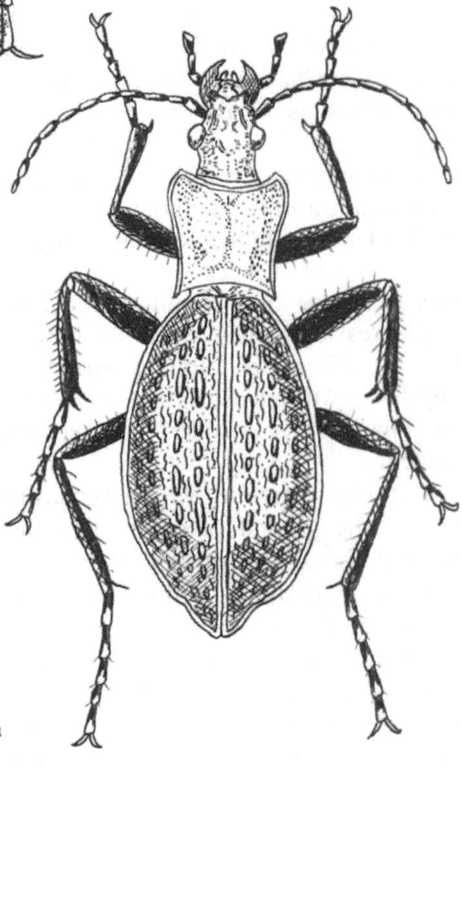

Fig. 2a

Fig. 2. (a) *Carabus intricatus*, moderate cychrization. Original size ca. 28 mm; (b) *Carabus creutzeri*, more developed cychrization. Original size ca. 25 mm; (c) *Cychrus attenuatus*, extreme cychrization. Original size ca. 15 mm. Fig. 2a drawn by H. U. Thiele

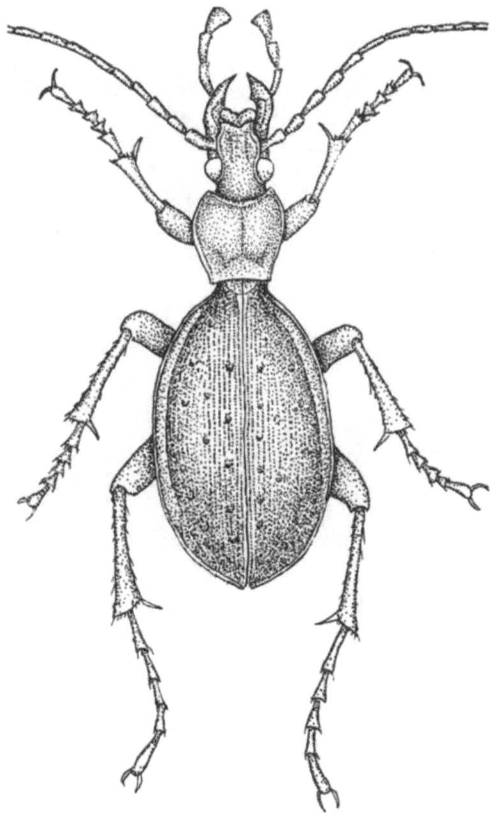

Fig. 2b

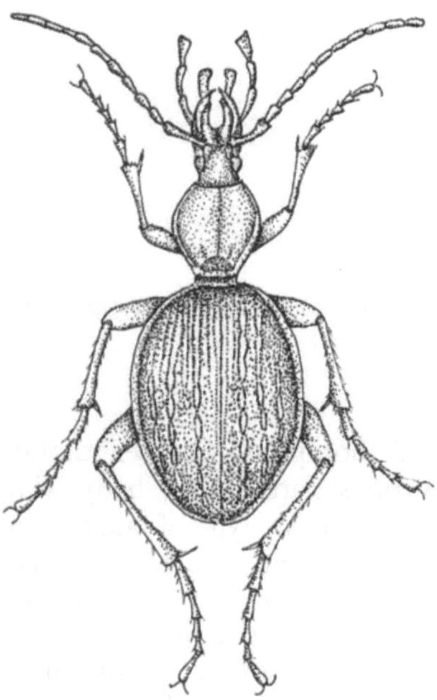

Fig. 2c

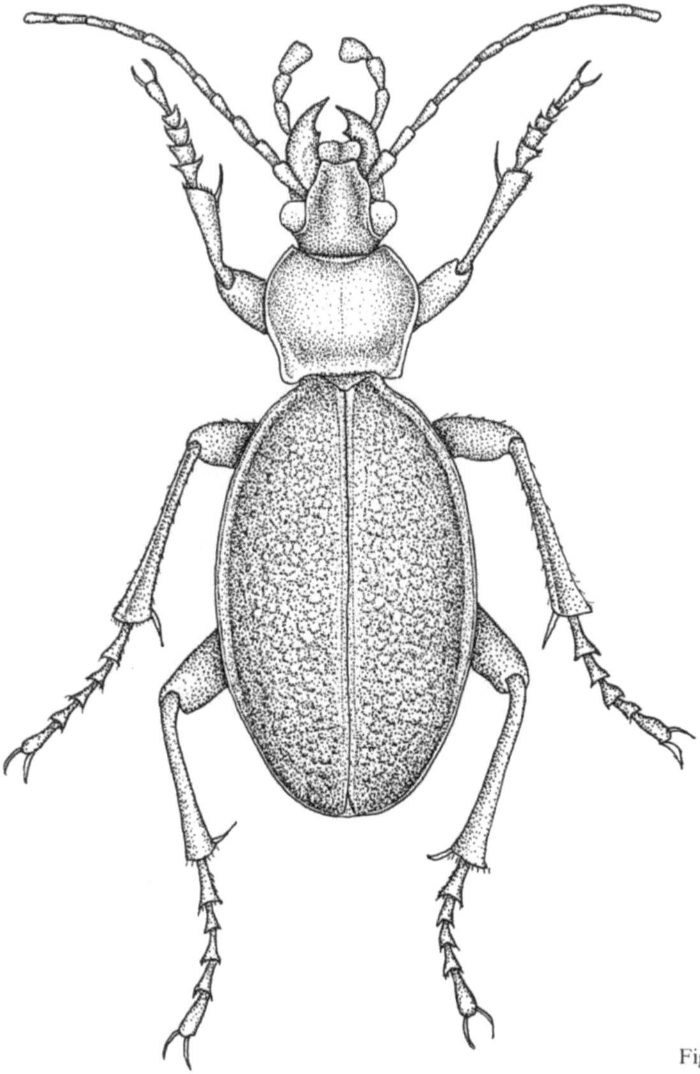

Fig. 3a

Fig. 3. (a) *Carabus coriaceus*, moderate procerization. Original size ca. 30 mm; (b) *Procerus gigas*, extreme procerization. Original size ca. 40 mm

body. The deep constrictions between head and thorax and between thorax and abdomen enhance the flexibility of the various parts of the body with respect to each other. Krumbiegel (1936) characterized this type of adaptation as follows: "An animal of this type is in a position to bite into its prey and follow its twistings and turnings, at the same time keeping its own body stationary ... species exclusively attacking snails are characterized by a very long and slender pronotum, a head of similar form and long mandibles. In the most extreme cases a pincette has developed, with the aid of which the animal is

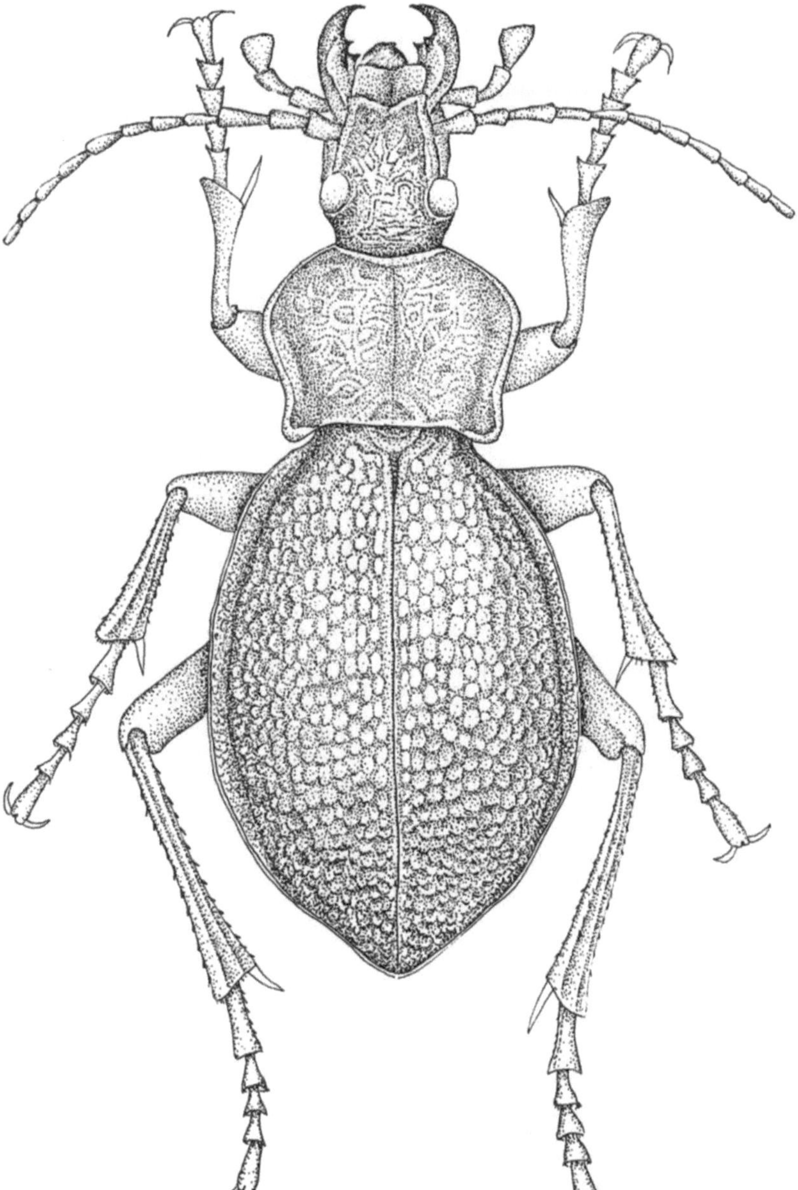

Fig. 3b

able to penetrate the snail shell by the normal route." This type is encountered in its most extreme form, as described above, in the genus *Cychrus*. The tendency to develop this complex of characters has appropriately been termed "cychrization" by Krumbiegel (1960). Members of the North American genus *Scaphinotus* fit this description, with their peculiar elongated, hook-like mandibles. Together with the spoon-shaped palpi they can penetrate deeply into the snail shell and extract the body of their prey.

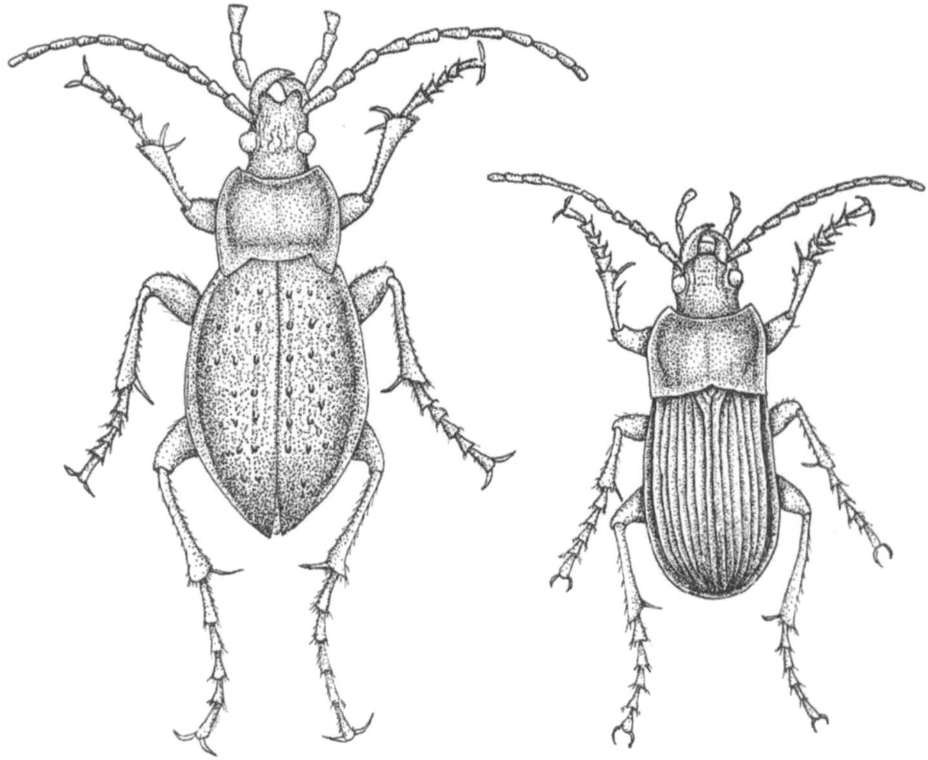

Fig. 4a Fig. 4b

Fig. 4. (a) *Carabus nemoralis*, moderate abacization. Original size ca. 22 mm; (b) *Abax ater*, extreme abacization. Original size ca. 18 mm

Fig. 5. *Mantichora latipennis*, a huge carabid of the subfamily Cicindelinae. Probably an extreme form of procerization. Photo Roer

The evolutionary tendency to develop into larger, heavier forms with broader heads, typical of the snail-shell breakers, has been designated "procerization" by Krumbiegel (from the genus *Procerus*). Cicindelinae of the genus *Mantichora* may also represent an extreme case of this type of modification (Fig. 5). These very large animals possess mandibles developed to an extreme not otherwise seen in carabids, and superficially reminiscent of the stag beetle. They live in the sandy Namib Desert of southwest Africa, in places where scorpions abound (Roer, personal communication). Unfortunately, we do not know whether the extreme development of the mandibles of *Mantichora* is an adaptation to particularly large and resistant prey, since no direct observations of its attacking scorpions have been reported.

A third type is described by Krumbiegel (1960) as "the group of borers that devour both living and dead prey. Apart from the effect of paralysing secretions, live prey is overcome more by the size and weight of the beetles than on account of manoeuvrability of the pronotum etc. Smaller and lighter species are as a rule much plumper and shorter in the leg as well. These feed upon carrion, preferably crushed snails or earthworms, but also on windfall fruit, berries, decaying fungi and the like. To this group belong all those species that in broad outline most obviously represent the genus *Carabus*, e.g. *glabratus, nitens, nemoralis*". The especially broad and ponderous form exhibited by species of *Abax* is due to the fact that notum and elytra are so flattened and widened that the body assumes the appearance of a kind of platform (Krumbiegel, 1960). Delkeskamp (1934) coined the term "abacization" for this type of development which is associated with rudimentation of the wings. In the case of *Abax* we are dealing with a predominantly carrion-consuming nutritional type.

The types shown in Figures 1–4 illustrate morphological tendencies in the evolution of carabids. They do not represent a series in which one beetle descends from another but only models for steps of evolution. Similar types are often the result of convergence (e.g. cychrisation in *Cychrus* and *Carabus creutzeri*).

In contrast to the above-mentioned kind of adaptations that are connected with a special *mode* of life, only a few cases of morphological adaptations to a special *environment* are known among the different carabid species. Some of the Cicindelinae of the genus *Tricondyla* in the tropical regions of southern Asia and New Guinea have abandoned the ground as an environment and have taken to the tree level of the tropical rain forests. A similar situation is seen in the Brazilian carabid *Agra tristis*. These species differ considerably from the typical ground dwellers. Squat extremities and strongly developed claws mark them as climbers. Some species appear to be convergent in their development with the Cerambycidae, which are the typical representatives of the characteristic life form of climbing beetles.

In temperate latitudes arboreal carabids are of little importance. Among the 361 species of the fennoscandian fauna Lindroth (1945) lists only eight species that spend their entire life on living or dead trees, seven of them belonging to the genus *Dromius*. No special morphological adaptation to an arboreal existence can be detected in these animals, nor in the *Calosoma* species that are known to climb trees in following their prey.

The bizarre flattening of the body of the ghost ground beetle, *Mormolyce phyllodes* from Southeast Asia, is an adaptation to seeking its prey below the cracked bark of trees. A convergent flattening in body shape is seen in the very much smaller species of the *Colpodes bromeliarum* group in Jamaica, an adaptation to life between the leaf sheaths of Bromeliaceae (Fig. 6; Darlington, 1970).

The best recognizable morphological differentiations confining a species to a particular environment are seen in carabids that have evolved into highly specialized diggers. The most striking example of this is provided by the Scaritini in which the foretibiae are flattened and widened and equipped with lateral teeth, thus providing the animal with an effective shovel. The deep constriction between thorax and abdomen enhances manoeuvrability for digging (Basilewsky, 1973; Fig. 7).

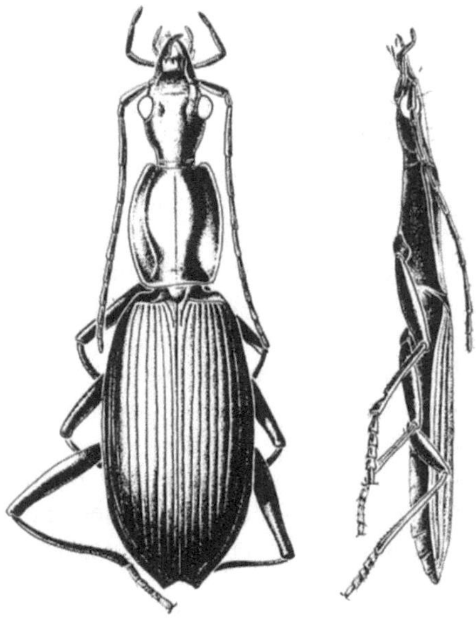

Fig. 6

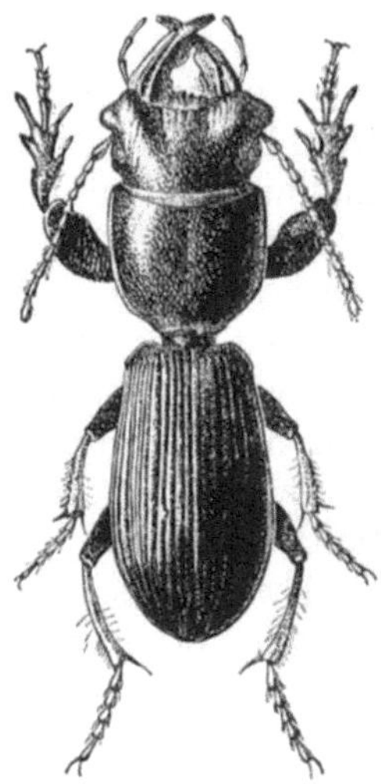

Fig. 7

Fig. 6. *Colpodes darlingtoni obtusior* from St. Andrew, Clydesdale, Jamaica. From above and from side to show body flattened for life between leaf sheaths of bromeliads. Actual size ca. 15 mm. From Darlington, 1971

Fig. 7. *Scarites biangulatus* from Madagascar. Typical adaptation of body for a digging habit of life. Actual size 28–34 mm. According to Jeannel from Basilewsky, 1973

Nevertheless, these few examples of adaptations of body form to *environment* still offer us no answer to the question concerning the underlying reasons for the occurrence of different species in adjacent habitats such as forest and open land.

Subsequent to completion of this section of the manuscript, several important contributions have come to my notice.

Sharova (1974) suggested a hierarchical system for the life forms of adult carabids. The main groups within this system are formed by the different modes of nutrition. In this respect it is merely an extension of the ideas of Krumbiegel and Delkeskamp (see above). The groups are subdivided according to the layer in which the carabids live (soil, soil surface, vegetation). In all, 29 morphoecological types were distinguished. Even a system of this nature, however, provides no explanation of the fact that closely related and structurally similar carabids live in quite different kinds of habitat.

Sharova (1960) also produced a valuable system of morphoecological types of ground beetle larvae. Here again, their body structure merely reflects an adaptation to a particular mode of nutrition, or the extent to which the animal's mode of life is epigaeic or hypogaeic (burrowing). The author draws attention to the fact that the larvae of species found in the Tundra are partially subterranean, and penetrate cracks in the ground. The larvae of the predominant species in the steppe regions of the Ukraine, on the other hand, live entirely in the ground and are morphologically adapted to burrowing. Aspects of this nature have so far been little investigated.

Erwin (1973) has recently published a short note in which he suggests dropping the name "ground beetles", since at least one-third of all tropical species live in the vegetation above ground. This arboreal element has large eyes, padded tarsi and elongated lateral setae of the elytra. The morphological adaptation of some tropical carabids to an arboreal existence undoubtedly warrants closer study, but the majority of tropical species are nevertheless genuine "ground beetles" and the species of the temperate zones are almost exclusively ground dwellers. For the large majority of carabids, therefore, the fact remains that their body structure in no way indicates the reason for their occupying a particular habitat.

B. Micromorphological Variations

Even if there are no detectable macromorphological differences in body structure between, for example, inhabitants of warm-dry and cool-moist habitats, the possibility remains that such animals might be distinguishable at a micromorphological level.

The stigmata play a very important role in the water balance. Bergold (1935) investigated whether the stigmata openings of carabids from dry habitats were smaller in relation to total body volume than those of species from moist habitats and found that this is, in fact, the case. The size of the stigmata openings and the total body volume were measured in nine species. A medium-sized species from a moderately wet habitat *(Carabus creutzeri)* was taken as a "reference beetle" and the stigmata area of the rest was calculated under the assumption that the surface area varies with the two-thirds power of the total body volume. If values calculated in this way were compared with measurements made on carabids from moister areas the latter were much higher than those anticipated from calculations (e.g. *Carabus clathratus*, that lives near water). Measurements on animals from dry habitats were relatively low as compared with the calculated values (carabids from the steppes of Turkistan and Africa: *Carabus staudingeri*, *C. tanypedilus*, *Tefflus hadquardi*). The stigmata of these carabids are situated in depressions.

For many years a discussion has been in progress concerning the concentration of colourful, shiny carabids in open fields and warmer habitats in general, as compared, for example, with woodlands. This might in some way be connected with the better protection from the sun afforded by greater powers of reflection.

Kirchner (1960), for example, found that 37% of the carabid individuals on warmer, sandy soil were coloured, as compared with 14% on a cooler clay soil. Krogerus (1948) showed, however, that the body of coloured, shiny carabids warms up just as quickly in sunshine as that of closely related black species.

The most interesting approach to this question was made by Lindroth (1974a). He demonstrated in electron microscopical studies that the iridescence of the surface of the elytra of carabids is due to their transverse microsculpture. The elytra of noniridescent carabids, on the other hand, possess a reticulate or irregular microsculpture, or even none at all. More iridescent carabids are found in warmer regions than in cold latitudes (almost none occur in the Arctic). Thermistor measurements showed, however, that *Abacidus fallax* (reticular structure) and *A. permundus* (iridescent, transverse structure) warmed up to the same extent. Some eyeless and unpigmented cave-dwelling carabids even exhibit a transverse elytral structure (*Neaphaenops tellkampfi* and *Pseudanophthalmus grandis*). The presence of large numbers of iridescent species on the margins of lakes and rivers cannot be connected with a particular elytral sculpture since the various types do not differ in their water-repellent properties. The different microstructures are of polyphyletic origin and, according to Lindroth (1974a), probably of no adaptive significance. They have only been preserved because they are "not harmful".

Chapter 2

Quantitative Investigations on the Distribution of Carabids

A. Methods

A large number of quantitative investigations on the distribution of carabids in different habitats has been carried out over the past 20 years, mostly in central Europe. The simple methods, common in the 20s and 30s, of collecting from the various habitats have been replaced in most cases by quantitative procedures. Quadrat determinations (collecting and sieving from areas of known size) are of only minor interest in the case of carabids because of their low population densities, but *Barber* traps (=pitfall traps), introduced about 1955, have come into widespread use for studying these insects in their habitats. The method utilizes collecting vessels let into the ground up to their rims: its manifold modifications have been dealt with by Heydemann (1956, 1958), Skuhravý (1956) and Tretzel (1955a). The collecting vessels may be empty (for live catches), or contain preserving fluid, either with or without bait.

A comparison between trapping and soil sampling for *Harpalus rufipes* and *Pterostichus vulgaris* in two different years revealed that the numbers caught do not run parallel with the size of the populations present in the habitat (Briggs, 1961). Since the number of animals trapped also depends on the animals' activity it does not, therefore, always reflect the abundance of the species in the habitat. For this reason Heydemann (1953) refers to the abundance resulting from trappings as the "activity density", and Tretzel speaks of the "activity abundance". These terms have been generally adopted since the parameters they represent provide a good estimate of the role of a species in an ecosystem, this not only depending on its frequency but also on its mobility (e.g. in catching prey).

The activity density depends, among other things, on the "environmental resistance" of a habitat (Heydemann, 1957) and on macro- and microclimate. These factors have different effects (sometimes concurrent) on various species. The activity abundances of different species are not necessarily comparable, as Bombosch (1962) correctly pointed out in a criticism of the trapping method. Nevertheless, it does offer a mechanical method of documenting in any one habitat species that escape simple catching, and gives an idea of the order of magnitude of their frequency, facilitates the simultaneous study of a number of habitats and allows an exact analysis of daily and annual rhythms in activity. Thus seen, it is superior to all other methods. Stein (1965) was able to show, in answer to certain points of criticism raised by Bombosch, that the habitat is not depopulated of carabids by the use of traps. He demonstrated that all dominant, subdominant and recedent species could be caught using only five vessels and that the quantitative ratios of the dominant species to one another did not change significantly if the number of traps was raised from five to 20.

That traps containing meat bait did not attract larger numbers was shown by Novák (1969) on extensive material consisting of field carabids. In various types of unbaited traps a total of 12,690 ground beetles was caught, whereas in an otherwise identical baited series in the same habitat at the same time 12,721 were caught. Of 33 species only one, *Pterostichus vulgaris*, was caught in significantly larger numbers: 2666 in baited traps, as compared with 1875 in unbaited traps. In the case of *Harpalus rufipes*, for example, the ratio in baited to unbaited traps was 1920:2039.

Certain of the disadvantages involved in the use of the usual small traps (e.g. 1 l glass jars) can be avoided by substituting large metal traps, several metres in circumference (Novák, 1969). Since the number of individuals trapped depends, among other things, upon the circumference available, the use of larger vessels brings results within a very short period of time. "Increased circumference of the trap definitely improves the possibility of making a quick survey of the nature and quantity of species inhabiting the sites under investigation." Only about half as many species were caught in the smaller glass vessels as in the larger metal traps. "The short-term use of the larger metal traps in a particular plant stock excludes many of the unfavourable circumstances hindering or even preventing qualitative and quantitative comparisons of trappings. The effect of variability in weather conditions, for example, is limited to a short period of time... The value of the results is to some extent enhanced by the fact that the adults are only disturbed for a very short period in their year-round population dynamics... Disadvantages of the larger metal traps are that the maintenance of their rims is time-consuming in bad weather and that prolonged periods of trapping affect population density".

According to Novák the results from smaller traps were influenced by how much of the surrounding area was cleared of vegetation. "Such areas proved to be attractive to adults of *Bembidion lampros*, *Harpalus rufipes* and *Trechus quadristriatus*, but were repellent to individuals of *Pterostichus vulgaris*". If employed for longer periods five to 10 traps suffice (223 days in the case in question) to show up all dominant carabids. If 10 traps are used reproducible figures for the relative frequencies of the dominant species can be obtained (Obrtel, 1971a). Although the number of species caught rose substantially with increasing number of traps (a total of 101 Coleoptera in 10 traps, 146 in 20 traps, 162 in 25 traps) this was due to an increase in the numbers of species represented by only one or a few individuals.

A surprising and not yet fully investigated finding of recent years is that a larger number of carabids was caught in traps containing 4–5% formol than in similar traps, in the same habitat, filled with water (Luff, 1968; Skuhravý, 1970).

Skuhravý's observations were made on *Pterostichus vulgaris* and *Harpalus rufipes*. The ratio of the catches in formol and water traps was 1055:671 for the former species and 676:501 for the latter. In the formol traps females of both species dominated although the two sexes were present in almost equal numbers in the water traps. Luff was able to show for a large number of carabids that more animals were caught in formol than in water traps (overall ratio 1798:864).

Whether or not formol attracts carabids can only be resolved decisively in laboratory experiments. Field investigations so far available permit other interpretations. Since formol is highly poisonous, perhaps fewer carabids can escape from it than from water. It was possible to demonstrate in very careful field studies that only 1.56% and 1.79% of the carabids (in two series of experiments) could escape from formol, whereby smaller species such as *Bembidion lampros*, *B. properans* and *Agonum dorsale* were more successful than larger ones (Petruška, 1969). Luff found a ratio of 1162:464 for the large *Pterostichus madidus* in formol and water traps respectively, but 205:181 for the small *Trechus obtusus*. The explanation might simply be that carabids are killed more rapidly in formol than in water. This view finds support in Skuhravý's observation that there were more females than males in the formol traps but equal numbers of the two sexes in the water traps. Assuming that there are no sexual differences in the attraction to formol, the males are probably better able to climb out of the traps since they are usually smaller and are equipped with bristles on the anterior tarsals, facilitating climbing. Similar findings by Luff for nearly all beetle families suggest that formol itself has no attractive effect (total catch 2737:1249).

Subsequent to the completion of this text I was given the opportunity of reading the unpublished thesis of Adis (1974), in which he compared the effect of formol, picric acid and water on the numbers of soil animals caught in traps of different size and pattern of distribution. It appears that, irrespective of the time of year, *Carabus problematicus* was invariably found in significantly larger numbers in formol than in water traps. In some cases *Pterostichus oblongopunctatus* was also found in larger numbers in formol but, depending upon the season and the order in which the traps were set up, water, too, was sometimes more attractive. The numbers of *Pterostichus metallicus* and *Trechus quadristriatus* were much the same in formol and water traps.

In the meantime, Adis and Kramer (1975) have shown in the laboratory that the attractive effect of formol on male and female *Carabus problematicus* is highly significant. Nevertheless, this could only be demonstrated consistently in animals caught in May, whereas only isolated individuals among those caught between June and October were attracted. When unbranched aldehydes up to undecanal were tested, it transpired that heptanal and pentanal were even more attractive than formol. Since formol is highly toxic and rarely occurs in nature, the authors suspect that the positive reaction to this substance when used as a baiting solution is due to erroneous perception by an organ that normally responds to another aldehyde.

The number of animals trapped in any one biotope is well correlated with the activity of the species in that biotope. The use of formol, however, necessarily involves an occasional slight distortion in the relative degree of activity exhibited by different species.

These recent results should not tempt us to discard results obtained from formol traps. In a forest region of western Germany parallel series, each of eight traps, were set up for 10 days each year over a six-year period, the one series filled with formol and the other empty (Thiele, unpublished), although the set-up and type of trap were identical in every other respect. The total catches are shown in Table 1 and again confirm that more carabids are caught in formol. It seems significant to me that the sequence of the carabid species when arranged according to individuals trapped, is almost the same in the two cases.

Table 1. Catches of carabids in formol (F) and empty (E) traps over seven 10-day periods in May or June 1970–1976, from two forest habitats near Cologne

	Habitat A		Habitat B		Both habitats	
	F	E	F	E	F	E
Nebria brevicollis	10	6	282	158	292	164
Several undetermined carabid species	58	51	57	54	115	105
Dyschirius globosus	74	70	2	0	76	70
Agonum assimile	6	7	59	55	65	62
Notiophilus (mainly *N. biguttatus*)	36	27	18	6	54	33
Abax parallelus	0	0	42	28	42	28
Abax ater	10	8	31	11	41	19
Harpalus (mainly *H. tardus* and *H. rufipes*)	36	16	0	3	36	19
Carabus (almost only *C. problematicus* and *C. nemoralis*)	4	‘8	3	3	7	11
Cychrus caraboides	1	1	0	0	1	1
Total	235	194	494	318	729	512

One indication of the value of traps in estimating abundance has attracted little attention (Schütte, 1957). Following a DDT campaign aimed at destroying *Tortrix viridana* 108 dead carabids were collected in a forest clearing where poisoning was so effective that it could be assumed that the entire carabid population had been exterminated. The percentage distribution of individuals per species was compared with that from baited traps without formol for the same spot over three years, and good agreement was found (Table 2).

Several authors have attempted to determine absolute abundances, in some cases using the Lincoln index method. Since few exact values for the population density of carabids are available in the literature some data will be given here. Dubrovskaya (1970) determined the abundance of carabids per m^2 in fields in Russia by direct counting and obtained mean values between 10.2 and 53.2 for the months of May to September. In mark, release and recapture experiments Frank (1971a) determined the abundance of 12 species on fields in Canada. For *Bembidion*, *Agonum*, *Pterostichus*, *Amara* and *Harpalus* species the values were usually in the region of $1/m^2$ and below. *Harpalus amputatus* and *Bembidion quadrimaculatum* showed the especially high values of 4.2 and 20 respectively.

Table 2. Composition of the carabid fauna in a forest in northern Germany. (According to Schütte, 1957)

Species	Relative abundances of the species (%)	
	Collected after poisoning with DDT	Total results of pitfall traps
Calosoma inquisitor	5.5	5.4
Carabus coriaceus	0.9	2.5
Carabus granulatus	10.2	9.3
Carabus nemoralis	11.3	14.6
Agonum assimile	12.9	3.2
Pterostichus niger	10.2	15.4
Pterostichus vulgaris	30.5	26.4
Pterostichus madidus	10.2	12.4
Abax ater	3.7	9.8
Abax parallelus	4.6	1.0
Total	100.0	100.0

In forests in England Frank (1967b) found higher values for some species: $11/m^2$ for *Pterostichus madidus*, $2.5/m^2$ for *Pterostichus vulgaris* and $6/m^2$ for *Abax ater*.

The density of *Harpalus rufipes* on arable land in England is given as 0.2–$6.0/yd^2$, whereas a higher abundance was measured for the same species on grassland, with $13.5/yd^2$ (Briggs, 1965, observations over a year). For *Nebria brevicollis* (adults) values below $1/m^2$ were recorded in England (Manga, 1972).

Scherney (1962b) made numerous estimations of the total population of larger carabids on fields in Bavaria (West Germany). Abundances of 0.2–$1.1/m^2$ were found on loamy sand or sandy loam and of 0.2–$0.6/m^2$ on loamy clay. On marshy ground the lowest abundance found was $0.08/m^2$. Using the Lincoln index method Kirchner (1960) found a population density of *Pterostichus vulgaris* on cultivated fields in western Germany of 1.4–5 individuals/m^2, a value of a similar order of magnitude to that found by Basedow (1973) using the quadrat method (6–$16/m^2$).

Few determinations have been made on larvae. With a few exceptions where the food supply was particularly good 1–3 *Harpalus rufipes* specimens/yd^2 were found by Briggs (1965). The maximum value for the population density of larvae of *Nebria brevicollis* was found to be $6.39/m^2$ in October (Manga, 1972).

It appears that a maximum of 1 specimen/m^2 is the general rule for larger carabids *(Carabus)*, and of the middle-sized species (e.g. *Pterostichus*) usually a few, and only in exceptional cases more than 10. A good idea of the total population is given by Heydemann (1962a; Table 3). His values were obtained using the Lincoln index method (details given in his publication). It can be seen that at the most favourable time of year a total of 43 specimens was found, giving a biomass of about 15 g/m^2 or 15 kg/ha. The value for numbers of individuals is in good agreement with those found by Dubrovskaya (1970). On an average the values for carabid populations seem to be higher on cultivated fields than in forests (Table 4).

Table 3. Number of individuals and weight of carabids per m^2 in August/September in the grassy regions of a polder. (According to Heydemann, 1962a)

Species	Average number of individuals per m^2	Average weight in mg/m^2
Pterostichus vulgaris	1.5	225
Pterostichus niger	2.0	400
Pterostichus strenuus	10.0	90
Clivina fossor	15.0	150
Amara convexiuscula	4.5	180
Amara apricaria	6.0	120
Harpalus rufipes	3.0	350
Harpalus aeneus	1.0	40
	43.0	1555 = approx. 15 kg/ha

Table 4. Numbers of carabid individuals in forests and various types of cultivated fields in the vicinity of Cologne (catches made over several months or a whole year—identical methods of trapping; from Kirchner, 1960; Thiele, 1964b; Thiele and Kolbe, 1962)

	Individuals per trap per month
Cultivated fields on clay	73–357
Cultivated fields on sand	31–77
Field hedges	17–23
Oak-hornbeam forests	4–22
Beech-sessile oak forests	13

A neglected but important question in determining abundance values is whether carabids are distributed at random or in patches. Reise and Weidemann (1975) found both alternatives in a beech wood: *Calathus micropterus* was dispersed at random, *Trechus quadristriatus* and *Notiophilus biguttatus* showed slight aggregations, and *Pterostichus oblongopunctatus* also differed significantly from random distribution.

B. The Carabid Fauna of Forests

The first clear distinction between field and forest carabids based on quantitative investigations resulted from the widespread catches made by Röber and Schmidt (1949) in the surroundings of Münster (in northern Germany). In the succeeding years many censuses were carried out on the populations of forest regions that were already exactly defined from the plant ecological (plant sociological) point of view, or at least floristically well characterized.

Investigations of this kind in certain forest plant communities of central Europe have progressed to such a degree that a comparative analysis of their carabid populations can be attempted. These plant communities can be classified as follows:

Order **Fagetalia Silvaticae** (Tall deciduous forests on neutral to slightly acid soil; equable, moist microclimate)
Associations: **Fagetum** *(mountain beech forest)*
Querco-Carpinetum (*oak-hornbeam forest of mountain regions*, partly closely related to the Fagetum or emergent from it as a result of succession or human interference; Thiele, 1956)
Querco-Carpinetum (*oak-hornbeam forest of the lowlands*)
Fraxino-Ulmetum (*water-meadow forest*; a small portion of the water meadow forests investigated here consists of moist Querco-Carpineta)

Order **Quercetalia Robori-Petraeae** (on silicaceous rock or sand, more acid soil; more fluctuating microclimate, drier and warmer)
Associations: **Querco-Betuletum** (*oak-birch forest*, mainly in lowlands)
Fago-Quercetum (*beech-sessile oak forest*, mainly in mountain regions)

The publications from which Table 5 was compiled and the regions investigated are as follows: Netherlands (van der Drift, 1959; den Boer, 1965). Deutsches Mittelgebirge (Rheinisches Schiefergebirge: Thiele, 1956, 1964b; Thiele and Kolbe, 1962; Paarmann, 1966; Neumann, 1971; Becker, 1975; Westphalian Sauerland: Thiele, 1956; Lauterbach, 1964; Kolbe, 1968b, 1970, 1972; Teutoburg forest: Giers, 1973; Weserbergland: Adeli, 1963–64; Knopf, 1962; Rabeler, 1962; Lohmeyer and Rabeler, 1965; Harz: Tietze, 1966a; Saxony: Beyer, 1964, 1972). Southwest Germany (Wutach gorge: Kless, 1961). Austria: Franz *et al.* (1959). North German lowlands (Rhineland: Lehmann, 1965; Jarmer, 1973; Thiele, unpublished; Westphalian Münster country: Stein, 1960; Wilms, 1961; Northwest Germany: Rabeler, 1957, 1969a; Lüneburg Heath: Niemann, 1963–64; Saxony: Tietze, 1966b; Mletzko, 1970, 1972). Czechoslovakia (Moravia): Obrtel, 1971b.

Most of the investigations cited in Table 5 are based upon catches with Barber traps and a few on meticulous quantitative quadrat counts. The symbols in Table 5 indicate "presence" in Tischler's (1949) sense of the word, i. e. the percent of those stands investigated in which the species was encountered. Differing abundances in the various types of forest community are therefore not shown. Only species that were represented with a "presence" of 50% or more in at least one of the plant communities appear in the table.

Characteristic societies of carabids can be assigned to some forest plant communities. This is particularly true of the larger units in the system of plant communities. Referring to carabids as well as to other soil arthropods, Thiele (1956) recognized "...the plant ecological order as the smallest biocenotic unit possessing characteristic species of plants and animals that are associated with it over wide geographical areas". Rabeler (1969a) elaborated further: "Although certain faunal differences are readily discernible between the larger groups such

Table 5. The occurrence of important carabid species in the plant societies of the forests of central Europe

	Fagetalia				Quercetalia	
	Fagetum (mountains)	Querco-Carpinetum (mountains)	Querco-Carpinetum (lowland)	Water-meadow forests (usually Fraxino-Ulmetum, lowland)	Fago-Quercetum (mountains) and Querco-Betuletum (mountains)	Querco-Betuletum (lowland)
Number of stands investigated	7	9	5	6	9	6
Peak in Fagetum						
Molops elatus	+++	o	–	–	o	–
Carabus auronitens	++	–	o	–	+	o
Pterostichus metallicus	++	o	–	–	+	–
Harpalus latus	++	o	+	+	–	–
Peak in mountain and/or lowland Fagetalia						
Pterostichus vulgaris	+++	++	+++	+++	+	o
Nebria brevicollis	++	++	++	++	+	o
Molops piceus	+++	+++	++	–	o	–
The same, but also in mountain Quercetalia (although in much reduced abundance)						
Abax parallelus	+++	+++	++	+	++	o
Abax ovalis	+++	++	–	–	++	+
Trichotichnus laevicollis (incl. *T. nitens*)	+++	++	–	–	++	–
Pterostichus cristatus	+	+	–	–	++	–
Peak in certain societies of the Fagetalia						
Pterostichus strenuus	+++	o	++	+++	+	+
Pterostichus madidus	+	++	++	o	o	+
Cychrus caraboides	+	+	+++	–	o	o
Peak in water-meadow forests						
Agonum assimile	+	+	++	+++	+	o
Carabus granulatus	o	o	+	+++	+	o
Leistus ferrugineus	–	o	–	+++	–	+
Clivina fossor	–	–	–	+++	–	–
Pterostichus anthracinus	–	–	–	+++	–	–
Agonum micans	–	–	–	+++	–	–
Pterostichus cupreus	o	o	o	++	–	–
Patrobus atrorufus	o	o	+	++	o	–
Trechus secalis	–	o	o	++	o	–
Asaphidion flavipes	o	–	o	++	–	–
Pterostichus vernalis	–	o	o	++	o	–

Table 5. (continued)

	Fagetalia				Quercetalia	
	Fagetum (mountains)	Querco-Carpinetum (mountains)	Querco-Carpinetum (lowland)	Water-meadow forests (usually Fraxino-Ulmetum, lowland)	Fago-Quercetum (mountains) and Querco-Betuletum (mountains)	Querco-Betuletum (lowland)
Number of stands investigated	7	9	5	6	9	6
Agonum viduum (incl. *A. moestum*)	–	–	o	++	–	+
Bembidion tetracolum	–	–	–	++	–	–
Agonum fuliginosum	–	–	–	++	–	+
Agonum obscurum	–	–	–	++	–	o
Peak in mountain Quercetalia						
Cychrus attenuatus	+	–	–	–	++	–
Peak in lowland Quercetalia						
Notiophilus rufipes	o	+	o	–	o	++
Notiophilus palustris	–	o	+	+	o	++
Calathus micropterus	o	–	o	o	–	++
Peak independent of forest type in mountains						
Carabus problematicus	++	++	+	–	+++	++
Carabus coriaceus	++	+++	+	–	++	+
No strict habitat affinity						
Abax ater	+++	+++	+++	++	+++	++
Carabus nemoralis	+++	+++	+++	++	+++	++
Pterostichus oblongopunctatus	+++	+++	+++	+	+++	+++
Pterostichus niger	++	++	+++	+++	++	++
Trechus quadristriatus	+	++	–	++	+++	o
Notiophilus biguttatus	++	o	o	+++	+	+++
Loricera pilicornis	+	+	o	++	o	++
Pterostichus nigrita	++	–	+	++	o	++
Carabus violaceus or *C. purpurascens*	++	o	o	o	+	++
Number of species	33	32	30	31	32	28
Number of species with a presence of ≥50 %	20	14	12	25	12	12

(The presence (see text) with which the species occurred in the different plant societies is indicated. –: absent, .: up to 24 %, +: 25 to 49 %, ++: 50 to 74 %, +++: 75 to 100 %). Only species which attained a presence of 50 % in at least one of the six habitat complexes investigated were included)

as the acidophilic and the mesophilic forests, a distinction becomes increasingly difficult with the attempt to characterize the individual forest communities (associations) ... The number of animal species that respond to the often only slight ecological differences existing between closely related smaller sociological units is relatively low".

However, the abundant material available from quantitatitive investigations permits, in some cases, a characterization of plant communities smaller than the plant sociological order, e.g. of societies ("Verbände") or even of associations, on the basis of their carabid fauna. As a rule, however, the characterization of the animal population of a particular ecosystem is not based upon species of one family alone, but draws upon many groups of animals just as the plant ecological system is not based upon only one species-rich plant family.

"It is not to be expected that a particular species be found exclusively in any one plant community. One-hundred per cent adherence to a particular community represents an ideal that is not encountered in nature; the isolated occurrence of a species elsewhere does not reduce its value as a characterizing species" (Thiele, 1956).

Bearing these limitations in mind, the carabid communities of the order Fagetalia (mesophilic deciduous forests on alkali-rich soils) can be well characterized. The typical species here are *Pterostichus madidus* (occurs in western Europe only), *P. strenuus*, *P. vulgaris*, *Molops piceus* and *Nebria brevicollis*. Numerous other species are also typical of the Fagetalia but exhibit further peaks in occurrence in certain communities within this order. *Molops elatus*, *Carabus auronitens*, *Pterostichus metallicus* and *Harpalus latus*, for example, show a very marked peak in occurrence in the Fagetum. Fifteen species clearly prefer the water-meadow forests. The majority of the central European riverside meadow forests so far investigated belong to the association Fraxino-Ulmetum, some, however, to the Querco-Carpinetum athyrietosum or moist oak–hornbeam forest, the carabid fauna of which shows a remarkable similarity to that of the water-meadow forests (Table 5).

Certain species which prefer the mountainous Fagetalia exhibit a second maximum in frequency in mountainous forests with acid soil (Fago-Quercetum), which belong to the completely different plant ecological order of Quercetalia robori-petraeae (on acid soil, with less ground cover and much more drastic fluctuations in microclimate).

The "presence" of these species in the mountainous Quercetalia is high although their abundance is greatly diminished. The following two examples are taken from the West German Mittelgebirge. In the first case soil-inhabiting beetles were caught over the course of a year in a scrub strip, the vegetation of which gradually changed from the oak–hornbeam type in the west to oak–birch in the east. A large drop can be seen in the abundance of those species exhibiting a maximum in the Fagetalia (Table 5), i.e.: *Pterostichus cristatus*, *Abax parallelus*, *Patrobus atrorufus*, *Agonum assimile*, *Trechus secalis* and *Nebria brevicollis* (Table 6).

A further investigation (Thiele and Kolbe, 1962) dealt with the carabids of an oak-hornbeam and a beech-sessile oak forest at the same altitude and with the same macroclimate. *Trichotichnus laevicollis* and *Pterostichus cristatus* were obviously the most frequent species in the oak-hornbeam forest whereas *Abax parallelus* and *A. ovalis* were restricted to this type (Table 7).

It has been shown experimentally for a number of species of the Fagetalia that they prefer moisture and in some cases also cold. Examples of moisture-

Table 6. Differences in abundance of carabids in two merging types of vegetation in a bushy strip

	Querco-Carpinetum		Querco-Betuletum	
Pterostichus cristatus	154	66	11	42
Patrobus atrorufus	63	51	26	14
Abax parallelus	69	7	—	3
Agonum assimile	24	34	3	5
Trechus secalis	43	4	2	—
Nebria brevicollis	6	19	—	1
Pterostichus vulgaris	9	25	2	12

(Results in each case from three traps over one year in four localities moving from West to East, 1956/57; from Thiele, 1964b)

Table 7. Differences in abundance of some carabid species in two forest societies identical as to altitude and macroclimate

	Querco-Carpinetum	Fago-Quercetum
Abax ovalis	40	—
Abax parallelus	19	—
Trichotichnus laevicollis	59	18
Pterostichus cristatus	328	60

(Results in each case from 10 traps in one vegetational period, 1958; from Thiele and Kolbe, 1962)

and cold-preferring species are *Molops elatus*, *M. piceus*, *Abax ovalis* and *Pterostichus metallicus*. The following species are also moisture-loving, although less dependent on a cool microclimate: *Pterostichus cristatus*, *P. madidus*, *Patrobus atrorufus*, *Nebria brevicollis* and *Abax parallelus*.

The microclimatic requirements of the carabids of the Fagetalia explain why, if they occur in Quercetalia, they prefer those of mountain regions with their cool, moist macroclimate. Their numbers are reduced, however, owing to the sparser ground vegetation and hence more variable conditions of temperature and humidity than in the Fagetalia. The importance of microclimatic requirements for their distribution among plant communities is well illustrated by the following observation (Thiele and Kolbe, 1962; Table 8): *Pterostichus cristatus* and *Trichotichnus laevicollis*, which in experiments prefer wet or cold conditions, occur more frequently in moist than in dry places within a mountainous Fago-Quercetum, whereas the warm-preferring *Carabus problematicus* and warm- and

Table 8. The abundance of some carabid species in moist and dry localities in a Fago-Quercetum

	Moist habitats	Dry habitats
Pterostichus cristatus	42	9
Trichotichnus laevicollis	15	1
Carabus problematicus	48	59
Pterostichus oblongopunctatus	12	18

(Results in each case from four traps in one vegetational period, 1958; from Thiele and Kolbe, 1962)

dry-preferring *Pterostichus oblongopunctatus* show no such differences. It is interesting that *C. problematicus* (prefers moisture) shows a peak in occurrence in the mountains which is not shown by *P. oblongopunctatus* (Table 5), a species preferring dry conditions.

The latter species is extremely eurytopic in forests, as are other climatically eurypotent forest carabids: *Carabus nemoralis* is eurythermic and euryhygric, *Pterostichus niger* is fairly thermophilic, *Pterostichus nigrita* and *Carabus purpurascens* are eurypotent with respect to temperature and moisture. Microclimatic requirements will be dealt with in more detail in Chapter 5.

Compared with the many species preferring the Fagetalia, only a small number choose the Quercetalia. An obvious peak in the occurrence of *Cychrus attenuatus* is seen in the Fago-Quercetum. Under experimental conditions this species prefers relatively low temperatures and a high degree of moisture and darkness (Lauterbach, 1964), thus resembling the characteristic species of the Fagetalia. Too little is known about the biology of the species, however, for its specific habitat affinity to be understood. This also holds for the three species of carabids in Table 5 which show peaks in occurrence in lowland Quercetalia.

That the Fagetalia provide a more suitable habitat for carabids than the Quercetalia is clearly seen from the following data. Table 5 lists 29 species which prefer some or all communities of the Fagetalia and only four species with a marked preference for the Quercetalia (11 species with no fixed habitat). Although the total number of species found is similar in all forest communities, the number of species with a high "presence" (occurring in 50% or more of the stands investigated) varies considerably. The water-meadow forests obviously head the list with 25 species. They are followed by the Fageta with 20 and the Querco-Carpineta of the mountains with 14 species. The two associations of the Quercetalia comprise only 12 species each with high "presence", just as the Querco-Carpineta of the lowlands, which are drier than the water-meadow forests. There is a very strong suggestion that the number of species of carabids runs parallel with the degree of moisture connected with the particular type of forest.

The Quercetalia with acid soil not only harbour the smallest number of carabid species but also less individuals. Thus, in the two populations of Querco-Carpinetum and Fago-Quercetum in the West German Mittelgebirge already mentioned (Thiele and Kolbe, 1962, p. 22), the number of species was 55:35 and of individuals 1942:756.

The carabid fauna of other types of forest have not been sufficiently well investigated to permit a complete characterization. In the submontane spruce forests of the Harz mountains and in the Bayrischer Wald typical species that also occur in the submontane Fagetum are *Pterostichus aethiops* and *Carabus silvestris* (Klein, 1965; Tietze, 1966a; Schmidt *et al.*, 1966).

A birch marsh in northwest Germany was found to be remarkably poor in carabids, although *Agonum* species (*A. obscurum* among others) were much in evidence (Rabeler, 1969b).

In the Ukrainian steppe a type of forest termed ravine forest ("Schluchtwälder") by Ghilarov (1961) is encountered. Floristically, they are similar to the Querco-Carpinetum of central Europe, but their carabid population consists mainly of species that inhabit arable land and dry grassland in central Europe, but are not encountered in forests: *Brachinus crepitans* and *B. explodens*, *Pterostichus coerulescens* and *P. cupreus*, numerous *Harpalus* and *Amara* species and many others. Of the typical forest animals of central Europe *Pterostichus oblongopunctatus*, a species that is both eurytopic and eurythermic, is the most abundant. It is tempting to assume that these thermophilic elements of the forest-steppe spread with the warmth-loving mixed oak forest at the height of the Atlanticum and that, in the succeeding phase which brought a cooling down of the climate in central and western Europe, they migrated into the open dry grasslands (Becker, 1975). Numerous species that inhabit forests in central Europe are found in the larger oak forests of the forest-steppe of southern Russia: *Abax ater*, *Pterostichus oblongopunctatus*, *P. niger*, *P. nigrita*, *P. vulgaris*, *P. strenuus*, *P. diligens*.

On the whole, the carabid fauna of European forests has little in common with that of the adjoining fields (Chap. 4.A.I.3.). It is therefore surprising that 13 of the 45 common forest species of central Europe listed in Table 5 are also to be found among the 50 most widespread field carabids of this region (see also Table 9). But of these 13 species almost half, i.e. the six species *Clivina fossor*, *Carabus granulatus*, *Trechus secalis*, *Pterostichus vernalis*, *P. cupreus* and *Asaphidion flavipes*, almost exclusively inhabit water-meadow forests and five more are very eurytopic species lacking a well-developed affinity for any one particular forest community: *Carabus nemoralis*, *Pterostichus niger*, *P. nigrita*, *Loricera pilicornis* and *Trechus quadristriatus*. Water-meadow forests often comprise sparsely wooded, meadow-like habitats, harbouring, as would be expected, littoral animals, and the carabid fauna of the central European agricultural land originated largely in littoral habitats. Five of the 13 species mentioned in this context are only found more frequently in meadows, and rarely on tilled fields.

Löser (1972) carried out investigations involving montane carabids on the edges of the West German Mittelgebirge, where the animals attain the limits of their area of distribution. By means of quantitative catches and climatological recordings, and by calculating correlations, he showed for the first time that soil moisture and the humidity of the air layer near the ground are very important factors in limiting the distribution of montane carabids in the direction of the lowlands. The number of individuals was shown to rise sharply for even small increases in altitude: using the same number of traps over the same period of time 7524 individuals were trapped below 100 m, more than 9900 between 100 and 150 m and 10,494 above 150 m.

Although moisture (water table) and not the composition of the vegetation is held by den Boer (1963) to be the most important factor determining the distribution of carabids, this does not conflict with the above observations: distribution is not directly dependent upon the vegetation. Those European forest communities which can be distinguished on the basis of their carabid fauna also differ as to microclimate, especially with respect to water balance. In forests on acid soil the air layer near the ground is subject to greater fluctuations in moisture and temperature than in the Fagetalia (Thiele and Kolbe, 1962). The forests investigated by den Boer in Holland (province of Drenthe) nearly all belong to the order of the Quercetalia robori-petraeae and, as would be expected from the foregoing, reveal no differentiation within the carabid population which could be related to a plant ecological subdivision into associations or sub-associations. In this case, it is much more evident that the distribution of carabid species will be found to depend on microclimatic factors.

Van der Drift (1959) found that the frequency of many forest species in Holland differed from one microhabitat to another despite an apparently homogeneous milieu (uniform plant community). Although the total frequency may fluctuate from year to year the pattern of distribution within a very small area is maintained over a longer period of time.

C. The Distribution of Carabids in Open Country

I. The Carabids of Cultivated Land

1. The Carabid Fauna of Arable Areas

Only about 25 years have elapsed since the fauna of these habitats first became the subject of more profound investigations. Today, carabid populations of these habitats of arable land are among the best studied in some parts of Europe.

It was long assumed that the frequent upheavals involved in cultivation and the repeated artificial changes in vegetation accompanying crop rotation must exclude any but a nonspecific fauna of ubiquitous species, consisting of relics of the original forest fauna with a few additional immigrant species from the steppes.

In the meantime it has transpired that the arable areas of large regions of Europe, despite very different climates, harbour a distinctly characteristic animal population, including a typical carabid fauna. For the following survey results taken from 32 publications[3] covering 29 arable habitats in a belt stretching from England across central Europe to Belo-Russia, have been evaluated. A particularly large number of the investigations originate in Germany, Czechoslovakia and Poland (Table 9).

The basic fauna is surprisingly homogeneous. 26 of the species found in the arable areas were encountered in at least one-third of the regions investigated. Of these, the following eight species were particularly frequent and characteristic,

[3] *Arable land:* England: Pollard, 1968; Western Germany: Tischler, 1958; Fuchs, 1969; Heydemann, 1964; Hossfeld, 1963; Kirchner, 1960; Thiele, 1964b; Weber, 1965a; Röber and Schmidt, 1949; Prilop, 1957; Lücke, 1960; Becker, 1975; Scherney, 1958, 1960a; German Democratic Republic: Müller, 1968, 1972; Geiler, 1956/57; Czechoslovakia: Skuhravý, 1959, 1970; Skuhravý and Novák, 1957; Novák, 1964, 1967, 1968; Petruška, 1967, 1971; Louda, 1968, 1971; Poland: Kabacik, 1962; Kabacik-Wasylik, 1970; Gorny, 1968a, 1968b; Russia: Dubrovskaya, 1970.
Meadows and grazing-land: Western Germany: Lehmann, 1965; Stein, 1965; Rabeler, 1953; Gersdorf, 1965; Boness, 1953; German Democratic Republic: Müller, 1968; Tietze, 1968, 1973; Hempel *et al.*, 1971; Hiebsch, 1964; Czechoslovakia: Doscocil and Hůrka, 1962; Bílý and Pavlíček, 1970.
Clover and alfalfa fields: Western Germany: Kirchner, 1960; Boness, 1958; German Democratic Republic: Geiler, 1967; Czechoslovakia: Obrtel, 1968.

occurring in more then 20, i.e. two-thirds, of the areas investigated: *Pterostichus vulgaris*, *P. cupreus*, *Harpalus rufipes*, *H. aeneus*, *Agonum dorsale*, *A. muelleri*, *Bembidion lampros* and *Trechus quadristriatus*. (All of them are also found on land which is not exposed to changes, such as meadows, clover and alfalfa fields, see below.) A few species are found only in western Europe *(Carabus auratus)* or are more frequent only in arable fields of this region *(Pterostichus niger*, *Nebria brevicollis)*. A larger number of species occurs either exclusively in eastern Europe or only there constitutes a considerable portion of the field fauna (see below).

Table 9. Occurrence of important carabid species in the agricultural regions of Europe

	Arable land	Meadows and pastures	Clover and alfalfa
Number of areas investigated	29	12	4
Pterostichus vulgaris	28	11	4
Harpalus rufipes	27	8	4
Bembidion lampros	25	7	4
Pterostichus cupreus	24	8	4
Agonum dorsale	23	4	3
Agonum muelleri	21	+	3
Harpalus aeneus	21	5	4
Trechus quadristriatus	21	5	3
Loricera pilicornis	19	7	3
Carabus cancellatus	17	4	–
Calathus fuscipes	17	5	4
Carabus granulatus	16	9	+
Synuchus nivalis	16	+	+
Clivina fossor	14	9	–
Calathus melanocephalus	14	5	+
Amara familiaris	13	4	4
Carabus auratus (western Europe only)	12	+	+
Asaphidion flavipes	12	+	4
Pterostichus lepidus	12	+	+
Broscus cephalotes	12	–	3
Dyschirius globosus	11	8	–
Stomis pumicatus	11	+	3
Pterostichus niger	11	9	+
Bembidion quadrimaculatum	11	+	+
Bembidion obtusum	10	–	+
Pterostichus coerulescens	10	11	+
Trechus secalis	+	7	–
Amara communis	+	7	3
Amara aenea	+	7	+
Pterostichus vernalis	+	7	+
Pterostichus nigrita	+	5	+
Pterostichus diligens	+	5	+
Amara aulica	+	5	+
Amara lunicollis	+	5	–
Carabus nemoralis	+	5	+
Nebria brevicollis	+	4	+
Amara plebeja	+	4	3

Table 9. (continued)

	Arable land	Meadows and pastures	Clover and alfalfa
Number of areas investigated	29	12	4
Only common in agricultural country in Eastern Europe			
Dolichus halensis	9	−	+
Amara consularis	8	+	+
Calosoma auropunctatum	8	−	+
Harpalus griseus	8	−	−
Amara bifrons	6	+	+
Acupalpus meridianus	6	−	+
Brachinus crepitans	5	−	−
Brachinus explodens	5	−	+
Zabrus tenebrioides	5	+	−
Calathus ambiguus	5	+	+
Pterostichus punctulatus	5	−	+
Harpalus distinguendus	5	−	+
Carabus scheidleri	5	−	−

The columns show the number of areas investigated in which the species occurred. + in column 1 indicates: in less than ten areas, in column 2: in less than four areas, in column 3: in less than three areas. Only species that inhabited at least a third of the populations in at least one of the three habitat complexes investigated were included

Comparative investigations on the occurrence of Carabidae were made by means of pitfall trapping mainly in fields of winter wheat from the end of May to the beginning of August, during 1973 and 1974, in Belgium, West Germany, the Netherlands, and Sweden. Trapped beetles were listed in tables according to their frequency. With one exception, the 15 most frequent species of both years are also included here in Table 9. Where a comparison was possible, no distinct differences were discernible between numbers of carabid species found in recent investigations and in those of about 20 years ago (Basedow *et al.*, 1976a).

2. *Carabid Fauna and Crop Plants*

No case is known in which any one species of carabid is characteristically linked with a particular crop plant. A complete changeover of the carabid fauna from year to year with the changing crops is most unlikely.

Nevertheless, "quantitative indicators" do exist for certain types of crop (Heydemann, 1955). All investigators agree that winter crops (especially winter grain) and root crops such as beet, potatoes, etc., differ particularly with respect to their carabid fauna. Examples of this were given by Heydemann as far back as 1955 (Table 10). In Poland, rye and potato fields (Kabacik-Wasylik, 1970) were found to differ widely, both quantitatively and qualitatively, in dominance as well as in the ratios of spring to autumn breeders. The dominant species in rye fields were: *Carabus cancellatus*, *Pterostichus lepidus*, *P. cupreus* and *Loricera pilicornis* whereas in potato fields *Broscus cephalotes*, *Bembidion lampros*, *Calathus*

Table 10. Carabids as quantitative indicators for winter and root crops on different types of soil. (From Heydemann, 1955)

a) Quantitative indicators for winter crops on heavy soil:	
Agonum dorsale	*Carabus cancellatus*
Agonum muelleri	*Pterostichus cupreus*
Carabus auratus	*Stomis pumicatus*
b) Quantitative indicators for root crops in general:	
Calathus fuscipes	*Trechus quadristriatus*
c) Quantitative indicators for root crops on heavy soil:	
Pterostichus niger	*Synuchus nivalis*
Pterostichus vulgaris	
d) Quantitative indicators for winter crops on sandy soil:	
Carabus convexus	*Microlestes minutulus*
Harpalus distinguendus	*Pterostichus coerulescens*
Harpalus tardus	*Pterostichus lepidus*
Metabletus foveatus	
e) Quantitative indicators for root crop fields on sandy soil:	
Broscus cephalotes	*Calathus melanocephalus*
Calathus erratus	*Harpalus griseus*

erratus, *C. ambiguus*, *Amara bifrons*, *A. fulva*, *Harpalus rufipes* and *H. griseus* were dominant. The species listed below were found by Petruška (1971) in barley and sugar beet in Czechoslovakia. The following were more frequent in barley fields: *Pterostichus cupreus*, *Agonum dorsale*, *Harpalus rufipes*, *Stomis pumicatus* and *Carabus scheidleri*. In the beet fields *Pterostichus vulgaris*, *Bembidion lampros* and *Trechus quadristriatus* dominated. A reciprocal relationship between the frequency of *P. vulgaris* and *P. cupreus* in cereal and root crop fields has also been found in western central Europe (e.g. Thiele, 1964b; Kirchner, 1960). In this case *P. vulgaris* is an autumn breeder and more frequent among root crops, whereas *P. cupreus*, which is more often encountered among cereals, is a spring breeder. Autumn breeders have often been found more frequently among root crops. Near Cologne, Kirchner (1960) caught 9712 autumn individuals on root crop fields but only 1905 spring breeders, although the number of species represented among the latter was larger. The data of Kabacik-Wasylik (1970), mentioned above, also point to a predominance of autumn breeders among root crops.

The different composition of the carabid fauna of the two groups of crops is in part due to the fact that the methods of cultivation employed in spring for the root crops affects the developmental stages of carabids to varying degrees (see. Chap. 8.A.). Kirchner (1960) points out that, primarily, the root crops offer much more shade than cereals so that the microclimate of the two kinds of field differs. A far higher proportion of species among the spring breeders prefers warmth and dryness, a factor which no doubt contributes largely to the differences in the distribution of the two types of species. *Pterostichus vulgaris*, for example, is hygrophilic and eurythermic whereas under experimental conditions *P. cupreus* prefers dryness and warmth.

Furthermore, Heydemann (1953) found fewer carabid individuals among root crops. The ratio for the different types of field was 2:1. Similar results were obtained by Scherney (1955) in Bavaria. He caught an average of 85 carabids/100 m^2/day on wheat, barley and mustard fields, and 66/100 m^2/day on potato and beet fields (but only 31/100 m^2/day on clover fields and grassland). On the other hand, Petruška (1971) found almost equal numbers of carabids in barley (5310 individuals) and sugar beet (4962 individuals).

The value of carabids as indicators of the ecological conditions prevailing in various types of crops is not to be underestimated. "The differences between the carabids in root crops and in cereals on one and the same type of soil are much more marked than those between the plants (i.e. the weed communities, author's comment)... The adaptations and migrations made necessary by crop changes on cultivated fields can be accomplished more readily by animals than plants and as a consequence the fauna provides an accurate picture of the ecological situation at any time" (Heydemann, 1955).

In comparison, little is known about the carabid fauna of arable land *on continents other than Europe.* Preliminary investigations have recently been reported from *North America,* but even in 1971a Kirk wrote: "Little is known about the biology of North American carabids or their role in cultivated fields...". Following four-year studies from South Dakota Kirk reported finding 127 species and gave data concerning their frequency. Just as in Europe, species of the

Table 11. The most common carabid species of cultivated fields in Ontario (Canada) and South Dakota (USA). (According to Rivard, 1964 and Kirk, 1971a)

Species (in order of decreasing frequency)	Ontario	South Dakota
Pterostichus vulgaris	+	−
Harpalus erraticus	+	+
Harpalus pennsylvanicus	+	+
Harpalus compar	+	−
Harpalus lewisi	+	−
Amara obesa	+	+
Pterostichus lucublandus	+	+
Harpalus caliginosus	+	(+)
Amara latior	+	(+)
Bembidion quadrimaculatum	+	+
Agonoderus comma	+	+
Tachys incurvus	+	(+)
Agonum placidum	(+)	+
Amara carinata	−	+
Chlaenius platyderus	−	+
Pterostichus alternans	−	+
Harpalus herbivagus	−	+
Microlestes nigrinus	−	+
Pasimachus elongatus	−	+
Pterostichus chalcites	(+)	+

+ : the species occurs regularly and frequently
(+): the species occurs less frequently
− : the species is absent

genera *Amara*, *Harpalus*, *Pterostichus* and *Agonum* are especially prominent, but *Chlaenius* species are more frequent on cultivated fields than in Europe. Of the *species* also occurring in Europe Rivard (1964a) found *Carabus nemoralis*, *Loricera pilicornis*, *Pterostichus vulgaris*, *Amara familiaris*, *A. aenea*, *Bembidion quadrimaculatum* and *Harpalus affinis* on fields in Canada (with the exception of *Loricera*, all introduced into America). Rivard's catches in the vicinity of Belleville, Ontario, comprised a total of 159 species. He lists 12 particularly frequent species and Kirk mentions 14. Of the 12 most common species in Ontario, six were also dominant in South Dakota, three others were found there in smaller numbers and only three were completely absent. Of the total of 20 species which occurred most frequently in one or the other region, 11, more than half, were common to both, which indicates a large degree of correspondence (Table 11).

Almost nothing is known concerning the ecology of the carabid fauna of *tropical* agricultural land. Some studies have been carried out in Surinam, employing the same methods as in Europe (van der Drift, 1963). Primary and secondary forest and a variety of cultivated areas on which groundnuts, tomatoes, melons and citrus fruits were grown, as well as grazing land were investigated.

Here again, the fauna of the cultivated areas had little in common with that of the forests. The carabid genus *Brachinus* was almost completely confined to the primary forest whereas *Taeniolobus* and *Pheropsophus* were limited to cultivated land. In places, the latter genus was found in spectacular numbers (a maximum of 2500 individuals in 12 traps in two weeks). Taken as a whole, however, the carabid fauna proved to be relatively poorly developed.

3. The Influence of the Preceding Crop on the Carabid Population

In western Germany an investigation was carried out on the carabid fauna of fields that had borne different crops such as clover or potatoes prior to being planted with vegetables in June/July (Kirchner, 1960). The large quantitative difference found in the populations of these fields was maintained for several weeks subsequent to the planting of the vegetables. The fact that clover harbours far fewer carabids than does arable land was reflected in the correspondingly lower numbers found (Fig. 8). Kabacic-Wasylik (1970) concluded that in the course of the changeover from one type of crop to another the spectrum of dominance of carabids undergoes a corresponding shift.

4. The Carabid Fauna of Permanent Forms of Cultivated Land

Meadows and pastures or clover and alfalfa fields have in common that the ground is left untilled for several years in succession. In order to come to conclusions as to the carabid fauna of land of this type in Europe, 12 publications concerning grassland and four dealing with clover and alfalfa fields have been consulted (see footnote p. 26).

As mentioned above, the eight species most common on arable land also occur in the undisturbed areas, where they, as a rule, occupy a dominant position (only *Agonum muelleri* is much less frequent in meadows). The species found

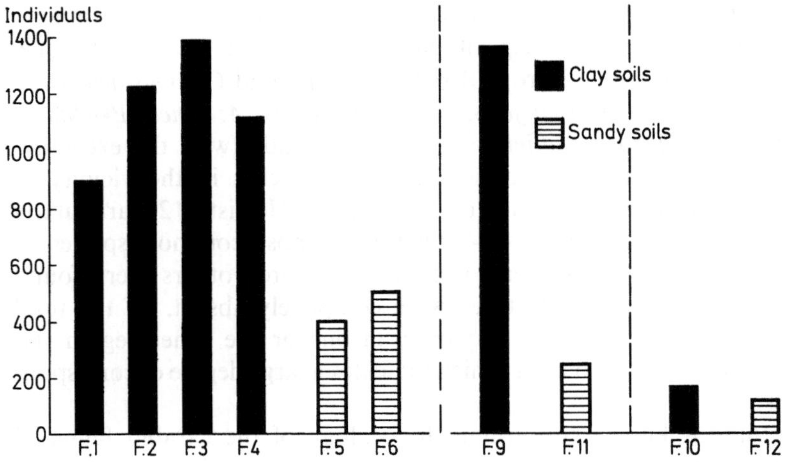

Fig. 8. Total catch of carabids from clay and sandy soils bearing different crops (*F1* to *F6*: vegetable fields; *F9* and *F11*: potato fields; *F10* and *F12*: clover fields). From Kirchner, 1960

in clover and alfalfa fields differ but little from those in fields that are tilled annually (Table 9), although the number of individuals is considerably lower (Kirchner, 1960, Fig. 8). According to Scherney (1960) the ratios of the number of individuals found on different types of cultivated land are approximately as follows:

Wheat	Barley	Potatoes	Clover	Meadow	Natural wasteland
10 :	8 :	7 :	5 :	3 :	1

The quantitative differences are probably due to the lower availability of space for carabids in the permanent cultures (see p. 216ff.).

Considerable qualitative differences also exist between the carabid populations of grassland and arable land, but not between the latter and those of clover and alfalfa fields. A large group of species that are frequent in meadows may also occur on arable land, but in a minor role. These are animals showing a strong preference for moisture, and choosing the meadows on account of their microclimate, e.g. *Trechus secalis*, *Pterostichus vernalis*, *P. nigrita*, *P. diligens*, *Nebria brevicollis* and others (Table 9). Of the carabids common in fields some are even more frequently encountered on meadows and may be considered to prefer the latter, e.g. *Dyschirius globosus* or *Pterostichus niger*. *Carabus granulatus* can justifiably be termed *the* meadow *Carabus* (see Chap. 6.A.3.). Other species of the genus play a minor role in meadows. A particularly interesting aspect is presented by the different distributional patterns exhibited by the two closely related species *Pterostichus cupreus* and *P. coerulescens*. Whereas *P. cupreus* was found in a large proportion of the arable areas investigated, and in far higher numbers than in meadows, *P. coerulescens*, on the other hand, was encountered in only one-third of the arable areas investigated but in 11 out of 12 of the meadows.

5. *The Influence of Soil Type on the Carabid Fauna of Agricultural Land*

It has been observed repeatedly that the differences, both qualitative and quantitative, between the carabid population of agricultural land with clay (heavy) soil and that with sandy (light) soil, are larger than the differences observed between the fauna of the various crops. This fact was first pointed out by Heydemann (1955, see Table 10). In western Germany the carabid fauna of fields with sandy and with clay soils bearing the same crops was compared (vegetable, potato and clover fields: Kirchner, 1960). "The carabid spectrum on different types of soil exhibits large qualitative and quantitative differences, scarcely any one species being equally distributed. Most species are more common on clay soil than on sandy soil, only a few species preferring the conditions offered by the latter". Kirchner, for example, caught 1353 carabids in a potato field with clay soil but, in the same period of time, only 235 individuals in a similar field with sandy soil. The species ratio was 17:11. Similar ratios were found on vegetable fields. "What are the reasons underlying the differences observed in fields with various types of soil? Clay and sand differ in the grain size of their particles, and particle size governs the water content which, in turn, is responsible for differences in plant cover. A root crop on a field with sandy soil never achieves anywhere like the complete cover found on a field with clay soil. Both water content and plant cover have a combined effect on the microclimate, which does, in fact, exhibit correspondingly large differences depending upon the soil type. Many carabid species obviously prefer fields with clay soil to those with a sandy soil... Clover fields even on sandy soils, however, are densely covered, so that the microclimatic differences between fields with different soil types are not so large as for root crop fields" (Kirchner, 1960). Kirchner found a smaller proportion of spring breeders, which on the whole prefer warmth and dryness, on clay soil than on sandy soil. The ratio of individuals of autumn to spring breeders on clay soil was about 6:1 and on sandy soils 2:1. Species that occur on both clay soil and sandy soil in the more continental type of climate in central Germany with its drier summer are confined to sandy soil in the Atlantic climate of Schleswig-Holstein, i.e. *Broscus cephalotes, Calathus erratus, Pterostichus punctulatus, P. coerulescens* and *Harpalus distinguendus.*

The following differences have been reported from Schleswig-Holstein between rye fields with heavy soil and those with light soil (Tischler, 1955): on heavy soil carabids accounted for 33% of the individuals among the animals of the soil surface, but only 13% on light soil. The corresponding species ratio was 15:13, and that for their contribution to the biomass of the epigaeic fauna 70:40. Investigations carried out by Scherney (1960a) in Bavaria also revealed a higher frequency of carabids on clay soil, and his quantitative indicators for clay soils agree on the whole with those given by Heydemann.

The higher frequency of carabids on clay soils is probably not only a result of a more favourable microclimate but is in part also due to the generally higher productivity of organic substances which, in turn, ensures a better food supply. An interesting observation in support of this idea is that the average weight of carabids from heavy soil is much higher than that of those from light soils (Heydemann, 1964).

The average weight of his trapped carabids was:

from light soil		from heavy soil	
in root crops	in winter cereals	in root crops	in winter cereals
42.5 mg	91.5 mg	127.8 mg	376.1 mg

6. *Concerning the Origin of the Carabid Fauna of Cultivated Land*

Since agricultural land is a habitat of recent anthropogenic origin the question arises as to the natural habitats from which its fauna, including the carabids, derives. Far from being a random collection of elements thrown together by chance, it has been shown to be a completely characteristic fauna.

The recognition that this fauna originated in Litoraea biotopes, that is to say in the periodically flooded coastal- and shore-zones of the sea and inland waters, and especially the drift line with its wrack layers, is mainly due to Tischler (for a comprehensive discussion see Tischler, 1958). The idea is at first surprising, but closer inspection reveals many similarities between littoral regions and cultivated fields. Both are subject to drastic changes in climatic factors, in the one case where silting-up leads to catastrophic disturbances, and in the other due to soil cultivation. The recurrent addition of self-decaying organic matter, on the one hand drift and seaweed and on the other manure, is a factor common to both. Of the 36 species of field carabids listed by Tischler for Schleswig-Holstein, 29 are familiar on the sea coast and 24 on river banks, and 35 were encountered in at least one of the two habitats. *Carabus auratus* was the only species not found in either habitat but Tischler regards it, too, as originally endemic to the littoral zone.

In West Germany Kirchner (1960) found that 90% of all carabid individuals on fields with clay soil and 87% on sandy soil were natives of beach drift. Carabids of the fields bear hardly any relation to the fauna of even closely neighbouring forests (see p. 153). In the more continental regions of Europe, however, the situation is somewhat different. On the basis of material gathered by Geiler (1956/57) from central Germany (Saxony) Kirchner (1960) calculated that only about 60% of the field carabids originate in beach drift, which is small when compared with Tischler's figures. In fact, in eastern central Europe a considerable proportion of the species occurring in fields are east European, and some, such as *Zabrus* and the *Brachinus* species or *Dolichus halensis* (see Table 9) can even be termed typical steppe elements.

II. The Carabid Fauna of Heaths and Sandy Areas

A comparison with the fauna of moors in northern Germany led to the recognition of carabid species that can at least be considered as characterizing species for sandy heaths (Mossakowski, 1970a). The following species were confined to heaths: *Calathus melanocephalus*$^+$, *C. erratus*, *C. fuscipes*$^+$, *Pterostichus lepidus*$^+$,

Harpalus aeneus⁺, *H. tardus*, *Amara famelica*, *A. equestris*, *Bembidion nigricorne*, *Notiophilus germinyi* and *Broscus cephalotes*⁺. Especially prominent on peaty areas and heaths were: *Cicindela campestris*, *Dyschirius globosus*⁺, *Carabus arcensis*, *Pterostichus coerulescens*⁺, *Amara lunicollis*⁺, *Olisthopus rotundatus*, *Harpalus latus*, *Bradycellus harpalinus* and *B. collaris*. A preference for heather-covered peat was shown by: *Bembidion lampros*⁺, *Trichocellus cognatus*, *Anisodactylus nemorivagus* and *Carabus convexus*. Of these species, those marked with + are also regularly encountered on arable land, some of them particularly or exclusively on fields with sandy soil in Atlantic western Europe. Many of them are thus not specifically confined to heaths nor are they characteristic species for this type of formation.

The species *Carabus nitens* is held by Mossakowski to be typical of *Calluna*, and was also found by Rabeler (1947) to be characteristic of the Calluneta of northern Germany.

Especially typical of dry, sandy grass areas, according to Mossakowski, are *Amara fulva* (with reservations), *A. infima*, *Cicindela hybrida*, *Harpalus rufus* and *H. neglectus*. Some of these species have also been caught elsewhere, particularly on sand. Schjøtz-Christensen (1957) found 28 carabid species on dry, sandy grass areas (Corynephoretum) in Denmark, including two Cicindelinae. The most important dominant species were *Amara infima*, *Harpalus neglectus* and *H. anxius*. In addition, *Cicindela hybrida*, *Carabus cancellatus*, *Harpalus tardus*, *H. smaragdinus*, *H. rubripes*, *Amara familiaris*, *Calathus erratus*, *C. melanocephalus*, *Pterostichus lepidus* and *Metabletus foveatus* were frequently encountered. In a further investigation in 1965 Schjøtz-Christensen also found particularly large numbers (usually more than 100 individuals of each) of *Cicindela silvatica*, *Carabus nemoralis*, *Leistus ferrugineus*, *Notiophilus germinyi*, *Broscus cephalotes*, *Harpalus aeneus*, *Bradycellus similis*, *B. collaris*, *Amara communis*, *A. aenea*, *A. apricaria*, *Pterostichus niger*, *Calathus fuscipes*, *Olisthopus rotundatus* and *Metabletus foveatus*.

Amara fulva was seldom met with in this investigation, although it was encountered by Lehmann (1965) on the banks of the Rhine near Cologne in West Germany and considered by him to be the characteristic species of the shifting sand zone.

The following "psammophilic" species were found in the yellow dune region in the Netherlands by van Heerdt and Mörzer-Bruijns (1960): *Amara spreta*, *A. quenseli*, *Harpalus neglectus*, *H. servus*, *Calathus mollis*, *C. melanocephalus*, *Dromius linearis* and *Metabletus foveatus*. In addition to these species, already familiar from similar habitats, three other species were common: *Trechus quadristriatus*, *Broscus cephalotes* and *Calathus erratus*.

Many of the species apparently confined to sandy areas are possibly attracted by the prevailing warmth and dryness, being thermo- or xerophilic. This is suggested by the observation that some of the species of sandy regions also show a preference for steppes and dry grassland, even on other substrates (see Chap. 2.C.V.). Schjøtz-Christensen (1957) found, for example, that *Harpalus tardus*, *anxius*, *smaragdinus* and *neglectus* preferred relatively high temperatures, particularly large numbers being found at 25–29° C, thus confirming experimental data obtained by Lindroth in 1949 using these species. The optimum for *Amara infima*, on the other hand, was shown to be in the low region between 15 and 20°C, although it is apparently a common inhabitant of sandy areas: a spectacular total of more than 10,000 individuals was caught in the Corynephoretum in Denmark (from a total of 35,000 carabids from three stations in five years, Schjøtz-Christensen, 1965).

It appears to be of greater significance that many sand-preferring species of carabids are xerophilic than that they are thermophilic. Sand has only a limited capacity for retaining water as compared with clay and dries out readily. Lindroth (1949) investigated experimentally the temperature and moisture requirements of 15 *Harpalus* species (see p. 212ff.). He then arranged the species according to their degree of thermophily on the one hand, and their degree of xerophily on the other. The four species mentioned above as being exceptionally frequent on sand were quite far down the list for thermophily although as regards xerophily they occupied the positions 3–6 among the 15 species.

Lindroth (1949) assumes that *Amara fulva*, noted for its striking preference for sand, is not found in western Norway on account of excessive humidity. For a whole series of species the preference for sandy ground increases from central to northern Europe *(Bembidion nigricorne, Bradycellus harpalinus, Demetrias monostigma, Dromius longiceps, D. melanocephalus, D. nigriventris)*. This also suggests that climatic factors play a role in "psammophily" and that the affinity for sand may partially at least be due to the fact that the thermophilic animals choose sand because of its high conductivity of heat.

In a classical study on the arthropods of the shifting sand regions on the coast of Finland Krogerus (1932) mentioned seven carabids as being stenotopic species for such habitats: *Cicindela maritima, Dyschirius obscurus, D. impunctipennis, Bembidion argenteolum, Amara fulva, A. silvicola* and *Dromius longiceps*. Krogerus had already recognized that the influence of sand on psammophilic species is partly direct and partly indirect. Grain size is of considerable importance: a grain diameter of about 0.2 mm marks an important limit between sands with a weaker and those with a stronger capillary effect. Fine sands have a greater capacity for retaining water and since the heat capacity "of water is greater than that of the air, warmth is held longer in fine-grained sand than in coarse-grained". Fine-grained sand therefore is a more suitable substrate for thermophilic organisms.

Many psammophilic carabids, such as *Harpalus* and *Dyschirius*, are digging species. Krogerus is of the opinion that from this point of view as well, fine sand provides a better substrate for the typical sand species. He found a positive correlation between increasing fineness of sand-grain size and increasing population density in *Dyschirius obscurus*. The question was approached experimentally by Lindroth (1949; Fig. 9). In seven species of *Harpalus* investigated, those species that have repeatedly been termed psammophilic, i.e. *H. tardus, anxius* and *smaragdinus*, exhibited a marked preference for fine sand if all other factors were kept constant. Only *H. neglectus* showed scarcely any preference. The diffuse distribution of *H. serripes* in these preference experiments agrees well with the observation that this species normally lives on coarse gravel.

Although psammophily, as far as can be judged from observations and experiments so far recorded, seems to depend largely upon a combination of the

Fig. 9. The distribution of seven *Harpalus* species on sands of different grain size in a substrate apparatus. Types of sand: *1:* 2–1 mm; *2:* 1–0.5 mm; *3:* 0.5–0.25 mm; *4:* 0.125 mm; *5:* 0.125–0.075 mm; *6:* <0.075 mm; *a: Harpalus serripes*; *b: H. neglectus*; *c: H. hirtipes*; *d: H. tardus*; *e: H. anxius*; *f: H. smaragdinus*; *g: H. rufitarsis.* From Lindroth, 1949

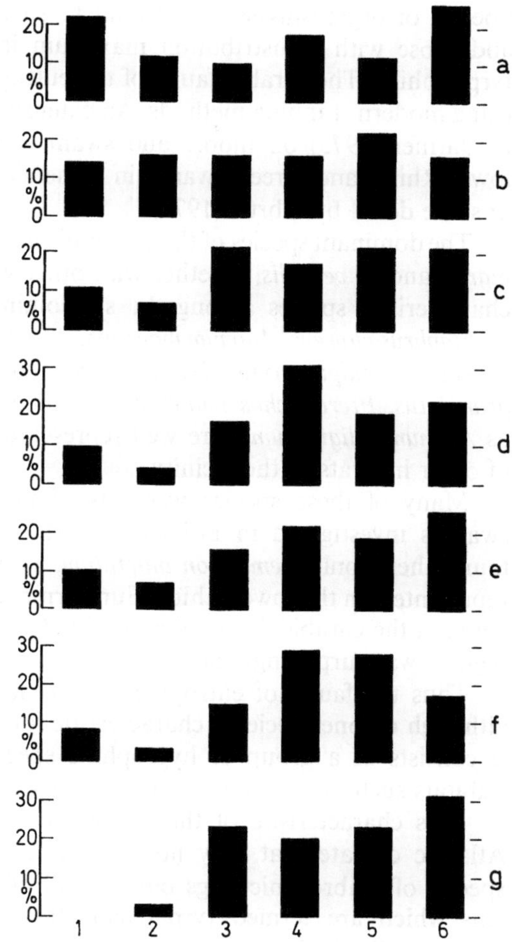

microclimatic requirements of the species concerned and the climatic conditions peculiar to sandy substrates, the nature of the substrate also exerts an immediate influence.

III. The Carabid Fauna of Moors

We shall now consider the differences between the fauna of the various types of moors, distinguishable as follows: (1) eutrophic, minerotrophic fens; (2) oligotrophic, minerotrophic fens which are connected with a mineral substrate poor in nutrients and the water borne by it. A specialized form of the latter type is presented by the heather moors (Ericetalia), which are frequently found in parts of western Europe influenced by the Atlantic climate. (3) Oligotrophic bogs, which receive neither nutrients nor water from the mineral substrate but are fed solely by rain (=ombrotrophic). They occur mainly in central, northern and eastern Europe and, in a specialized form, in the European mountains.

Species of organisms strictly bound to such moors are termed tyrphobiontic, and those with a distribution maximum in such habitats are referred to as tyrphophilic. The carabid fauna of the eutrophic fens has been little investigated using modern trapping methods. An extensive study, however, has been reported by Jarmer (1973) on moors and swamps bordering the old branches of the lower Rhine, and a reed swamp in Bohemia (Czechoslovakia) has been studied in some detail by Obrtel (1972).

The dominant species of the eutrophic fens on the lower Rhine are *Pterostichus nigrita* and *P. vernalis*, together with other widely distributed species. Especially characteristic species among the subdominants and recedents were found to be *Elaphrus cupreus*, *Agonum moestum*, *A. thoreyi*, *Bembidion assimile*, *Pterostichus gracilis*, *Acupalpus mixtus*, *Trechus discus* and *Oodes helopioides*. Although *Carabus granulatus*, *Pterostichus minor*, *niger*, *diligens*, *anthracinus* and *vulgaris* as well as *Agonum fuliginosum* were well represented they were also found in a variety of other habitats in the vicinity.

Many of these species were also found to be the dominant carabids of swamps investigated in Bohemia (Obrtel, 1972). Of all the dominant species found there, only *Bembidion inoptatum*, a southeast European species, was not represented on the lower Rhine. Furthermore, the degree of ecological agreement between the carabid fauna of eutrophic fens in these two widely differing faunal regions was surprisingly large.

Thus the fauna of eutrophic fens seems to have quite a typical make up, although no one species is characteristic for, or exclusive to, the habitat. Rather, it consists of a group of hygrophilic species that also occur in other moist habitats such as meadows, shores and meadow forests.

It is characteristic of the oligotrophic heather moors of regions with an Atlantic climate that they not only lack the tyrphobiontic and tyrphophilic species of ombrogenic bogs but also all of the typical species of the eutrophic fens which are termed tyrphoxenic (avoiding oligotrophic bogs): i.e. many *Elaphrus*, *Bembidion* and *Agonum* species, *Oodes helopioides* or *Chlaenius nigricornis* (Jarmer, 1973). Eurytopic species of wet habitats dominate: *Pterostichus minor*, *nigrita*, *diligens* and *niger* alone accounted for 92% of the individuals found in the sphagnum-covered moor regions. The same species have also been found in neighbouring *Molinia* associations and birch swamps in addition to most of the other species common in sphagnum regions, i.e. *Dyschirius globosus*, *Agonum fuliginosum*, *A. gracile*, *P. vernalis*. Only *Agonum thoreyi* and *A. muelleri* were confined to the sphagnum zone. *A. thoreyi* and *P. vernalis* are the only species among those mentioned which are seldom encountered on true oligotrophic bogs. The most characteristic species of the oligotrophic bog region studied was *Trichocellus cognatus*, which, although almost exclusive to oligotrophic bogs in central Europe, is much more eurytopic in northern Scandinavia and Great Britain, according to Lindroth (1945).

Although oligotrophic bogs became a subject of interest to zooecologists earlier than forests or cultivated land, almost all investigations until very recently were of a qualitative or semiquantitative nature (Mossakowski, 1970a). This is true even of the particularly comprehensive Finnish studies of which those of Renkonen (1938) and Krogerus (1960) warrant special consideration.

According to Krogerus, 11 of the 51 species collected on Finnish oligotrophic bogs can be considered as eucoenic species: i.e. *Dyschirius nigricornis, Elaphrus lapponicus, E. uliginosus, Notiophilus aquaticus, Bembidion humerale, Trechus rivularis, Pterostichus aterrimus, Agonum ericeti, A. livens, A. consimile* and *A. mannerheimi.* But by no means all of these species are either tyrphobiontic or tyrphophilic over their entire range of distribution. Lindroth's (1945) data, at least, indicate no such strict habitat binding in the case of *A. livens* and *P. aterrimus.*

Krogerus reported finding *Pterostichus nigrita, minor, strenuus* and particularly *P. diligens*, besides *Agonum viduum, fuliginosum* and *gracile* regularly accompanying the above-mentioned species which are, as stated, more or less confined to oligotrophic bogs. (The frequent occurrence of *P. diligens* on central European moors poor in nutrients was observed by Mossakowski, 1970a and Jarmer, 1973.)

In the course of his studies on oligotrophic bogs in north Germany Mossakowski (1970a) found the tyrphobiontic species *Agonum ericeti* as the sole representative of species typical of the habitat, and devoted a special investigation to its habitat affinity (Mossakowski, 1970b). According to a classical paper by Peus (1932) the tyrphobiontism of many moor species is a case of regional ecological limitation ("regionale Stenökie"). Many such species are eurytopic in the continental regions of Eurasia: only in Europe, particularly in the west, are they associated with oligotrophic bogs. The microclimate on bare peat areas, with its extreme diurnal and annual fluctuations, especially with respect to temperature, is similar to the macroclimate of the more continental parts of Eurasia.

Investigations were carried out in the German Mittelgebirge (Harz: Sonnenberger Moor) by Mossakowski (1970b) on mosaic-like complexes of oligotrophic bogs and eutrophic fens. He found *Agonum ericeti* to be fairly narrowly associated with oligotrophic bogs, whereas its counterpart on eutrophic fens was *A. fuliginosum.* Competition does not seem to play a role (this holds for *Pterostichus nigrita* as well, which is confined to eutrophic fens) since on other moors where all three species occur their numbers decline in parallel with one another.

A. ericeti prefers dryness and warmth (Krogerus, 1960), its temperature preference maximum lying between 25 and 30° C (mean 26.6° C). In a one-factor experiment *A. ericeti* chose warmth, dryness and an acid substrate. In combination experiments it chose dry/warm, acid/warm or dry/acid in preference to the opposite combinations. If the beetles could choose between dry/alkaline or wet/acid, the choices varied widely, as would be expected from the previous data. Microclimatic requirements of this nature fit in well with the views of Peus. In June 1957 daily amplitudes in temperature of up to 36.2° C were measured on the Sonnenberger moor, and a maximum of 43.5° C on a site with dry, dark peat (maxima of over 60° C have been recorded on similar sites in Bavaria). However, microclimatic requirements alone do not completely account for the affinity of certain species to oligotrophic bogs since *Agonum ericeti* is limited to this type of habitat over its entire range of distribution, including the Soviet Union and western Siberia. Krogerus found experimentally that the tyrphobiontic species *A. consimile* prefers moderately warm, very wet and fairly acid substrates, thus providing an example of a tyrphobiontic species by no means extreme in its climatic requirements. The possibility exists that the occurrence of this species is directly connected with an acid substrate (see Chap. 6.B.VI.). However, Mossakowski (1970b) justifiably demanded information as to the ecological requirements of the larvae.

IV. Carabids of the Litoraea Zones (Coastal and Shore Habitats)

The carabid fauna of river banks and lake shores has been the subject of several investigations, the more recent ones using traps.

Eutrophic fens and bare zones on the *banks of old stretches of the lower Rhine* have been investigated by Jarmer (1973). He reported finding a series of characteristic carabids strictly confined to this region: *Agonum viduum, A. viridicupreum, A. marginatum, Elaphrus riparius, E. uliginosus, Acupalpus dorsalis, Clivina collaris, Agonum gracile, Panageus cruxmajor*. Krogerus (1948) has already shown for some of these species that their preference for moist soil is the same in experiment and in nature. Some species that are dominant on the banks of the old Rhine are otherwise mainly associated with cultivated land: e.g. *Pterostichus cupreus, Harpalus rufipes, Carabus granulatus*. Apart from these dominant species, other species from cultivated land were frequently encountered: *Pterostichus niger, P. vulgaris, Carabus monilis, Bembidion lampros, Trechus quadristriatus, Amara communis, Nebria brevicollis* and others. This clearly illustrates the partial overlapping of the fauna of cultivated land and of the littoral region, already mentioned on p. 34. Exhaustive lists of the animals found in shore drift, including many carabids, can be found in a publication by Dürkop (1934).

A comparison of the shore carabid fauna found by Jarmer (1973) with the results of other authors shows that it is indeed typical (six publications on stagnant and flowing inland waters). Of his 63 shore species only three had not been found on shores by any other author.

On the other hand, certain species typical of the banks of the flowing Rhine, as for example *Amara fulva*, were completely absent from the banks of the old Rhine. *Amara fulva* itself is confined to the shifting sand regions of the former.

The shore fauna of the main Rhine has been thoroughly investigated in the vicinity of Cologne (Lehmann, 1965). Starting at the water's edge and working inland five zones could be distinguished: (1) a coarse gravel zone without vegetation; (2) a fine gravel zone with isolated grasses; (3) a thinly covered grassy zone; (4) a sandy zone almost devoid of plant cover (the ground surface dries out and the fine sand is kept in constant movement by the wind); (5) a meadow forest zone (in most places only remnants).

The numbers of species and individuals varied characteristically from zone to zone (Table 12). The number of species rose progressively from the water's edge whilst the dominance spectrum showed an increasing diversity of the common species. The number of individuals increased on the whole; the small rise in the gravel zone can be explained by the presence of animals that had strayed from other zones and wandered along the water's edge.

Table 12. Numbers of species and individuals of carabids in the shore zones of the Rhine. (According to Lehmann, 1964)

		Number of species	Dominant	Subdominant	Number of individuals	Spring breeders
Zone 1 and 2:	Coarse gravel and fine gravel	26	4	5	603	80%
Zone 3:	Grassy patches	32	3	12	434	93%
Zone 4:	Moving sand	40	5	10	543	75%
Zone 5:	Water-meadow forest	44	5	9	1496	17%

The dominant species of the shore zones, outside of the water meadow forests were: *Bembidion femoratum, B. tetracolum, Trechus quadristriatus, Amara similata, A. fulva* (in the sand zone only), *Pterostichus vulgaris* and *Clivina collaris.* Especially constant and frequent subdominants were: *Agonum marginatum, A. ruficorne,* and *Bembidion punctulatum.* The following completely different species dominated in the forest zone: *Nebria brevicollis, Pterostichus madidus, P. vulgaris, Abax ater* and *Calathus piceus.*

According to Lehmann carabids are unable to survive the winter in zones 1–4, which are affected by floods. This means that larval overwinterers (autumn breeders) cannot exist in these zones (save for *Amara fulva*, which already inhabits the upper regions of zone 4). This type of breeder is not capable of existing nearer to the river than in meadow forests, which are not regularly inundated by the winter high water. Since the source of the Rhine is high up in the mountains there is often a summer high water as well, so that even the broods of spring breeders cannot always develop on its banks. In any case, the adults have to fly away in autumn (all species inhabiting river banks are capable of flight), and Lehmann was able to demonstrate that the same species fly in to take up their old positions once more after the winter high water has subsided.

Lehmann compared his results with those of four other publications concerning the carabids of river banks, three of them from Scandinavia (Palm and Lindroth, 1937a, b; Palmén and Platonoff, 1943; Karvonen, 1945; Kless, 1961), none of them, however, employing traps. On the basis of this comparison and other ecological data it was possible to recognize the 18 species characteristic of the Rhine banks as typical of river banks in general.

In the above-mentioned investigations spring breeders were found to dominate over autumn breeders: the ratio of spring to autumn *species* was according to

Palm and Lindroth:	33:3
Palmén and Platonoff:	58:4
Karvonen:	14:3

Kless found no autumn breeders on river banks.

In northern Norway the situation on the river banks is obviously different (Andersen, 1968). Larval overwinterers account for 25% of the *species* living there, which may be connected with the absence of a winter high water. The "spring flood" in May–June is probably of shorter duration as a rule than the summer high water of the Rhine. In contrast to Lehmann's findings, Andersen is of the opinion that adults as well as other stages are able to survive floods of this nature. It should be taken into consideration that at high water the flood water in northern Norway is much colder than that in the Rhine (see Chap. 5.C.II.).

It is impossible to offer any general statements about the *drifting of carabids down-river as a means of dispersal*: Lehmann considers it to be of minor importance on the Rhine. Although he did in fact catch a living specimen of each of the two foothill species *Lionychus quadrillum* and *Bembidion fasciolatum* on the banks of the Rhine near Cologne, it should be borne in mind that they were part of a total catch numbering 3862 carabids. Furthermore, it is unlikely that montane species of carabids washed down to the flat Rhineland by accident would be capable of permanent existence there. Lindroth (1949) emphasized that cold-preferring species from the alpine regions and high-lying forests would hardly be carried down to the lowlands by rivers. On the other hand, he attributed an important role in distribution to rivers flowing through a mountain barrier between two lowland regions. *Bembidion virens* probably crossed the Scandinavian Fjeld chain from west to east in this way, and it is possible that rivers aided the distribution of *Agonum piceum* and *Pterostichus minor* from Sweden to Norway. It is most unlikely that the beetles swim down river but rather that clumps of earth falling into the water with adherent vegetation provide the small animals with a means of transport.

Carabids of the Salt Marshes: a Biological Land–Sea Boundary

Extensive investigations on this habitat have been carried out by Heydemann (1962a, b, 1963, 1964, 1967a). On the North Sea coast of Germany, salt marshes are the first formations on the freshly reclaimed polders. A comprehensive survey of the carabid fauna of newly reclaimed land in comparison with that of agricultural land of long-standing is given by Heydemann (1964).

Smaller carabid species are much more in evidence on the polders than on older land: the number of individuals is high but the biomass is small, a situation which is even more pronounced on exposed foreland. On the North Sea coast *Carabus* accounts for 0.09% of all individuals, for 14.2% inland in Schleswig-Holstein, and for 40% near Munich. In contrast, the genus *Bembidion* makes up 23.9% of the trappings on the North Sea coast, 8.8% near Leipzig and 0.92% near Munich.

These figures probably reflect to some extent the availability of food. Winter cereal fields on heavy soil in inland regions yielded an activity biomass of carabids 20.5 times higher than that of foreland reclaimed 1–2 years previously, although the activity density was only 44% of that of the coastal biotope. The average weight of an inland carabid (376.1 mg) was 46.4 times that of a coastal carabid (8.1 mg), according to typical trap results at the time of maximum activity. In Schleswig-Holstein the foreland had the largest density and smallest biomass of all habitats investigated. Five years after land reclamation, however, the biomass had increased to 15.6 times this value and thus achieved the level found inland.

V. Carabids of the Steppes, Dry Grasslands and Savannas

On the open primaeval steppes of the eastern Ukraine 23 carabid species were found by Ghilarov (1961) and 36 in the so-called ravine forests ("Schluchtwälder") although not a single species was common to both habitats. Characteristic species of the former habitat that were absent from both agricultural land further to the west in Europe and from the ravine forests of the steppes, were, for example, *Notiophilus laticollis, Amara scytha, Carabus errans, C. hungaricus, Olisthopus rotundatus, Pterostichus sericeus, Harpalus caspius* or *Zabrus spinipes*. Some of these species were also encountered in the montane steppes of the northwestern Caucasus above the forest zone (Ghilarov and Arnoldi, 1969). On the other hand, in the Caucasus other species, i.e. *Amara communis, bifrons, equestris, Harpalus rufipes, Calathus fuscipes, Dyschirius rufipes*, widespread on cultivated land and some even common in central Europe, are found on the primaeval steppes.

Thorough investigations have been made of the carabids and other arthropods of the *forest steppe zone* of southern Russia (Arnoldi and Ghilarov, 1963). A large proportion of the individuals (Table 13) are carabids.

Table 13. Carabid adults and larvae/m^2 in steppe forests and forest steppes in southern Russia. (From Arnoldi and Ghilarov, 1963)

	Adults	Larvae	Total
Oak forest	6.5	1.0	7.5
Groups of trees	3.7	1.1	4.8
Forest meadows	5.4	1.6	7.0
Mown steppe I	6.9	0.8	7.7
Mown steppe II	6.2	0.4	6.6
Unmown steppe I	8.4	1.7	10.1
Unmown steppe II	7.0	0.7	7.7

An interesting case is presented by the carabids of dry grasslands in the Atlantic regions of western Europe. For zoo-geographical reasons no large degree of similarity to the steppe fauna of southern Russia can be expected. Nevertheless, of 34 species showing a peak in distribution in the dry grasslands of the Bausenberg in the Eifel in western Germany (Becker, 1975) the following nine species also occurred consistently in the forest steppes of southern Russia, according to comprehensive tables published by Arnoldi and Ghilarov (1963): *Badister bipustulatus, Harpalus vernalis, Bradycellus collaris, Amara similata, Amara tibialis, Brachinus crepitans, Bembidion properans, Amara communis* and *A. aenea*. Of the species with a definite peak in distribution on cultivated fields on the same mountain, only *Pterostichus vulgaris* also occurred in oak *forests* of the forest steppe zone of southern Russia.

Little agreement exists between the fauna of west European dry grasslands and that of the above-mentioned primaeval steppes of southern Russia, from which the species found in the dry grasslands of the Eifel are missing. "None of the species confined to the steppes was encountered in the region under investigation. In the steppes as well as in their shrubby ravines Ghilarov (1961) found *Harpalus vernalis, Brachinus crepitans* and *B. explodens. Leistus ferrugineus, Harpalus distinguendus* and *H. rubripes* are encountered in the shrubby ravines only. The shrubby ravines of the steppes and the xerothermic mixed oak forests corresponding to the ravine forests are inhabited by *Badister bipustulatus, Harpalus azureus, Amara aulica* and *Pterostichus coerulescens. Panagaeus bipustulatus, Bembidion lampros, Harpalus quadripunctatus, Amara ovata, A. communis* and *Pterostichus cupreus* are limited to the ravine forests. All of the species mentioned are in the Eifel confined to open land, particularly the dry grasslands" (Becker, 1975). The conclusion to be drawn from this is that the climatic demands of these species remain constant over the whole area of distribution. In the continental climate of southern Russia they are confined to warm dry forests since the open steppes are too hot and dry. In western Europe they probably achieved dispersal as inhabitants of warm, dry, mixed oak forests under the optimum climatic conditions of the post-glacial period, and only as the climate gradually became cooler were they forced onto the dry grasslands.

Modern quantitative investigations of the carabid fauna of the *African savanna* using the quadrat method have been reported from the Ivory Coast (Lecordier and Pollet, 1971). An adjacent gallery forest and the boundary line between the two was included in the investigations. The number of species, number of individuals and biomass rose from the savanna to the edge of the forest but sank rapidly inside the forest (Table 14). Each biotope had its own characteristic carabid fauna, the following species attaining a dominance of 6% or more in the individual habitats:

Table 14. Annual means for number of species, biomass and number of individuals of carabids in a gallery forest and the adjoining savanna on the Ivory Coast. (According to Lecordier and Pollet, 1971)

	In the savanna		Forest edge	Forest interior
	40 m from the forest	20 m from the forest		
Number of species	47	46	68	36
Biomass: g/100 m^2	2.51	2.51	4.08	1.16
Number of individuals	76	86	119	33

Open savanna:	*Abacetus tschitscherini, A. iridescens, Neosiopelus nimbanus, Abacetus ambiguus, Dichaetochilus rudebecki.*
Savanna near forest:	*Abacetus iridescens, A. tschitscherini, Neosiopelus fletifer, N. nimbanus, Abacetus ambiguus.*
Forest edge:	*Styphlomerus gebieni, Hiletus versutus, Stenocallida ruficollis, Laparhetes tibialis.*
Forest:	*Abacetus flavipes, Hiletus versutus, Styphlomerus gebieni, Abacetus amaroides.*

Only 31.6% of the species were common to both forest and forest edge whilst the open savanna and forest had almost no species in common. The forest edge was in this case the best populated habitat. It is noteworthy that the carabid population of the savanna was lower than that of the south Russian forest steppe by a factor of about 10 (see Table 13). The maximum biomass of 0.04 g/m^2 was far below the 1.5 g found in the grassy regions of northern Germany even under the most unfavourable of conditions (Table 3, p. 18). The number of individuals recorded here on 100 m^2 could be found in favourable situations in temperate latitudes on 1 m^2 or at least on 10 m^2. The remarks concerning the carabids of tropical cultivated land on p. 31 and the general comments on p. 100 and Chapter 9.B. should be referred to in connection with this relatively low number of carabid individuals in the tropics.

D. General Remarks Concerning the Distribution and Structure of Carabid Communities in Different Habitats

The introduction of pitfall traps has done much to clarify our quantitative and qualitative picture of the distribution of carabids in different habitats and plant communities.

Many cases are known in which the affinity of a species for a particular type of habitat is extremely narrow. For example, deciduous forests belonging to different plant communities may vary considerably with respect to their carabid populations. If a species is strictly bound to one particular plant community its numbers change even over small distances from a very high frequency to zero with alterations in vegetation.

In a well-defined faunal region such as central Europe it is possible to predict, with some degree of certainty at least, which species of carabids can be expected to occur in a particular plant community. Conversely, from a quantitative census of carabids conclusions can be drawn as to the nature of the habitat from which they originated. Not that this should be taken to mean that each carabid species is only capable of existing in one habitat. The water-meadow forests provide a good example of this, since they are inhabited by certain carabid species that distinguish them from other types of forest, although the same species can be found in other, quite different, types of non-woodland habitat.

Carabids are thus often narrowly adapted to particular habitats. Some of the critical factors involved, such as microclimate, have been touched upon briefly in various places, but one of the main objects of this book is a thorough consideration of factors of this nature. It will become apparent that an autecological analysis of the habitat selection of individual species is indispensible for elucidating the reasons underlying the limitation of a species to certain habitats.

Attempts have recently been made to describe the diversity of populations of organisms in different habitats with the aid of mathematical indices (e.g. with the Shannon-Wiener formula). The value of these methods for characterisation, and the possibilities of interpreting such index values are subjects of controversy (Hurlbert, 1971). Some examples will be given further on (see. Chap. 4.B.VI.).

The method of calculating diversity represents a further step in elucidating the pattern of dominance shown by groups of organisms within an ecosystem. Results obtained from traps are well suited for establishing a scale of dominance for carabid species in various environments with different living conditions. If the catches are arranged in order of percent frequency (=dominance) of the individuals of the species, differences between habitats in which certain factors exert an extreme or a less extreme effect are shown up clearly.

Thienemann, as early as 1920, formulated his basic principles concerning biocenoses as follows: the more variable the conditions in a particular environment the greater the number of species occurring at this site. The more extreme the conditions become, the more impoverished is the biocenosis with respect to number of species, but the greater the number of individuals occurring within each species.

An example of this is given by an analysis of carabids trapped in a thicket and a cultivated field (a large, uniform vegetable field, see Fig. 10). There are fewer dominant and subdominant species in the field, with few species definitely playing a leading role. In the thicket, however, which offers more variation in living conditions, the dominance gradient is less steep, with more species in the higher categories of dominance.

The same situation can be observed if comparisons are made between different extreme habitats within one and the same type of vegetation, e.g. deciduous forest. In two forests investigated simultaneously in West Germany, using identical methods, a much steeper gradient of dominance was found in the Fago-Quercetum on acid soil with a more extreme microclimate, than in the Querco-Carpinetum on neutral soil with a more equable microclimate (Table 15). The four most frequent species in the Fago-Quercetum accounted for more than 90% of all individuals whereas this total was only reached by the eight most frequent species in the Querco-Carpinetum. The total number of species in Querco-Carpinetum is nearly double that in Fago-Quercetum.

Similarly, in the sphagnum zone of a dystrophic body of water the dominance values showed a much steeper rise than in the marginal zones of eutrophic waters (Fig. 11). The extreme character of the margins of the dystrophic waters consisted mainly in much more drastic daily fluctuations in microclimate (Jarmer, 1973).

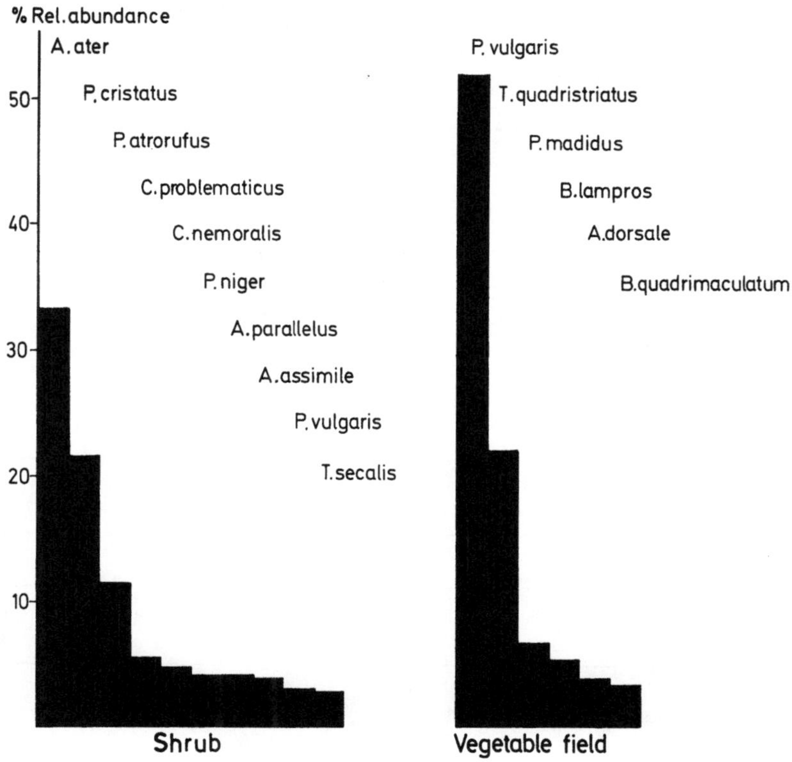

Fig. 10. Dominance structure of the carabid fauna in bushes *(left)* and a vegetable field *(right)* in western Germany

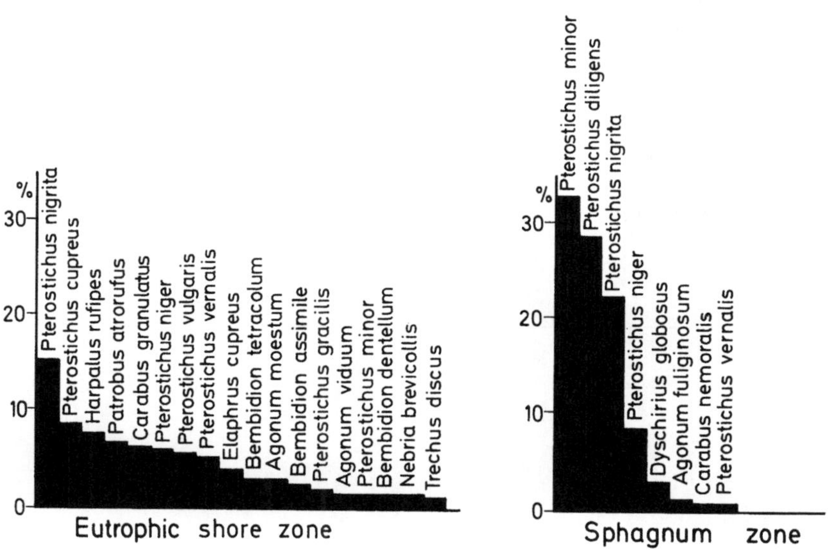

Fig. 11. Dominance structure of the carabid fauna in the shore zones surrounding a eutrophic body of water and in the sphagnum zone of a dystrophic body of water in western Germany. From Jarmer, 1973

Table 15. Carabids from two different forest societies in West Germany, arranged in order of abundance and dominance. (According to data from Thiele and Kolbe, 1962)

Fago-Quercetum		
	Abundance	Dominance
Abax ater	360	56.5%
Carabus problematicus	130	20.4%
Pterostichus cristatus	60	9.4%
Pterostichus oblongopunctatus	40	6.3%
Trichotichnus laevicollis	18	2.8%
Cychrus attenuatus	12	1.9%
Trechus quadristriatus	7	1.1%
Carabus nemoralis	6	1.0%
Agonum assimile	4	0.6%
	637 9 species	100.0%

Querco-Carpinetum		
	Abundance	Dominance
Abax ater	488	39.8%
Pterostichus cristatus	328	26.7%
Molops piceus	76	6.2%
Trichotichnus laevicollis	59	4.8%
Molops elatus	41	3.3%
Abax ovalis	40	3.3%
Nebria brevicollis	38	3.1%
Carabus problematicus	34	2.8%
Pterostichus oblongopunctatus	27	2.2%
Pterostichus niger	22	1.8%
Carabus coriaceus	20	1.6%
Abax parallelus	19	1.5%
Pterostichus madidus	14	1.1%
Pterostichus strenuus	12	1.0%
Pterostichus vulgaris	4	0.3%
Trechus quadristriatus	3	0.3%
Carabus nemoralis	2	0.2%
	1227 17 species	100.0%

Chapter 3

The Connections Between Carabids and Biotic Factors in the Ecosystem

A consideration of the ways in which carabids are linked with biotic factors in their ecosystem involves three questions:

1. To what extent do biotic factors govern the distinctive *pattern of distribution* of carabids in different habitats? This requires, on the one hand, an investigation of their distribution in macrohabitats corresponding to particular ecosystems and, on the other hand, their distribution among microhabitats within the ecosystem itself, i. e. characteristic manner of aggregation and dispersal.

Relevant factors include interspecific competition, parasites and predators, and, additionally, the social behaviour of the carabids.

2. How far do biotic factors determine the *population density* of carabid species within an ecosystem?

A central question in the field of population ecology is that relating to the factors regulating the population density of an animal species. In contrast to most of the abiotic factors, to be dealt with later, the effect of biotic factors such as competitors, predators and parasites upon a population *depends upon density:* not only do they influence the population density of, for example, their prey and host populations, but this in turn exerts a reciprocal effect in the form of a regulatory mechanism on the factors themselves.

Besides the factors mentioned already, an additional factor, namely intraspecific competition between the individuals of one species, plays a part in the regulation of population density.

Since the same factors are involved in both distribution and population density of a species these individual factors will be dealt with separately. The defence mechanisms employed by carabids in dealing with their enemies, parasites and other unfavourable environmental factors will also be discussed.

3. Carabids themselves represent biotic factors within their ecosystems, not only in the kind of feed-back relationship already mentioned which links them with their competitors, enemies and parasites. We also have to take into consideration the significant influence which carabids as predatory organisms exert upon their prey within the ecosystem.

Judging from their world-wide distribution and the large numbers in which carabids occur, it seems likely that they play an important part in the nutritional chains of the ecosystem. We feel justified, therefore, in devoting special attention to these questions.

For the above reasons, carabids as entomophages have attracted interest as potential tools in biological pest control. Whether, in fact, they would prove to be of value in this field remains open to question. Inseparably linked with this is the possible influence exerted on carabid populations by man himself as a factor in the ecosystem.

A. Inter- and Intraspecific Competition

I. Interspecific Competition

In an extremely large number of animal groups it is a well-recognized fact that closely related, sympatric species exhibit different peaks in ecological distribution. This phenomenon is particularly common among species of one and the same genus, and leads to a more or less clear-cut "intrageneric isolation". It was described by Monard in 1920 and termed "Monard's principle" by Illies, although it had already been established by Steere as early as 1894 (cf. Tretzel, 1955b). In his investigations on 55 animal and 27 plant societies in Great Britain Elton (1946) found 86% of the genera in the animal societies and 84% of those constituting the plant societies to be represented by only one species. On an average, 1.38 animal and 1.22 plant species per genus are present in any one society. This in no way reflects the average number of species belonging to one genus, which in fact amounts to 4.23 for 11 large groups of insects occurring in Great Britain. At a rough statistical estimate, the percentage of monotypic genera is only 50. A certain element of doubt is introduced by the inconsistency encountered in the allocation of a species to a particular genus and the strong tendency to subdivide genera.

From Monard's principle it has often been deduced that the existence of competition between species of a genus is a hindrance to their coexistence in one and the same habitat. Only this aspect will be dealt with in the present chapter. Further, it has been taken to follow from "intrageneric isolation" that competition, resulting in "niche segregation", has contributed to the phylogenetic differentiation of species. The part played by competition in the speciation of carabids will be considered further in the chapter on species diversity in the carabids (see Chap. 10.A.I.).

As early as 1949 Lindroth observed a permanent state of coexistence between species of the genera *Harpalus*, *Ophonus*, *Calathus*, *Dromius* and *Amara*, even within one and the same microhabitat. He found large numbers of different species of *Bembidion* running about together on river banks in North America (written communication), some of them taxonomically quite closely related.

In 1947 Williams had already noted that the number of genera with 1, 2,... species forms a logarithmic series. In samples poor in species, therefore, a considerable reduction in the number of species per genus in comparison to the total fauna is to be expected from purely mathematical considerations alone. Lindroth's (1949) investigations on the Scandinavian carabid fauna confirmed this in all respects.

Intrageneric isolation is nevertheless known to occur in carabids and was confirmed quantitatively by Thiele (1964a, 1964b) for a series of species pairs using catches made in Barber traps. Species of the same genus seem to be able to exclude one another nearly or entirely from a habitat complex, as for example *Agonum assimile* and *dorsale* or *Pterostichus cristatus* and *vulgaris* (Figs. 12, 13). The same was found by Paarmann (1966) for the species pair *Pterostichus angustatus* and *P. oblongopunctatus*. It is also possible for two species of the

Fig. 12. Ecological vicariation of two species of the genus *Agonum*. *Above:* Arrangement of the traps in a bushy strip *(6)* of the oak-hornbeam type *(left)*, gradually changing into oak-birch type *(right)*. Two rows of traps along the edges, one in the middle. *1* and *2:* narrow field hedges; *1:* oak-hornbeam hedge; *2:* oak-birch hedge. Each with two rows of traps along the edges, traps *1 h, i, k* displaced towards the middle. Centre and below: results of a one-year catch of the field species *A. dorsale* and the forest species *A. assimile*. From Thiele, 1964a

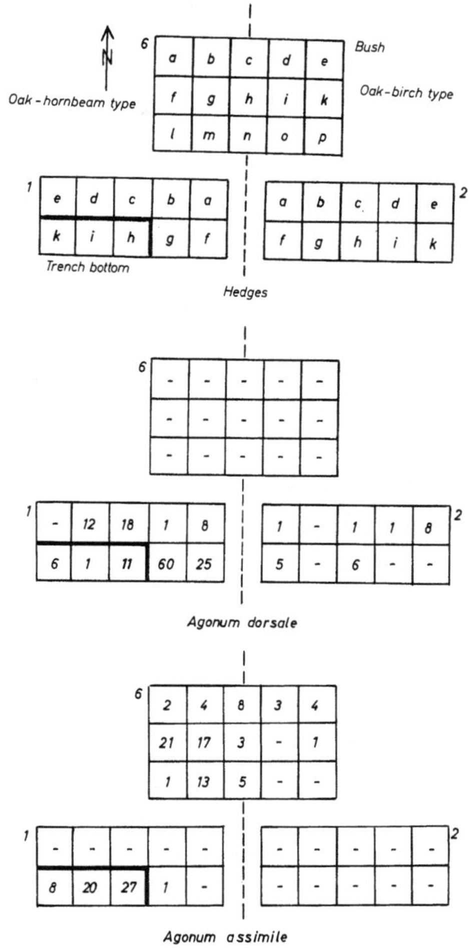

same genus to coexist in a habitat, whereas in another, immediately adjacent habitat, one of the species is present in large numbers but the other is rare or completely absent (*Abax ater*, *A. ovalis*, Thiele, 1964a). Observations of this kind strongly suggest that a state of competition exists between the members of the pairs of species involved and that one of the pair is better suited to the conditions prevailing in a particular habitat, so that the other, less well-adapted of the two, is supplanted. Competition has been defined by Schwerdtfeger (1968) as the common demand for, or common utilization of, something that is of limited availability, or, more concisely, "Competition is the common demand for a requisite of limited availability".

The assumption that closely related species have similar ecological niches and that for this reason species of one and the same genus compete particularly strongly with each other is often made without a closer consideration of the facts. The ecological separation of two species of a genus is then rashly stated to be due to competition. Such an assumption is no more than a working hypothesis as long as it remains unsupported by experimental evidence.

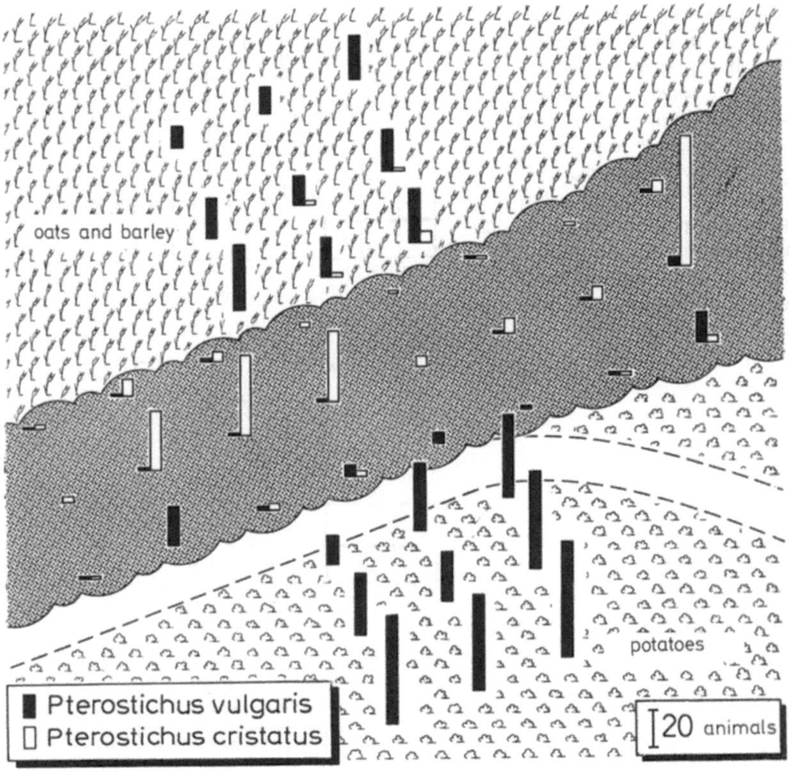

Fig. 13. Occurrence of a field-dwelling carabid *(Pterostichus vulgaris)* and a forest-dwelling species *(P. cristatus)* in a hedge and the neighbouring fields. Each double column represents the catch from one trap. Period of catch 4. July–8. August 1957. From Thiele, 1964a

The supposition of interspecific competition is often reached on the basis of quantitative investigations on distribution. That this is not permissible was quite unmistakeably pointed out by Tretzel (1955b): "A state of competition existing between two species in their natural biotope ... can ... not be proved unequivocally. If ... in supposedly identical biotopes, large differences in abundance become noticeable and the relative frequency of the species changes reciprocally, it still remains open to question whether in fact it is not abiotic factors that are responsible... Convincing evidence that two natural biotopes are identical seems to be unattainable".

The difficulties presented by competition experiments, especially those involving predators with a large radius of activity like the carabids, have been discussed by Thiele (1964a). In order to render such experiments technically feasible at all, a population density far above that prevailing in the field has to be taken into account, a fact which reduces the value of the results. Conclusions can best be drawn in cases where a separation of two species is observed in the field but where no sign of a competitive effect, or at most only a slight one, is observed in laboratory experiments even at excessively high population densities.

Experiments of this type provide the evidence necessary to refute the idea that ecological separation (intrageneric isolation) is due to competition.

Next, some experiments will be discussed in which adults of two, in most cases, very closely related species were kept together over lengthier periods of time. Any changes in population density were recorded. As controls, members of the individual species were kept alone under identical conditions and at the same population density.

Experiment 1: *Pterostichus vulgaris* and *P. cristatus* (Fig. 14). These two species are ecologically vicarious. In the hills of the Bergisches Land, east of Cologne, West Germany, *Pterostichus vulgaris* is almost exclusively a field

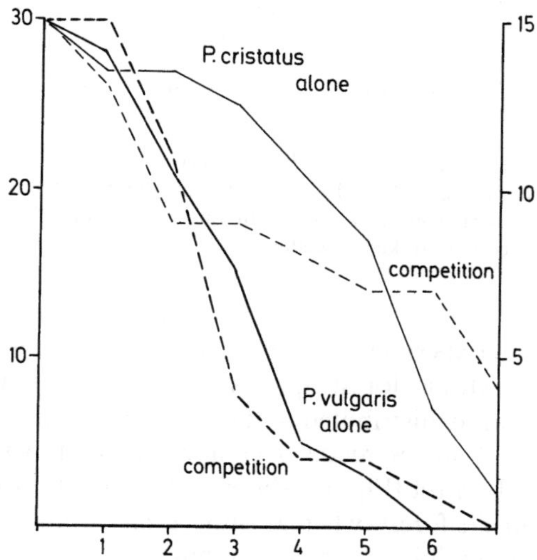

Fig. 14. Death rate of *Pterostichus cristatus (thin continuous line)* and *P. vulgaris (thick continuous line)*, when 30 individuals of only these species were used in each experiment *(left ordinate)*. *Broken lines* show the death rate in a competition experiment in which 15 individuals of each species were kept together *(right ordinate)*. *Abscissa:* months after commencement of experiment

species, whereas *P. cristatus* seldom occurs elsewhere than in forests and scrub. This was clearly illustrated by catches from a scrub strip and the neighbouring cultivated fields (Fig. 13). The population density curves for *P. vulgaris* in competition experiments and control take an almost identical course. *P. cristatus*, on the other hand, decreases more rapidly in the competition experiment than in the control, but only up to the point at which *P. vulgaris* has almost died off as a result of its shorter life span. This experiment reveals a competitive effect in which *P. cristatus* is the loser.

Experiment 2: *Pterostichus vulgaris* (field species) and *Abax ater* (forest species) (Fig. 15) belong to two different genera, both widespread in central Europe, but are of equal size.

The numbers of *Abax ater* sink more rapidly in competition experiments than in the controls, until *P. vulgaris* begins to die off naturally as a result of its shorter life span. The curves for *P. vulgaris* are again almost identical in experiment and control. In this experiment *A. ater* is the losing partner in competition.

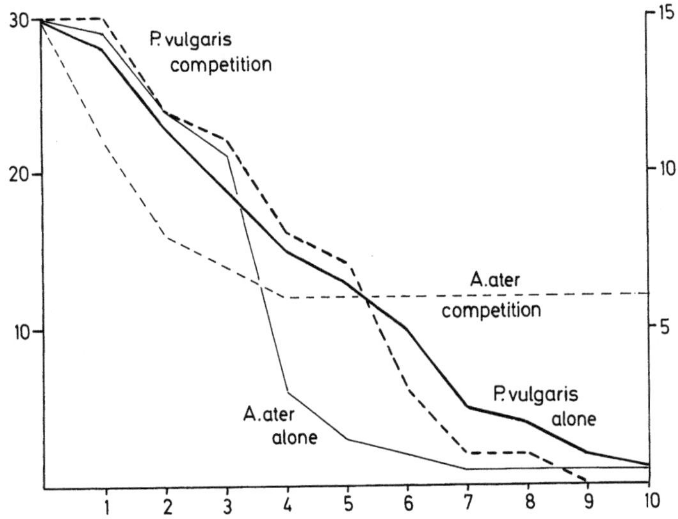

Fig. 15. Death rate of *Pterostichus vulgaris (thick continuous line)* and *Abax ater (thin continuous line)*, when 30 individuals of only these species were used in each experiment *(left ordinate)*. *Broken lines* show the death rate in a competition experiment in which 15 individuals of each species were kept together *(right ordinate)*. *Abscissa:* months after commencement of experiment

Experiment 3: *Abax ater* and *A. ovalis* (Fig. 16a–c). *Abax ater* is a highly eurytopic forest species but is also found in clearings, whereas *A. ovalis* is mainly distributed in the moist tall forests of the hills.

Catches made over a whole year with Barber traps on the margin of a tall forest (Fig. 17) show that the distribution of *A. ater* and *A. ovalis* is similar in the forest, whereas only *A. ater* occurs in the immediately adjoining clearing. A possible interpretation might be that *A. ovalis* can only withstand competition

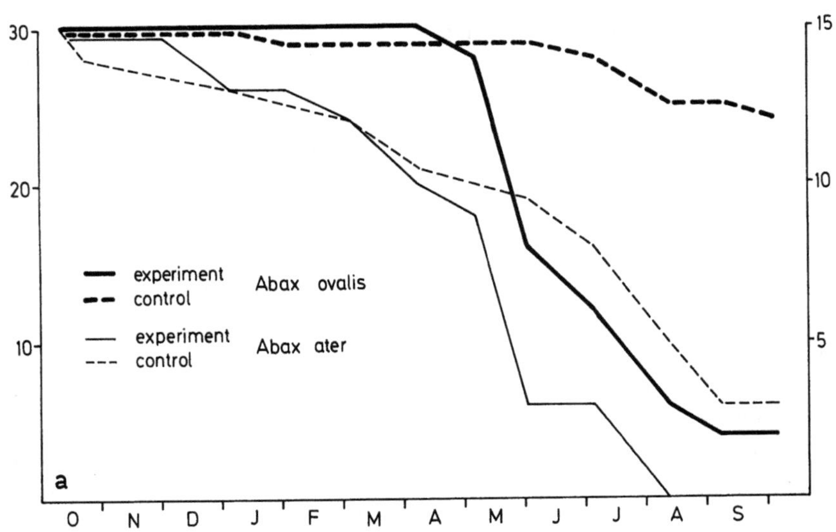

Fig. 16a.

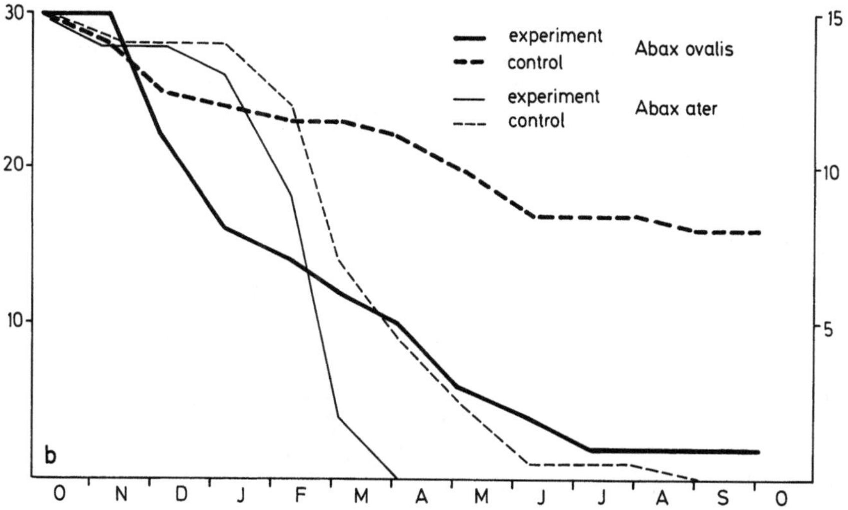

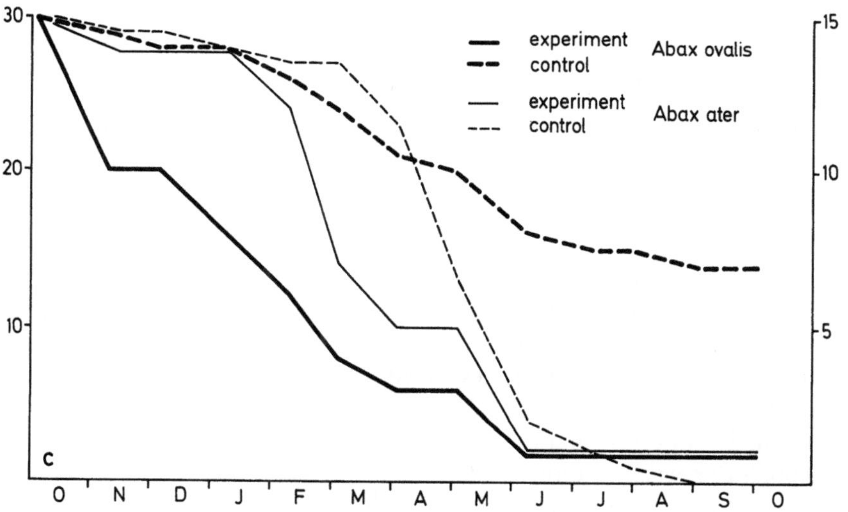

Fig. 16a–c. Three repetitions of a competition experiment with *Abax ater* and *A. ovalis.* In the controls only 30 individuals of one species were used alone *(left ordinate)*, whereas in the experiments 15 individuals of each species were placed together *(right ordinate)*. Course of the mortality curves over the months of the year *(abscissa)*. From Thiele, 1964a

from the larger *A. ater* under optimum habitat conditions and is for this reason ousted from the clearing.

In three similar experiments a more rapid decrease in numbers of *A. ovalis* took place in the face of competition as compared with the controls consisting of pure populations of *A. ovalis*. With *A. ater*, on the other hand, the curves for competition and control experiments are very similar in shape, the speed of depletion being only slightly higher in competition experiments than in controls.

In experiments 3B and 3C the available substrate volume was smaller than in 3A and a uniform increase in speed of depletion was recorded in both these experiments. This points to the involvement of inter- as well as intraspecific competition. *A. ovalis* proved to be the weaker competitor in this experiment.

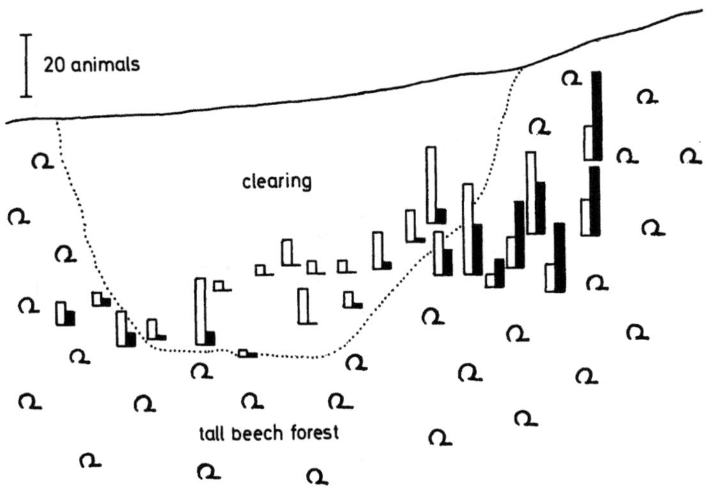

Fig. 17. Occurrence of *Abax ater (white columns)* and *A. ovalis (black columns)* in a tall beech forest and in a forest clearing. *Each double column* represents the catches from one trap. Period of catch: May–September 1959. From Thiele, 1974

For all three series of experiments, therefore, it appears that competition can be demonstrated under the experimental conditions prevailing. The nature of the competition effect could not, however, be elucidated. One possibility might be that the stronger species effects some form of alteration in the material constituting the substrate, in a way unfavourable to the weaker species. It is more likely, however, that the stronger prevents the weaker species from feeding, although this was not directly observed.

The application of experimental results to the situation in the field presents a number of difficulties. In all of the experiments described the substrate surface amounted to 1/20 m^2 and each experiment involved 30 beetles, corresponding to a population density of 600/m^2. Using the quadrat method (Thiele, 1956) in a Fagetum (mountain beech forest), however, a natural population density of 0.07 *A. ater* and 0.43 *A. ovalis*/m^2 was found, or 0.5 *Abax* per m^2. Thus the experiments described above involved a population density 1000 times that existing under natural conditions. A density of 1.5–5 individuals/m^2 on fields was reported by Kirchner (1960) for *Pterostichus vulgaris*, using the Lincoln-index method. Thus even in the other experiments the population density was 100 times higher than that normally prevailing in the field.

Despite the higher population density the effect of competition only shows up clearly after several months. The partner is by no means eliminated in the experiments with *P. vulgaris*. Some further experiments will now be described to provide examples in which no competitive effect was observed despite large increases in population density.

Experiment 4: *Pterostichus cristatus* and *P. nigrita.* A year-round faunal analysis of two forest habitats (small woods) in the hills of the Bergisches Land near Wuppertal (West Germany, habitat 1) and in the adjoining lowlands near Cologne (habitat 2) revealed similar spectra for the carabid fauna of the two habitats (Fig. 18). Of 15 species described, 12 are common to both habitats. The three most common species are

Habitat 1	Habitat 2
Abax ater	*Abax ater*
Pterostichus cristatus	*Pterostichus nigrita*
Patrobus atrorufus	*Patrobus atrorufus*

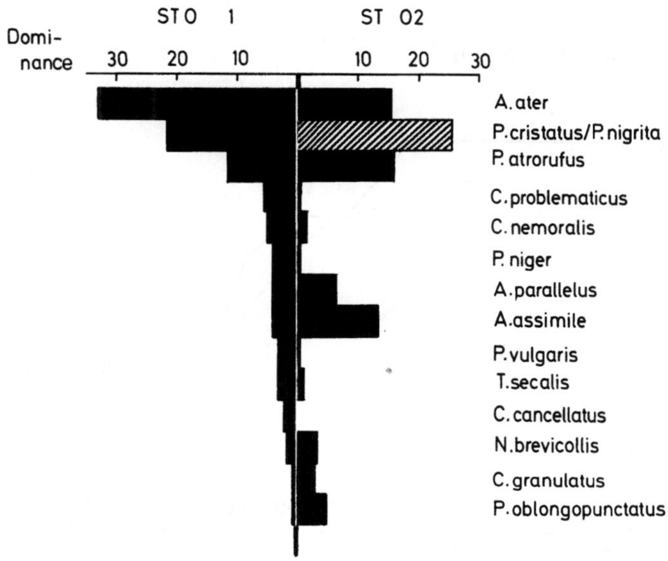

Fig. 18. Comparison of the dominance structure of the carabid societies in two oak-hornbeam habitats in the West German Schiefergebirge near Wuppertal *(ST O 1)* and in flat country near Cologne *(ST O 2)*. The carabid fauna of both forests is very similar but instead of *Pterostichus cristatus* which occurs in the hills *(second column, black, left)*, *P. nigrita* is found in the lowland *(second column, hatched, right)*

In an otherwise very similar society it appears that in habitat 2, *P. nigrita* replaces *P. cristatus*, a montane species not found in the plain (habitat 2). *P. nigrita* is, nevertheless, to be found in all altitudinal belts. The question is whether it seems plausible that *P. nigrita* can be lacking in habitat 1 on account of competition due to *P. cristatus*, although well able to fill its niche in habitat 2.

4a. Competition Experiments on *Pterostichus cristatus* and *P. nigrita* (Fig. 19). *P. cristatus* is relatively short-lived. It is an autumn breeder, the adults of which mostly die off following the reproductive period. *P. nigrita*, in contrast, is a

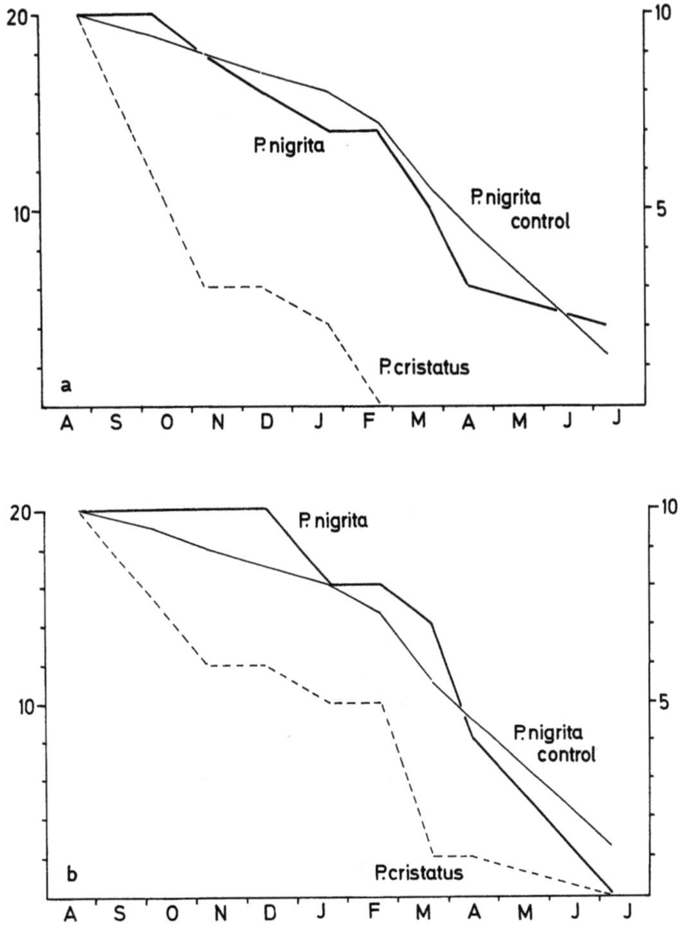

Fig. 19a and b. Two competition experiments with *Pterostichus nigrita* and *P. cristatus*. Mortality curve of *P. nigrita (thin continuous line)* in a control experiment involving 20 individuals of this species alone *(left ordinate)*. The *thick continuous line* is the population curve for 10 *P. nigrita* together with 10 *P. cristatus (broken line, right ordinate)*. *Abscissa:* 12 months of the year from beginning of experiment

spring breeder, and its adults overwinter and live longer. But again, a high population density of *P. cristatus* seems to have no influence on the *P. nigrita* population.

4b. Competition Experiments with *P. nigrita* and *Abax parallelus* (Fig. 20). These experiments were performed because the species are known to live together and exhibit the same type of annual rhythm. *A. parallelus* is larger and is a potential competitor of *P. nigrita.*

In the experiments *A. parallelus* survived *P. nigrita.* Their population curves are the same both in experiment and controls and, here again, there is no evidence of competition between the two species.

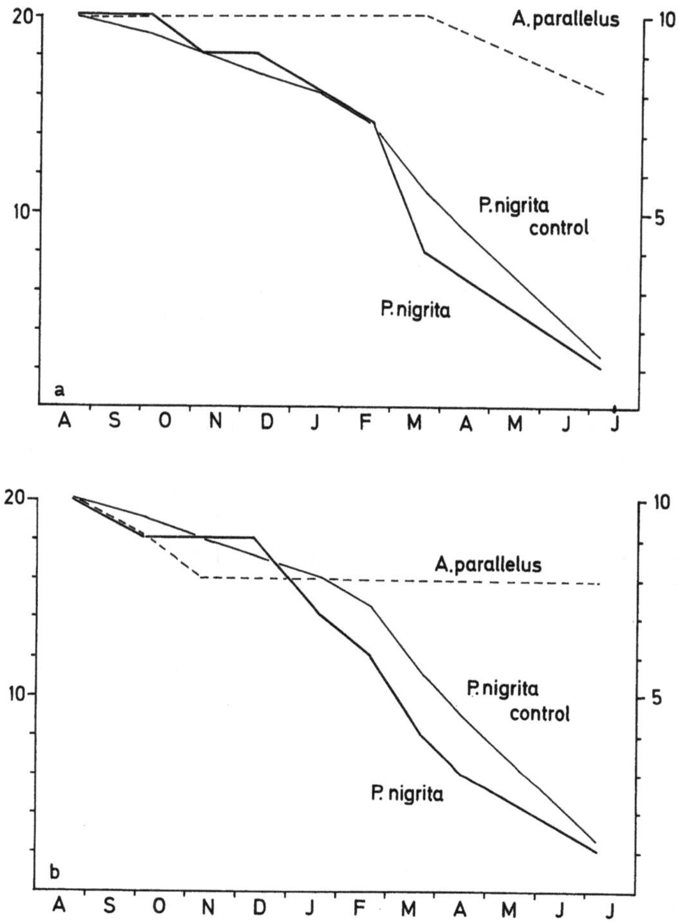

Fig. 20a and b. Two competition experiments with *Pterostichus nigrita* and *Abax parallelus.* Mortality curve of *P. nigrita (thin continuous line)* in a control experiment involving 20 individuals of this species alone *(left ordinate). Thick continuous line:* population curve for 10 *P. nigrita* together with 10 *Abax parallelus (broken line, right ordinate). Abscissa:* 12 months of the year from beginning of experiment

In summing up it can be said that the experimental population curves for *P. nigrita* are always the same, irrespective of whether it lives alone or with other species of carabids (Fig. 21). The experiments demonstrate, therefore, that *Pterostichus nigrita* is in no way affected by the presence of a second, related species.

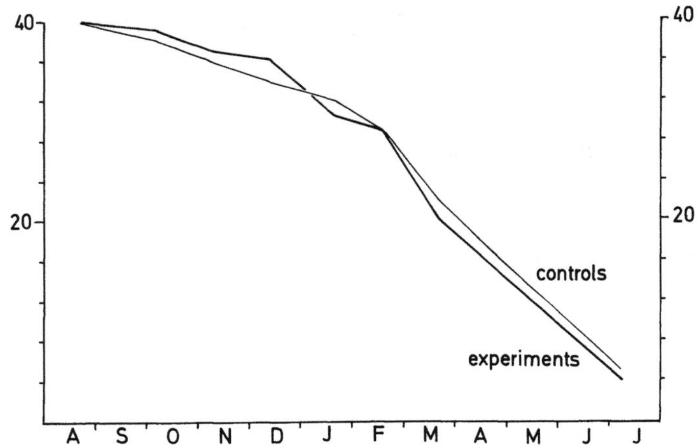

Fig. 21. Summary of the results of the competition experiments with *Pterostichus nigrita.* Average curves of the mortality rates in the experiments in which *P. nigrita* was kept alone (*thin curve:* 40 animals in all, two experiments) and in which they were kept with another species (*thick curve:* 4 experiments with in all 40 animals). *Abscissa:* 12 months of the year from beginning of experiment. The two curves are almost identical

Experiment 5: *Pterostichus oblongopunctatus* and *P. angustatus* (according to Paarmann, 1966; Fig. 22). The two species are ecologically vicarious. *P. oblongopunctatus* is a eurytopic forest species whereas *P. angustatus* lives on the margins of forests in warm, dry clearings, on bare felled patches and in burned areas. The two species thus occur as immediate ecological neighbours but they are mutually exclusive. Both belong to the subgenus *Bothriopterus* Chaud.

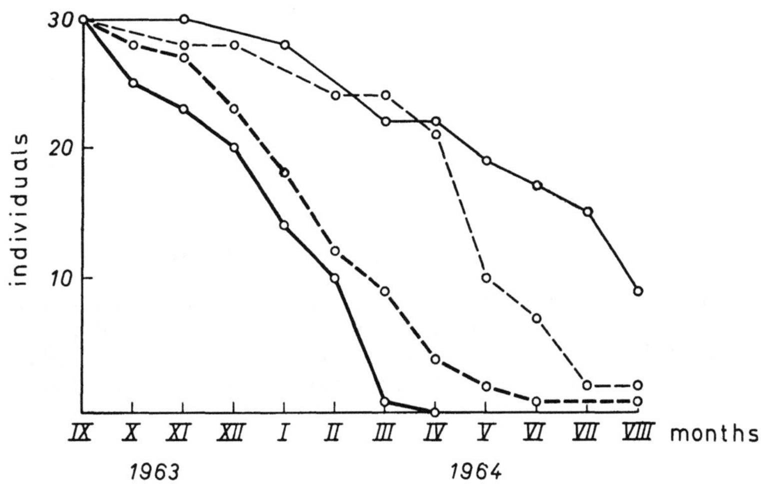

Fig. 22. Mortality rates of *Pterostichus angustatus (continuous thin line)* and *P. oblongopunctatus (continuous thick line)*, in each case 30 specimens of one of the species alone. *Broken curves* show the mortality rates in a competition experiment in which 15 individuals of each of the two species were kept together (results summed up from two equal experiments). *Abscissa:* 11 months of the year from the beginning of the experiment. From Paarmann, 1966

In experiments performed by Paarmann using the method described earlier, no mutual influence of the two species could be detected from the shape of the population curves.

The overall results of the above experiments lead to the following conclusion: in natural surroundings a competitive effect between the adults of similar and generically related species appears to be possible but unlikely. It would, at the most, lead to a reduction in numbers of the weaker partner, but not to its complete dislodgement from a habitat in which it could live in the absence of competitors. In no case would it result in the phenomenon of ecological separation of two species of the same genus.

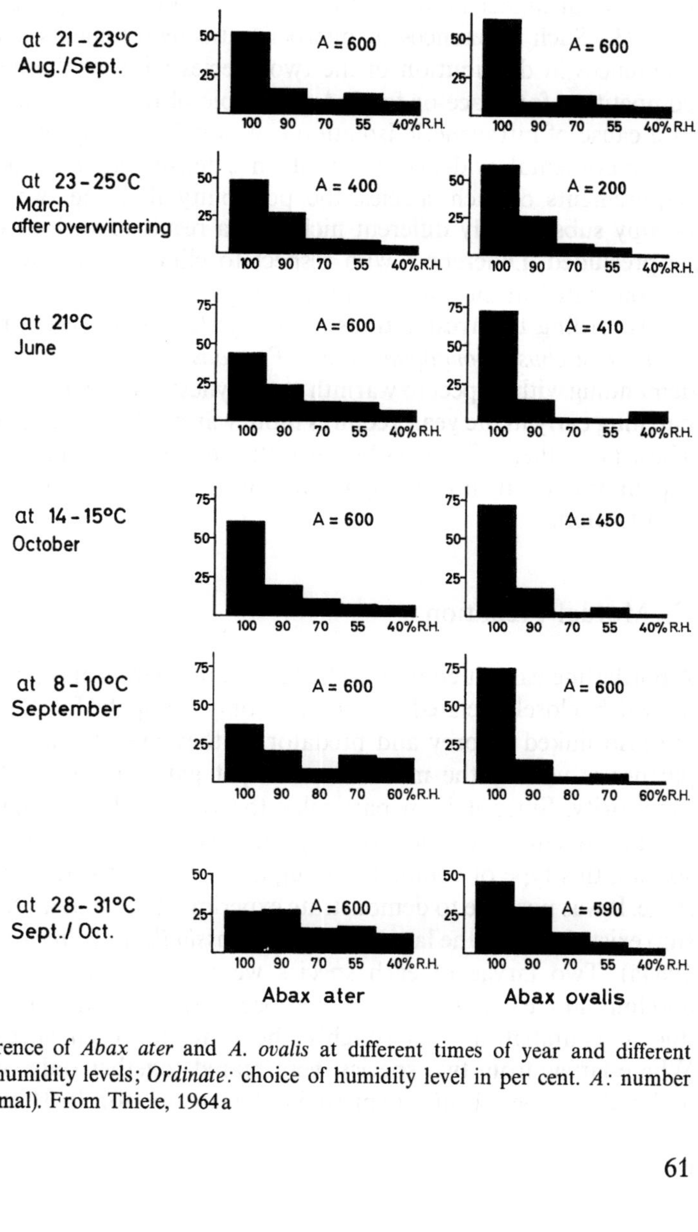

Fig. 23. Humidity preference of *Abax ater* and *A. ovalis* at different times of year and different temperatures. *Abscissa:* humidity levels; *Ordinate:* choice of humidity level in per cent. *A:* number of recordings (10 per animal). From Thiele, 1964a

It is surprising that interspecific competition is so lightly assumed wherever the phenomenon of intrageneric isolation is observed, without resorting to experimental proof. The food requirements of many species of animals are more familiar than their demands as to climate, since the latter have not been investigated experimentally. This situation can easily lead to an overestimation of the role of similarities in the ecological niches of two generically related species. Let us take, as an example, the case of *Abax ater*—*Abax ovalis*. As adults, both are scavengers and predators. Their larvae live exclusively on earthworms, so that the nutritional niches of the two species actually do overlap. Experiment has shown, however, that their microclimatic requirements differ. *Abax ater* is eurythermic but *Abax ovalis* is strongly stenothermic, with a maximum at 15° C (Thiele, 1964a). Although both species prefer moisture *A. ovalis* exhibited a somewhat higher need for it than *A. ater* in many experiments (Thiele, 1964a, Fig. 23). Such differences in microclimatic requirements suffice to explain the differences in distribution of the two species without the necessity of assuming competition for space or food. An example of this kind shows that competition as a cause of intrageneric isolation can only justifiably be assumed if supported by experimental evidence. Without an extensive analysis of the environmental requirements of each species, the possibility that the two species in question occupy substantially different niches as a result of various biotic needs cannot be precluded. Differences with respect to climatic requirements can readily be demonstrated in the case of the species pair *A. ater*—*A. ovalis*.

According to investigations made by Paarmann (1966) the same holds true for *Pterostichus oblongopunctatus*—*P. angustatus*. The latter appears to be more demanding with respect to warmth and dryness. The first larval stage in particular, hatching early in the year, requires more warmth than that of *P. oblongopunctatus*. These and other differences between the species (see Chap. 5.A.I.6.) are sufficient explanation for their differing habitat preferences, without the need for assuming competition.

II. Mutual Predation

A borderline case (it could also be termed interference) is presented by situations in which closely related species not only compete for food or territory, but are also linked as prey and predator. If they are of unequal strength and one partner is usually the more successful, it gains a specific type of competitive superiority. Since it is in particular the carabid larvae that are so aggressive towards members of their own species, as well as to larvae of other carabid species, this type of competition ought, above all, to be detectable at the larval stage. It was possible to demonstrate experimentally that a prey-predator relationship exists between the larvae of *Agonum assimile* and *Pterostichus nigrita* (Thiele, 1964a). Two larvae of each species were reared together in the experiments, and four larvae of one and the same species in the controls (Fig. 24a, b). Mortality due to cannibalism was roughly the same for each species in the controls. Where larvae from two species were reared together only one species survived, and in this respect *P. nigrita* proved to be far superior. In 31 out of 40 experiments

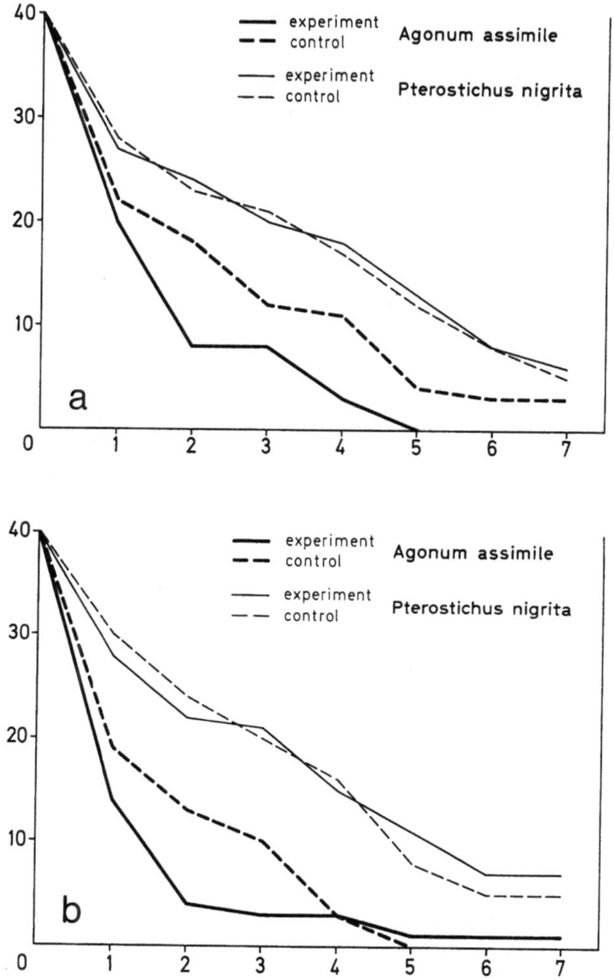

Fig. 24a and b. Experiments on mutual predation with larvae of *Agonum assimile* and *Pterostichus nigrita*. (a) Mortality curves of the two species under constant temperature conditions (17° ± 1° C). *Curves* represent total values from 20 experiments each with two larvae of each species and ten controls with four larvae of only one species. (b) The same, but with fluctuating temperature (13°–25° C). *Abscissa:* 10-day periods from beginning of experiment (duration of experiment 70 days). From Thiele, 1964a

it was the only survivor, whereas in only nine cases did *A. assimile* survive. Trap catches made over several years in a forest area with a wet floor near Cologne revealed that *P. nigrita* was always present in larger numbers than *A. assimile* (Fig. 25), both with regard to total numbers and in the individual traps. Although such results are perhaps due in part to the predatory superiority of the larvae of *P. nigrita* this type of "competition" can on no account be taken as the sole decisive factor in determining the pattern of distribution. In its adult form *A. assimile* prefers dryness whereas *P. nigrita* is much more strongly hygrophilic. The former is also very resistant to drying-out as an adult,

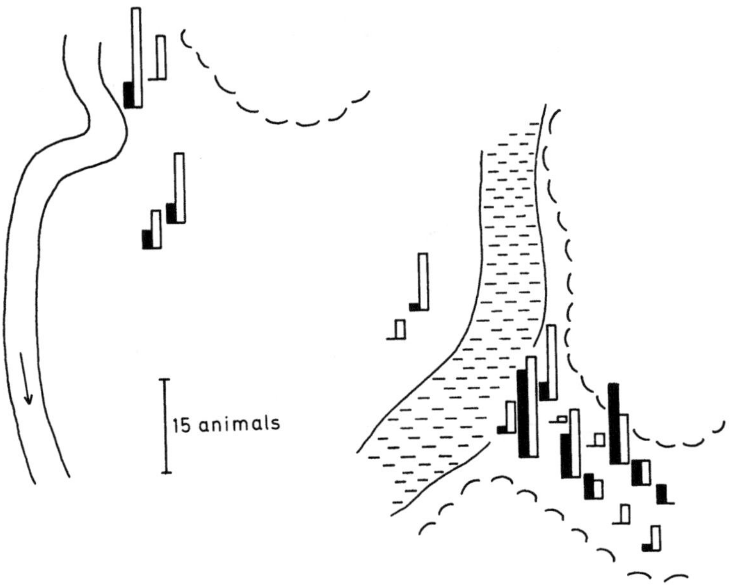

Fig. 25. Occurrence of *Pterostichus nigrita (white columns)* and *Agonum assimile (black columns)* in a West German water-meadow forest. *Each double column* represents the catch from one trap. *Left* of the diagram the course of a stream; *horizontal hatching*, a swampy stream. Catches made from April to September 1960

in contrast to the extreme sensitivity of *P. nigrita* in this respect. That *P. nigrita* dominates in the wet forest mentioned above, whilst at a distance of only a few kilometres, in a forest with only a moderately moist floor, *A. assimile* occurs in spectacular numbers and *P. nigrita* is totally absent, is not necessarily due to competition or mutual predation, but may simply reflect the microclimatic needs of the two species. Although the experiments do not preclude the possibility of competition, they cannot be taken as unequivocal proof that it is in fact involved. The population density in the experiments was 600 larvae/m^2, which is presumed to be about 100 times greater than naturally occurring densities (not known exactly). It is in any case clear that the great superiority of *P. nigrita* as revealed experimentally does not suffice to exclude *A. assimile* from its habitat. In fact the two species can coexist for many years, even if their frequency may vary. Similar experiments by Paarmann (1966) revealed the superiority of larvae of *Pterostichus oblongopunctatus* to those of *P. angustatus*. The former species is superior at experimental temperatures of 20° C and 25° C, a property that is particularly well seen in the first larval stage. Since *P. oblongopunctatus* lays fewer eggs with more yolk than *P. angustatus*, the larvae of the former are larger and stronger at the first stage than those of *P. angustatus*.

In this case, too, the abiotic demands of the larvae of the two species are so different that, combined with very unequal powers of dispersal, they can account for the differences in distribution without the additional assumption of competition (Chapter 5.A.I.2.).

III. Intraspecific Competition

In the face of limited availability of requisites (essential environmental factors) an increasing population density must necessarily, after a certain point has been reached, lead to competition for these requisites between members of a population.

Intraspecific competition can be demonstrated experimentally, but, with some caution, it can also be deduced from observations made in the field.

Experiments on Intraspecific Competition (Interference)

Reduced Fertility Due to Crowding. Experiments carried out on *Pterostichus nigrita* in order to determine the number of offspring produced following different photoperiodic treatment, clearly show that the figure sinks considerably with rising density of population. Experiments were performed under otherwise identical conditions with single pairs and with groups consisting of five males and five females.

One particular type of experiment (long day followed by interrupted short day, see Chap. 6.B.V.) resulted in 16 larvae per group (n=7 groups), which means 3.2 larvae per female. In two parallel series of experiments with single pairs, on the other hand, an average of 38 larvae per female was obtained in one case (n=20) and 22 per female (n=5) in the other. The number of offspring of single pairs is thus seen to be higher by a factor of 10 than for animals kept in groups. Another experiment involving a different photoperiod, from which high numbers of offspring were expected, resulted in only 63 larvae/group (n=4 groups) or an average of 12.5 larvae/female, which is still definitely lower than in the experiments with single pairs.

The reason for the reduction in numbers of progeny in the group experiments can be sought in disturbances during egg maturation and deposition, as well as a rise in mortality of the eggs and young larvae due to their being eaten or disturbed by the burrowing of crowded adults.

Schwerdtfeger (1968) points out that although, just as with the interspecific factors, it is a simple matter to distinguish intraspecific competition theoretically from "interference", it is sometimes extremely difficult in practice. "Interference" refers to the immediate and direct detrimental effects exerted by individuals of one and the same species upon one another. Intraspecific competition, however, results indirectly from demands for a requisite of limited availability.

Interference between the individuals of a species can more readily be demonstrated experimentally than intraspecific competition.

Direct Quantitative Observations on Interference have been made solely by Novák (1959). He observed the encounter behaviour of two individuals of one species kept together in Petri dishes for five minutes. The investigations on *intra*specific behaviour involved *Pterostichus cupreus* and *P. vulgaris*. *P. cupreus* turned out to be the more aggressive of the two. Males of the same species displayed more aggressivity towards one another than the females. A clearly delineated "pecking-order" was established, the stronger animal putting the weaker to flight and pursuing it with more or less vigour. With the exception of some very light-weight animals superiority was not dependent on body weight.

Concerning the applicability of conclusions drawn from such experiments to the situation in the field the same reservations apply here as in all other experiments on competition. Interference can be shown to exist, but its significance in habitat affinity and in the regulation of population density in the field remain open to question since population densities are substantially lower under natural conditions.

Cannibalism is a particularly drastic form of intraspecific interference, especially well developed in the *larvae* of carabids, a fact which renders their breeding a matter of great difficulty. To ensure success the larvae have to be reared individually.

In controls for the interference experiments on *Pterostichus nigrita* and *Agonum assimile* (see Chap. 3.A.II.), in which four larvae of only one species were reared together, large losses of larvae were due to attacks on members of the same species (Fig. 24). Similar observations were made by Paarmann (1966) in his control experiments on interference between *Pterostichus oblongopunctatus* and *P. angustatus.*

The degree of cannibalism is shown by the following data taken from Thiele (1968). The numbers of adults emerging if larvae were isolated at the first larval stage are compared with those from communal broods each consisting of four first instar larvae.

	Number of tests	Number of larvae	Emerged adults
	Agonum assimile		
Tests with 1 larva I	27	27	7=26%
Tests with 4 larvae I	96	384	54=14%
	Pterostichus nigrita		
Tests with 1 larva I	349	349	124=36%
Tests with 4 larvae I	117	468	76=16%

If four larvae are kept together the population obtained is only half the size of that obtained when the larvae are reared in isolation. Frequent observations point to cannibalism as being responsible for this vast reduction.

Intraspecific competition, or at least interference between members of one and the same carabid species, can therefore be shown to exist in populations kept under experimental conditions. Results of this kind are subject to the same reservations regarding their applicability to conditions prevailing in the field as those concerning interspecific competition or interference (see Chap. 3.A.I.). Here, too, the experimental density exceeds that found in nature by several orders of magnitude.

IV. The Theory of Regulation of Population Density by Competition and Other Biotic Factors

The limitations involved in applying the results of laboratory experiments to conditions prevailing in the field force us to seek new methods for studying the influence of biotic factors on carabid populations.

In summary, the main disadvantage of the data so far available is that they have been obtained on populations of densities 2–3 orders of magnitude above those normally existing in nature. Even if competition or interference is established experimentally it may well be that the species involved are regularly found side by side in the field at high population densities. In other cases, closely related species that are more or less mutually exclusive in the wild, exhibit little or no competition or interference under experimental conditions.

Observations on the effect of predators and parasites on carabid populations have not revealed a significant role for them in habitat affinity or in regulation of population size. Thiele (1964a) concluded, for example, that biotic factors play a lesser role in the distribution of carabids than the abiotic factors.

Abiotic factors are to a large extent the influence of weather and microclimate and have the property that their effect varies independently of population densities. A conclusion such as the foregoing must necessarily provoke the opposition of at least some population ecologists. Nicholson (cited here from Schwerdtfeger, 1968) is convinced that factors that are not dependent upon density cannot be responsible for maintaining equilibrium in a population: this can only be effected by density-dependent (and these are exclusively biotic) factors. "Rising population density brings with it a higher consumption of limited requisites and an enhanced effect of hostile factors. This in turn steadily reduces the favourable effect of several factors until the lowered abundance permits them to increase once again. In this manner population density can be maintained at a level appropriate to the characteristics of the animals and of their environment". A basic idea of this nature leads to the conclusion that population density, if governed by density-dependent factors, ought to exhibit regular fluctuations, whereas if abiotic factors are of major importance then irregular fluctuations should be seen. Uniform fluctuations in populations consisting of prey-predator organisms were first calculated by Volterra (1931). In principle they ought to apply equally to the competition of a monophage herbivore for its food plant, or to a host-parasite relationship. In model experiments, mainly beginning with those of Gause (1934), fluctuations of this kind, conforming with theoretical expectations, were demonstrated in the laboratory. Schwerdtfeger (1968) remarks critically that: "Nicholson neglects the fact that a population is embedded in a community, presumably because he took laboratory animals reared under constant conditions as the empirical basis for his theory".

An attempt to unite theory, model experiments and field observations into one common system was made by Ohnesorge (1963). He constructed models simulating population movements upon which (1) chance factors alone, (2) density-dependent factors alone, or (3) a combination of both were allowed to act. In this way, purely theoretical model curves for population changes, with characteristic properties, were obtained. Empirical population curves obtained from field observations can be compared with the theoretical curves, and from their shape conclusions can be drawn as to the factors influencing population density. Ideally, it ought to be possible to compile a "key" to the types of regulation of population density (Figs. 26–28). Unfortunately, insufficient results from long-term experiments on changes in density of carabid populations are available. Since most carabids live for one year only, trappings made year by year at one and the same site provide some information on the changes in population density.

Taking one case of this kind the following tentative conclusions can be drawn (Fig. 29).

1. There is no conclusive evidence of a consistent pattern, for example sinusoid, in the fluctuations in population density curves.

2. The summer of 1964 was exceedingly dry and a synchronous drop in numbers of moisture-loving species was observable in 1965. Species preferring dryness, on the other hand, were in no way affected by the anomalous weather of 1964 (Table 16).

At the moment it seems that the gap between the two extreme points of view in population ecology, as formulated by Schwerdtfeger (1968), is gradually closing. An idea developed by Eidmann (1937) from the analysis of populations of destructive insects is undoubtedly valid. He held that even if the environmental

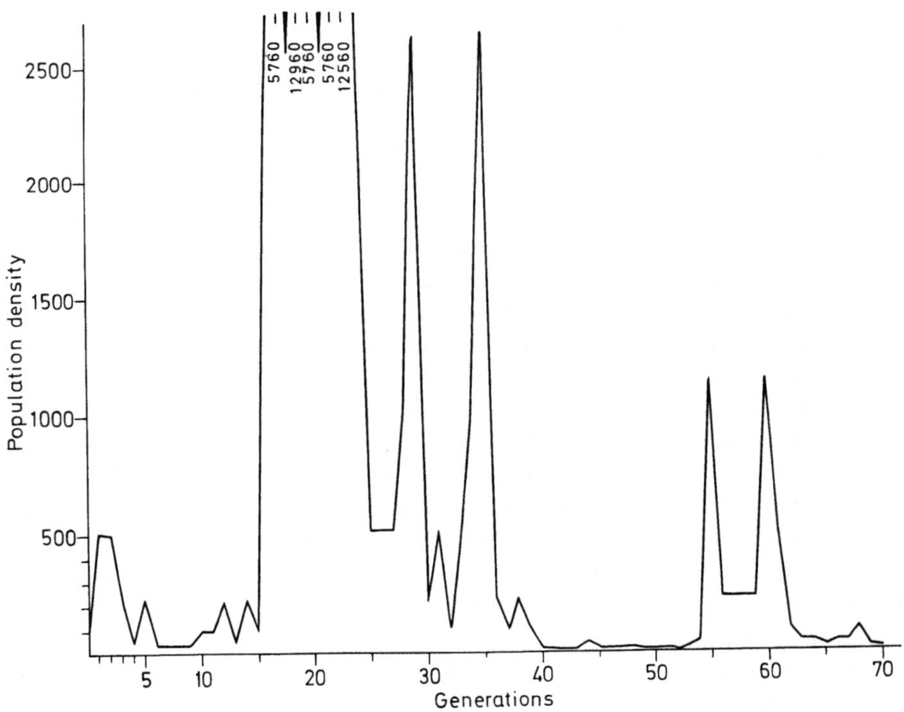

Fig. 26. Simulation of the changes in size of a population where its dynamics are governed only by chance factors. Initial population = 100. From Ohnesorge, 1963

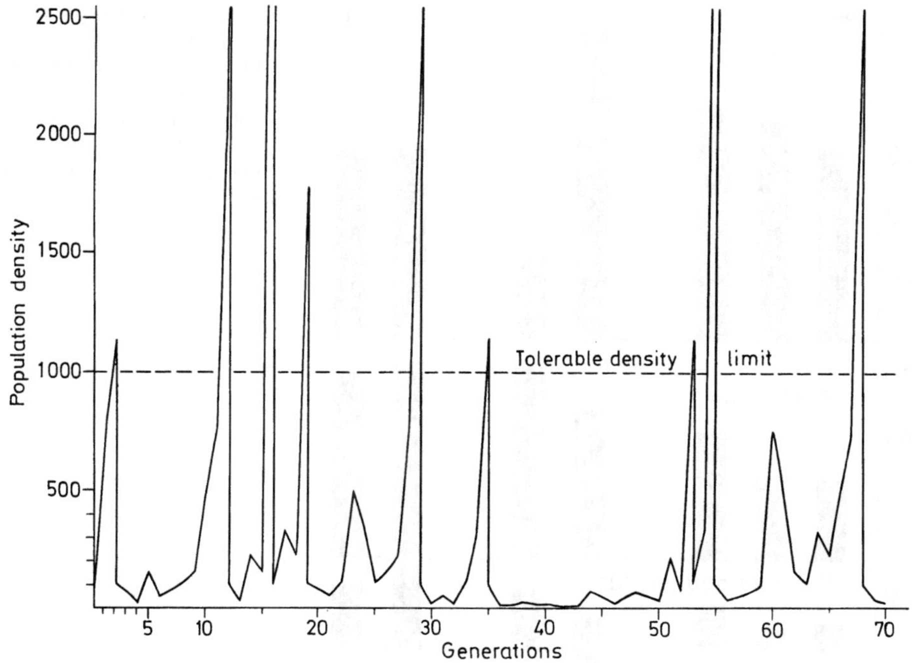

Fig. 27. Simulation of changes in size of a population where the dynamics are governed by a combination of chance factors and an overpopulation factor triggered by high density. Initial population = 100. From Ohnesorge, 1963

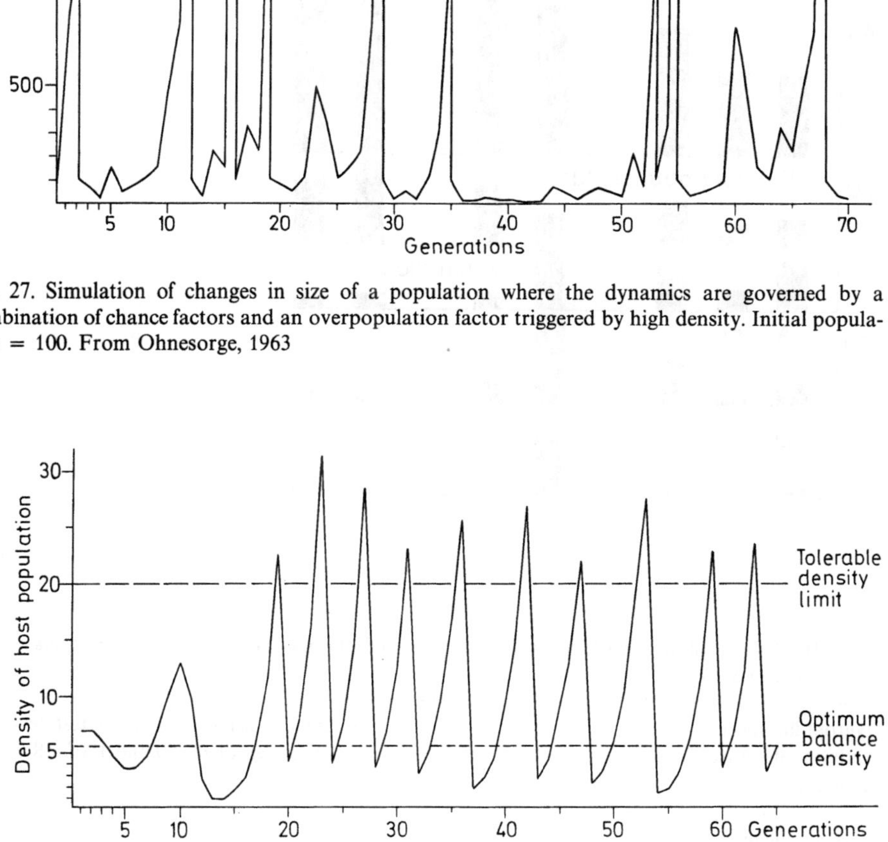

Fig. 28. Simulation of the changes in size of a population where parasites and an overpopulation factor triggered by high density govern the dynamics. From Ohnesorge, 1963

conditions of a particular organism only just suffice to maintain its coefficient of reproduction above 1, the population density must increase continuously. Limitations in the capacity of the environment prevent this from progressing infinitely and periodic collapses are observed whenever certain limiting levels

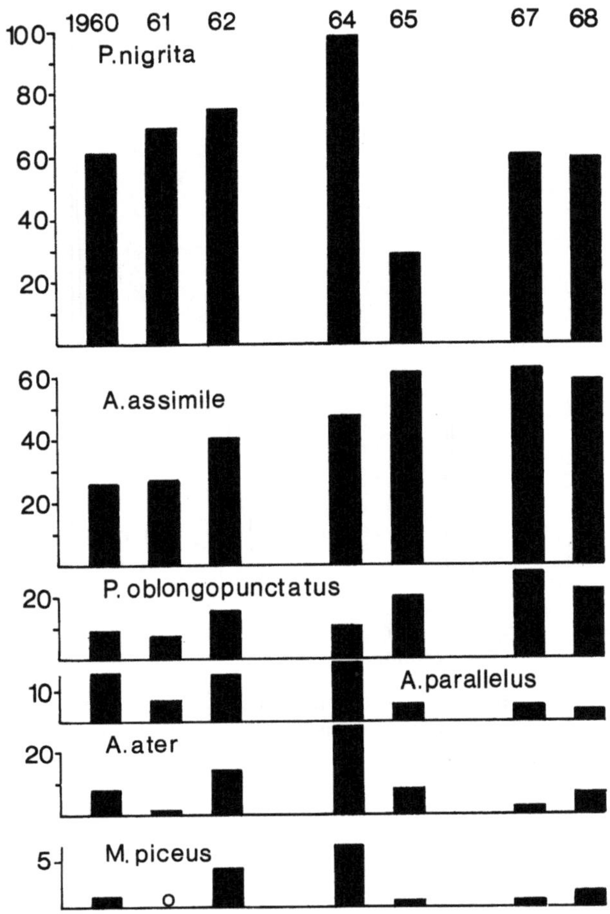

Fig. 29. Changes in population density of six carabid species based on trap catches made in spring from 1960–1968 in a West German water-meadow forest. *Ordinate:* Number of animals caught per 100 "trap weeks"

Table 16. The moisture requirements of some carabid species showing various reactions to the dry summer of 1964

Hygrophilic species exhibiting decrease in numbers	Percent choosing the two wettest levels of a humidity gradient (90 and 100% R.H.)
Abax parallelus	87%
Abax ater	69%
Molops piceus	61%
Pterostichus nigrita	45%
Xerophilic species exhibiting no decrease in numbers	
Pterostichus oblongopunctatus	26%
Agonum assimile	13%

of population density are attained. A simplified model of this nature may well hold for some phytophagous insects but is by no means of general applicability. Schwerdtfeger (1968) makes the following comment: "A population can only be prevented from exceeding its capacity limits and bringing about its own destruction by the presence of a limiting mechanism that is entirely dependent upon density and reacts immediately when the latter rises. The only factor able to do this is intraspecific competition, or in other words, competition and interference within the population itself". In addition, it should be mentioned that negative feedback systems involving predators and parasites can also play a role. Intraspecific competition is over-emphasized by Schwerdtfeger here. On the other hand, it is true of most organisms that, as Schwerdtfeger himself says, "a mechanism of this kind is only required in exceptional circumstances. In general, densities fluctuate around a mean value and a large rise in abundance is prevented by the combined action of density-independent and partially density-dependent factors" (Schwerdtfeger, 1968).

This implies that regulation of population density by means of density-dependent factors is not an omnipresent mechanism but "is only invoked as a safety mechanism or emergency brake in exceptional cases" (Schwerdtfeger, 1968). In simulation experiments Schwerdtfeger allowed "the 'density' of black or white stones in a game of chance to fluctuate within certain limits over 57 'generations', after which it inevitably began to rise. If, at this point, a simple regulatory mechanism such as increasing nutritional competition could be invoked... to decrease the 'density' the latter would once more begin to fluctuate within its old limits. The 'density curve' obtained from the model strongly resembles genuine abundance curves from investigations made in the field" (Fig. 30).

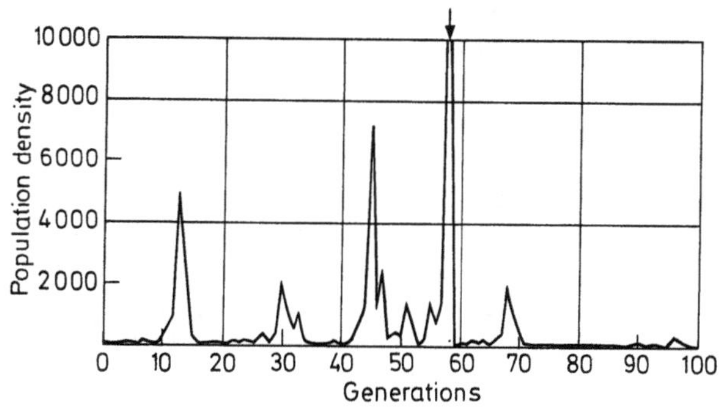

Fig. 30. Changes in population density in a simulation experiment. *Arrow:* regulation by nutritional competition. Explanation in text. From Schwerdtfeger, 1968

That populations of organisms usually fluctuate within narrow limits can be attributed to the fact that a considerable number of factors, some beneficial and some detrimental, act simultaneously upon the population density. In this way they balance each other out and stabilize the population density, thus avoiding the risk of overpopulation or of extinction which would result if the density were to sink below a critical level. Den Boer (1971a) calls this principle "spreading of risk" and describes the contributory factors as follows:

"Because the habitats of natural populations generally are very heterogeneous, there will be local differences in microweather, food, natural enemies, etc. Therefore, the chances of surviving and of reproducing will be different for individuals living in different sites within a natural habitat, even when they are phenotypically identical. Hence, for each generation in turn, the change in numbers may be expected to be different in different parts of the habitat of a natural population; this means that, for the population as a whole, the effect of relatively extreme conditions in one place will be damped to some degree by the effect of less extreme conditions in others. In other words, from generation to generation the risk of wide fluctuation in animal numbers is spread unequally over a number of local groups within the population; this will result in a levelling of the fluctuations in the size of the population as a whole. Moreover, the animals may move from one place to another within the habitat and such movements, even if they occur wholly at random, will contribute to this stabilizing tendency of spatial heterogeneity, since in this way very high or low numbers in some places will be levelled out more thoroughly. Although it is outside the scope of this paper, it must be noted here that heterogeneity within the population will also contribute to the relative stabilization of animal numbers; changes in numbers in one phenotypic or age group will be more or less counter-balanced by changes in other such groups. In this way, the effect of fluctuating environmental factors on the population as a whole is continuously damped to some degree by the phenotypic variation and/or by the variation in age composition (by this variation the range of tolerance of the population is increased as compared with that of the individual animals)". Den Boer's concept was developed for populations of carabids. In a meticulous series of long-term experiments in the field he was able to show, using *Pterostichus coerulescens* and *Calathus melanocephalus*, that fluctuations from year to year in the reproduction rate of a large but panmictic population remained within much narrower limits than those of the local populations of which it consists (Figs. 31, 32).

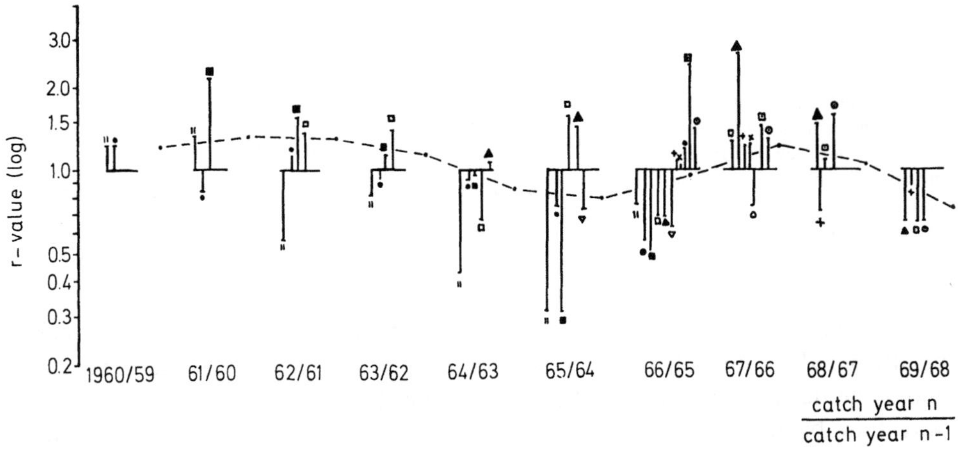

Fig. 31. Yearly changes in reproduction rate *r* (ordinate, logarithmic scale) in *Pterostichus coerulescens* in a Dutch heath region. *Vertical lines:* reproduction rates in parts of the populations; *broken curve:* mean value for the entire population. From den Boer, 1971 a

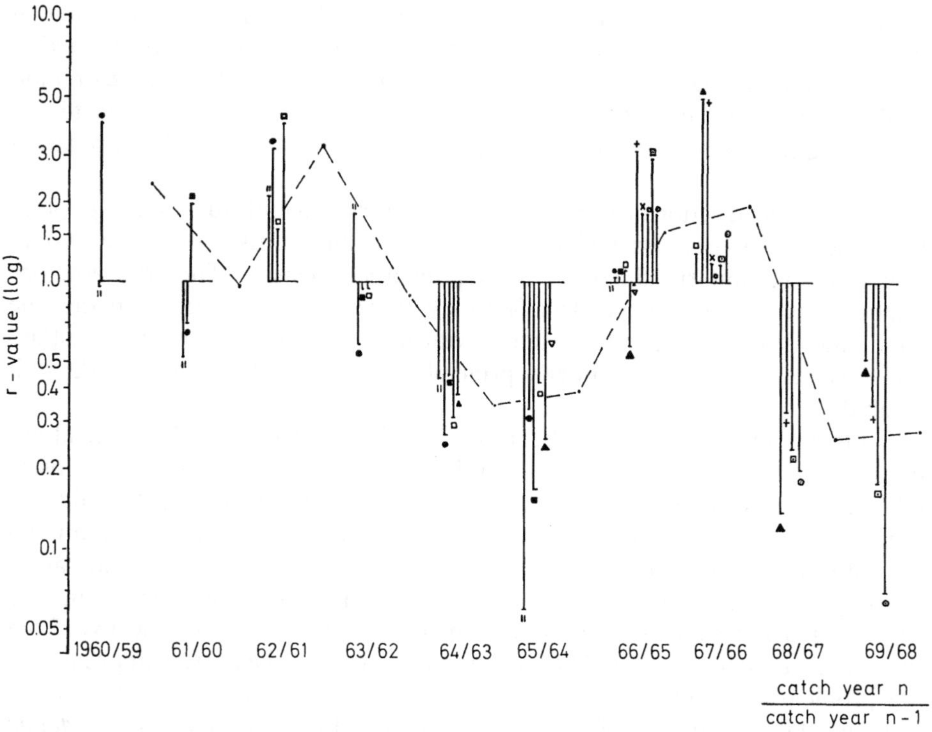

Fig. 32. Yearly changes in reproduction rate *r* in *Calathus melanocephalus* in a Dutch heath region. Explanation, see Figure 31. From den Boer, 1971 a

In computer simulations Reddingius and den Boer (1970) demonstrated that fluctuations in population density of insects are, in fact, greatly damped in accordance with the principle of the "spreading of risk". They formulated models for population development under the influence of a variety of factors. One of the innovations was that the authors did not permit the weather factor to fluctuate by chance (the usual practice is to simulate changes in the weather factors with values from random tables). Instead, they used published values from meteorological tables. This procedure is more realistic in that such data "may be serially correlated as well as correlated among themselves". Increasing the number of weather factors in this model (they were chosen for their feasibility in exerting an influence on the rate of reproduction) up to 10, progressively enhanced the stability of the population. A further stabilizing element was provided by the introduction of the assumption that the populations consist of nine sub-populations, between which migratory exchange is possible. Yet another stabilizing effect resulted from increasing the number of age groups within a population.

It is also possible to introduce density-dependent crashes into this model, starting when the capacity limit of the available space is reached. Nicholson (cited from Schwerdtfeger, 1968) suggests that they tend to stabilize the population density. On this point, the ideas of Reddingius and den Boer approach Schwerdtfeger's intermediate view, as described above. But, and this seems to be very important, population collapses at *irregular* intervals randomly distributed in time have a similar stabilizing effect, although here the danger of extinction would seem to be slightly larger than in density-dependent crashes. In the opinion of Reddingius and den Boer[4] it is impossible to draw conclusions as to the underlying causes from the more or less regular course taken by the population curves.

The combined action of a large number of factors on a multitude of sub-populations and phenotypes—as envisaged by the "spreading of risk" principle—could well result in uniform fluctuations in population density. The contradictory implications of density-dependent factors with regulatory function, on the one hand, and density-independent factors with no regulatory action on the other, are thus resolved by den Boer's principle, based largely on his work with carabids.

At this point, the possible role of interspecific competition in bringing about intrageneric isolation, or, in other words, the ecological vicariation of closely related species of one genus, warrants further brief consideration. We did not attribute any great importance to this factor (see Chap. 3.A.I.) in determining the pattern of distribution of carabids. Nonetheless, a theoretical consideration of the problem leads to the conclusion that even weak competition from the superior partner suffices in the long run to oust the weaker of the two from the space at their disposal. The first attempt at a mathematical approach to this problem was made by Volterra and d'Ancona (1935). Gause (1934) provided experimental confirmation of their calculations as follows: if *Paramecium caudatum* and *P. aurelia* are kept together only one species survived, depending upon the nutritional substrate used.

From the purely mathematical point of view this "competitive exclusion effect" does not come into play if the rate of increase of both species is equal. It follows that the larger the differences in growth rate the greater the competition, but if the coefficients of increase differ only slightly the elimination of one of the two species can take much longer. If one of the competitors becomes rare the competitive effect may be reduced. Further, the fact that laboratory experiments are usually carried out in a very homogeneous environment, a situation seldom encountered in nature, means that two species kept together are obliged to utilize the available space in a highly similar manner. But the more varied the environment the less the competition. *Paramecium bursaria* and *P. aurelia* can co-exist in unlimited numbers in a suspension of yeast, since the former feeds upon the sedimented particles, whereas the latter ingests those floating on the surface (Gause *et al.*, 1934). Crombie (1946, 1947) found that *Tribolium* eliminates *Oryzaephilus* in mixed cultures, but if small glass tubes are provided in flour, as refuges for the *Oryzaephilus* larvae, the two species are able to exist side by side. The same end can be achieved by introducing fragments of wheat grains, into which the *Oryzaephilus* larvae, but not those of *Tribolium*, can penetrate.

Experiments of this type represent attempts to simulate the situation prevailing in nature. They reveal the likelihood that potential competitors have a greater

---

[4] I wish to thank Dr. den Boer for his additional written and verbal information on this topic.

chance of living permanently side by side if they are provided with the opportunity of exploiting their common environment in different ways. In other words, they occupy different niches (see Chap. 10.B.).

Thus there is no contradiction in the finding that two species of carabids which compete with one another in laboratory experiments may well live side by side in large numbers in their natural environment.

B. Positive Intraspecific Relationships

Apart from the mutually detrimental effect exerted by one carabid upon another due to competition, there are also indications of a type of social behaviour in which the members of a species are mutually beneficial, or are at least tolerated by each other.

I. Aggregation

Nebria brevicollis and *Agonum dorsale* have long been known to aggregate. Allen (1957) observed aggregations of 30–40 individuals of the latter species on very small areas, but my own unpublished observations suggest that the number of individuals can be much larger. In this species aggregation is connected with the occupation of its winter quarters. Greenslade (1963b) describes similar aggregations of *Nebria brevicollis* and *Brachinus crepitans*. Lindroth (personal communication) has drawn attention to the large numbers of *Dromius* species aggregating in autumn, especially at the base of the trunks of fir trees.

The first experiments in connection with this phenomenon were carried out by Greenslade (1963b) on *Nebria brevicollis*. He invariably observed a clumping effect between individuals in a homogeneous experimental vessel. The deviation from a random distribution was highly significant.

Thorough studies on aggregation, however, have only been made in the two species *Brachinus sclopeta* and *B. explodens* (Wautier, 1971), both of which are long-lived (up to 4–5 years in the laboratory). Five populations of the former, each consisting of 25 individuals, were investigated over a period of 25 months, and three populations of the latter species over 13 months.

The results revealed an annual rhythm in aggregation. Aggregation is at its peak in winter, during the phase of sexual inactivity, but declines with the commencement of the reproductive period in spring. A daily rhythm is also detectable, with maximum aggregation in the morning, whereas nighttime brings an exodus of individuals in search of food. In laboratory experiments 77 and 199 specimens, respectively, were placed one by one into two terraria. They immediately formed small groups and within a few days the largest group had successively assimilated all of the smaller groups.

II. The Mechanism Underlying Social Behaviour: Pheromones

Wautier (1971) demonstrated that attraction between individuals of *Brachinus* depends upon olfactory stimuli which are perceived by the antennae and not by the palpae. Since she experimentally excluded recognition by means of optical and tactile stimuli the conclusion can be drawn that pheromones occur in *Brachinus*, although their chemical nature is so far totally unknown. A substance qualified to fulfil this role would have to be persistent and extremely stable, retaining its attractivity for the beetles over several years.

III. The Significance of Aggregation

This is a topic upon which little else but speculation has so far been published (Greenslade, 1963b). Since *Brachinus* possesses highly effective defence glands it might be postulated that a combined reaction to hostile predators would stand a much larger chance of success. In the case of *Nebria brevicollis*, Greenslade is of the opinion that aggregation not only protects the animals from losing too much water, but is also advantageous in ensuring that the sexes remain together during the summer dormancy of the, as yet, immature animals (see Chap. 6.B.III.).

Aggregation is apparently not governed by sexual pheromones, as the investigations of Wautier (1971) have shown. For one thing, it declines at the beginning of the reproductive phase. Further, not only could interattraction between *Brachinus* females be demonstrated, but also mutual attraction of males, as well as aggregation between immature members of both sexes. Goulet (1974) found that the females of *Pterostichus adstrictus* and *P. pennsylvanicus* did not attract the males if direct contact between the two sexes was experimentally hindered but the possibility left open for the males to perceive volatile substances, if any, given off by the females.

On the whole, aggregation seems to occur in only very few carabid species.

IV. Care and Provision for the Brood and Its Ecological Significance

The females of many species of carabids do not simply deposit their eggs on the ground but have been observed to dig a small hollow in the substrate with the apex of their abdomen into which they deposit the egg. Females of *Carabus* and of some other species dig themselves into the ground completely before egg deposition (von Lengerken, 1921).

Broscus cephalotes prepares a sloping shaft of up to 30 cm length for its eggs, which are then deposited in oval side chambers usually arranged more or less in pairs at intervals of 1–1.5 cm. After deposition of one egg in each chamber the connection between chamber and main shaft and the portion of the main shaft leading to the surface is tightly packed with sand (Kempf, 1954). Clausen (1962) described a similar form of behaviour in *Craspedonotus tibialis* in Japan.

Any further provision for the brood has only been described for a few other species. Using circular movements *Abax ater* females scrape off particles of moist clay with the styli of their telescopically-extended hind-quarters and cover the hindmost abdominal segments with a thin layer. The abdomen is then pressed onto the substrate and the female deposits an egg in the earthy sheath. "When the abdomen is withdrawn from the sheath the dorsal portion closes down on the lower half like a lid. The remaining gap is filled immediately by horizontal side to side movements of the posterior part of the body. The soil sheaths, constructed as they are of the finest particles, probably serve to prevent desiccation and fungal attack" and in addition protect the eggs from marauders (Löser, 1970, 1972). Similar cells are constructed by *Abax exaratus* (Brandmayr, personal communication) and *Abax springeri* (Brandmayr and Brandmayr, 1974).

According to Clausen (1962) the type of behaviour described in great detail by Löser has been seen in other carabids and was first reported by Riley in 1884. His studies, however, just as King's (1919; quoted in Clausen, 1962) findings on several species of *Brachinus*, *Chlaenius* and *Galerita*, also cited by Clausen, have found no mention in reviews on the topic of brood care in beetles.

Although the shape of the chambers varies from species to species the method of construction is the same as that employed in *Abax ater*. This is remarkable, in view of the fact that the occasional occurrence of the phenomenon in genera so little related to one another suggests a polyphyletic origin.

On the other hand, *Percus navaricus*, a near relative of *Abax*, moulds its sheaths, which look exactly like those of *Abax ater*, with its mandibles. The female first of all forms a bowl-shaped structure which she then lifts up to her adbomen with her legs for egg deposition. The "bowl" is then either left where it is or attached to a clump of earth (like *A. ater*) and finally provided with an earthen lid (Lumaret, 1971). *Agonum dorsale* has been observed to attach its eggs, each in its own earthen sheath, to the lower surface of strawberry leaves (Dicker, 1951). Kreckwitz (personal communication) confirmed this observation under laboratory conditions.

Another slight variation in behaviour was seen in the North American species *Tecnophilus croceicollis* (Larson, 1969). The female began by loosening particles of earth with her abdomen. The particles were "then picked up singly with the apex of the abdomen to which they adhered, probably by means of an adhesive substance produced by the accessory glands. When sufficient material was collected, the female climbed onto some object such as a twig... against which the abdomen was pressed and a drop of fluid released. The abdomen was then moved away from the twig, drawing the drop of fluid out until it hardened into a silk-like strand... When the thread reached a length of about 1 to 6 mm the female injected an egg into the ball of soil particles that she still carried on the apex of her abdomen. The egg and soil covering were then released and were left dangling in the air".

As early as 1874 Bargagli recorded that the eggs of *Percus passerini* were always encased in an earthen crust. He also claimed that the female guarded her brood, a form of behaviour not seen in the other species so far mentioned.

This brings us to the most advanced form of behaviour, involving a type of *brood care* in which the female guards her eggs up to the moment when the larvae hatch. So far, only a few cases of this nature have been reported.

As shown in Table 17 brood care has been observed in 14 species of the tribe *Pterostichini*, all of which behave in a very similar manner (Fig. 33). The females dig out a nesting hollow of a few cm diameter in rotten wood or often in the ground under stones, and remain on the eggs until the larvae disperse, soon after hatching[5]. During the three weeks that they remain on the nest the females consume no food. Löser noticed no defence reaction to other arthropods. Komárek (1954) and Löser (1970) assume that the females prevent the eggs from being attacked by fungi whilst in the nest.

This interesting form of behaviour, too, is apparently polyphyletic in origin. It is a striking fact that the majority of species exhibiting brood care are montane in distribution.

Table 17. Carabid species of the Pterostichini tribe, in which brood care has been observed (brood guarded by female)

Species	Observed by	Published by	Laboratory or wild
Pterostichus multipunctatus	Boldori, 1933	Boldori, 1933	Wild
Pterostichus anthracinus	Lindroth	Lindroth, 1946	Laboratory
Pterostichus metallicus	Weidemann	Weidemann, 1971b	Laboratory
Pterostichus metallicus	Neumann	unpubl.	Wild
Pterostichus metallicus	Brandmayr	Brandmayr, 1974	Wild
Pterostichus yvani	Vigna	Brandmayr, 1974	Wild
Pterostichus maurus	Vigna	Brandmayr, 1974	Wild
Abax parallelus	Thiele, 1960	unpubl.	Laboratory
Abax parallelus	Hůrka	unpubl.	Laboratory
Abax parallelus	Löser	Löser, 1970, 1972	Wild and Laboratory
Abax ovalis	Lampe	Lampe, 1975	Laboratory
Molops piceus	Jeannel	Jeannel, 1948	Wild
Molops piceus	Komarek	Komarek, 1954	Wild
Molops piceus	Thiele, 1970	unpubl.	Wild
Molops piceus	Brandmayr	Brandmayr, 1974	Wild
Molops austriacus	Weirather, 1926	Brandmayr, 1974	Wild
Molops ovipennis	Brandmayr	Brandmayr, 1974	Wild
Molops plitvicensis	Brandmayr	Brandmayr, 1974	Wild
Molops striolatus	Brandmayr	Brandmayr (pers. comm.)	Wild
Molops senilis	Ravizza	Leonardi, 1969	Wild
Molops edurus	Osella	Brandmayr, 1974	Wild

[5] An isolated observation of a *Molops piceus* female closely surrounded by about a dozen 2nd-instar larvae was reported by Jeannel. (Note at correction: Dr. P. Brandmayr kindly informed me that Jeannel's observation must have been erroneous. From Jeannel's description it could be deduced that these larvae were newly hatched first instars—see Brandmayr and Brandmayr, 1974.)

Fig. 33. Female of *Abax parallelus* in half-opened nest with eggs. From Löser, 1970

It seems that only relatively few carabids have developed the complicated type of behaviour connected with brood care. With one exception (to be dealt with later), it is confined to species of the Pterostichini tribe, in which it has developed convergently in species of several genera. These are predominantly species that lay a small number of eggs and take longer to develop. Brandmayr and Brandmayr (1974), who have dealt at some length with the subject, particularly emphasize that the embryonic development of these species takes relatively long. They regard this as an adaptation to a cooler climate, and consider the guarding of the eggs by the female during this lengthy phase as being of adaptive value.

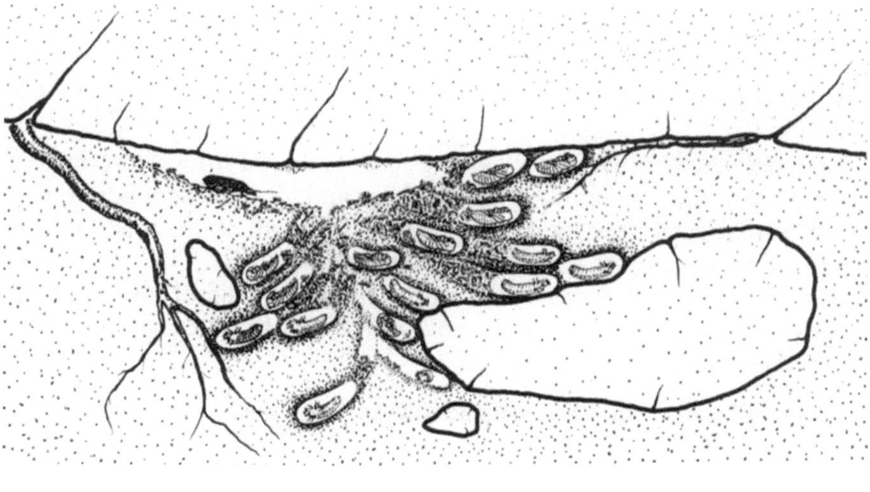

Fig. 34. Nest of *Carterus calydonius* at an advanced stage of development of the brood. The upper cells contain pupae, the lower prepupae or 3rd instar larvae. From Brandmayr and Brandmayr, 1974

Brandmayr's observed brood care not only in the Pterostichini but in *Carterus calydonius* as well, a species belonging to the Ditomina (Tribus Harpalini). Like other species of this subtribe, *Carterus* hoardes plant seeds (from *Daucus* in this case) in subterranean cells for aestivation in the dry summer of the submediterranean climatic region. The female lays its eggs in individual cells of the subterranean nests, where the larvae complete their entire development and feed, like the adult beetles, on the seeds. The maternal female guards the exit of the compound nest (Fig. 34). Not only, therefore, does the female *Carterus* lay up food stores for its own aestivation like other Ditomina, it also combines this activity with egg deposition and prolonged brood care. In Brandmayr's opinion such a coordination of adult and larval requirements represents an adaptation to the dry summer climate.

C. Parasites of the Carabids

Mention of parasitic organisms attacking carabids is scattered in the literature and hard to find, which also holds true for their predators. The present book seems to afford an appropriate opportunity for making a comprehensive survey of animal groups known to parasitize or prey upon carabids. It is hoped that those interested in the ecology of carabids will thereby be spared some of the labour involved in seeking out the relevant literature. The author is fully aware that, despite his efforts, the literature cited is not exhaustive. Nevertheless, the most important groups of parasites and predators (see Chap. 3.D.I.) have been mentioned, and it is hoped that this will provide a background for attempts at estimating the importance of parasites in the distribution and control of carabid populations.

I. A Survey of Parasites Found in Carabids Arranged Systematically

Bacterial and Viral Diseases. As far as can be judged from the review article by Krieg (1961) these diseases are unknown in carabids. This does not necessarily guarantee that such diseases do not occur in ground beetles since they have been found in many other beetles, but reflects, rather, the lack of attention so far devoted to carabids by insect pathologists. Histopathological investigations on prepupal and pupal stages of *Pterostichus nigrita* that had died without apparent cause revealed no such infections.

Fungi—Phycomycetes: Entomophthoraceae. *Empusa* infections have only been seen sporadically in *Pterostichus oblongopunctatus*. "This fungus was presumably introduced by a dipteran, *Sciara thomae* L. The dipteran developed on ground meat in the beetle stocks and was afflicted by the same fungal disease. In the field, however, this species of dipteran is in no way connected with *Pterostichus* species or with any other carabids" (Paarmann, unpublished report).

Fungi imperfecti: Hyphomycetes. Laboratory-bred carabids, especially *Pterostichus oblongopunctatus* and *P. angustatus*, were frequently found to be afflicted with *Metarrhizium anisopliae* (Metsch.) Sorokin. This is a nonspecific entomophagous fungus and can probably attack all carabids. It was found on every developmental stage except the eggs. "The mycelia break out of the dead animals and cover them with a thick white sheath. Older adults that had been kept for longer periods of time were the most frequently attacked, and from them the infection could spread to larval stock" (Thiele, 1968c). This is a very infectious fungus, but can be combatted by frequent changes of substrate at the first indication of attack. Cases of infection of carabids with this fungus in the wild are unknown.

Sturani (1962) reports frequent infection of his stocks of *Carabus* larvae with *Botrytis* (=*Beauveria*) *bassiana* (Bals.) Vuill. This species is not specific for carabids either, and is responsible for the much-feared silkworm disease "Calcino".

B. bassiana probably plays a role in checking the numbers of overwintering potato beetles. Under laboratory conditions the fungus brought about 40% mortality in the species, but was responsible for a death rate of only 3.4% in *Harpalus rufipes*. *Paecilomyces farinosus* (Dicks.) Brown and Smith brought about 70% mortality in potato beetles, but only 2% in *H. rufipes* and 8.3% in *Pterostichus cupreus* (Kmitowa and Kabacik-Wasylik, 1971). *B. bassiana* was frequently found in laboratory-bred *Amara ingenua* by Bílý (1975).

Fungi—Ascomycetes: Pyrenomycetales. The nonspecific insect-parasitizing species *Cordyceps militaris* Link and *Torrubia cinerea* have been found in *Carabus* species (Sturani, 1962).

Fungi—Ascomycetes: Laboulbeniales. This series consists of only one family, the Laboulbeniaceae, comprising about 1200 species (Dr. Müller-Kögler, personal communication), all of which parasitize arthropods and are for the large part specialized on beetles.

Thanks to the work of Lindroth (1948) and Scheloske (1969) we know more about the host spectrum and epidemiology of Laboulbeniaceae than of the other fungi parasitizing carabids. Lindroth found them in 21 species of carabids, Scheloske in 61, and almost all belonged to the subfamily Harpalinae. They have not been observed in Carabini, Cychrini and Cicindelinae. As a rule, only a few fungi were found on any one beetle. Lindroth found at the most 20 on animals caught in the wild, although much higher numbers can occur in laboratory-bred stock. The thalli are found especially on the posterior and lateral portions of the elytra, but in severe cases they occur on the lower side of the body as well. Neither Lindroth, Thiele (1968) nor Scheloske were able to detect a detrimental effect on the beetles, which did not appear to be shorter-lived when attacked by fungi.

In the course of his examination of seven Swedish species of *Harpalus* (subgenus *Ophonus*) Lindroth found that the parasitic fungi did not occur in the entire area of distribution of their hosts, but only in the southernmost and southeastern parts of the country. His experiments revealed that the carabids are more likely to be infected from earth contaminated with spores than by contact with infested members of their own species. Scheloske, however, showed that copulation plays a significant role in transfer of the parasites.

Since the spores of Laboulbeniales only remain infectious in the ground for a few weeks the fungi are obliged to spend the winter on the insect in order to survive. Monophagy of certain species of fungi on carabids that overwinter exclusively as larvae is therefore ruled out, because carabid larvae are never attacked by these fungi.

Protozoa: Sporozoa. The only protozoans known to parasitize on carabids are the gregarina. Sturani (1962) reported *Monocystis legeri* L. F. Blanch. in *Carabus auratus*, *C. punctatoauratus* and other *Carabus* species. Delkeskamp (1930) found gregarina in *C. nemoralis*, *hortensis* and *cancellatus* as well as in *Pterostichus niger*. The same species seems to have been involved in each case. The infection progresses very rapidly: Delkeskamp cites Wellmer as reporting that previously healthy individuals artificially infested with ripe spores on October 25th showed mature cysts by November 2nd.

Whereas gregarina are harmless in the intestinal tract they can lead to the death or at least castration of their host as parasites of the body cavities. This is not surprising, if it is borne in mind that Delkeskamp found as many as 220 cysts in the abdominal cavity of *Carabus nemoralis*. Puisségur (1972) found gregarina cysts in the body cavities of six out of 13 *Carabus* species investigated in the Pyrenees.

Nemathelminthes: Nematoda. Mainly members of the order Mermitoidea have been found as parasites on carabids. Rivard (1964b) summarized older observations and reported the occurrence of *Mermis albicans* Sieb. in *Amara similata* in Europe. He himself found an immature specimen of *Hexamermis* sp. in *Bembidion nitidum* in Canada. Isolated individuals of *Hexamermis* sp. were also found in *Nebria brevicollis*. Such mermithids can achieve a length of 50 cm. According to Tischler (1965) pupae of *Bembidion* species are attacked by *Heterotylenchus bovieni* and those of *Clivina fossor* by *H. stammeri*, although the damage done to the hosts is negligible.

Burgess (1911) records parasitization of *Calosoma sycophanta* in North America by undetermined nematodes and the occurrence of rhabditids in *Carabus monilis*.

Nemathelminthes: Nematomorpha. The larvae of members of this class develop in the body cavities of arthropods. Carabids are often mentioned as hosts in the literature: according to von Lengerken (1924) *Gordius* occurs in *Carabus auratus*, and Sturani (1962) reports its occurrence in *C. coriaceus*, *hortensis* and others. Tomlin (1975) reports one occurrence in *Pterostichus vulgaris*. Rivard (1964b) suspects that some of these reports are due to confusion with the very large mermithids (see above).

Acari (Mites). They are among the more important parasites of carabids although reports of their occurrence in ground beetles are widely scattered.

Acari: Parasitiformes. *Parasitus* species can occur (especially in cultures) in masses on the bodies of larger carabids (*Carabus* sp.). However, this is not a case of parasitism but rather of phoresia. The *Parasitus* species are predators and are transported by scarabaeids, in this way reaching the excrements of grazing animals where they find the large numbers of insect larvae which form their prey. Although the behaviour of *Parasitus coleoptratorum* and its symbiosis with *Geotrupes* is well known from the work of Rapp (1959) nothing has been published concerning the phoresia of *Parasitus* in carabids.

In the following groups true parasitism is involved:

Acari—Trombidiiformes: Podapolipidae. A monograph by Regenfuss (1968) deals with this family in great detail. The Podapolipidae are of particular interest because they are to a large extent specialized on Coleoptera, and even on carabids, with which they are very closely linked. The legs of the females of all genera, with one exception, have undergone reduction to three pairs or even one pair, and even these may be extremely small. In the males the legs are generally reduced to three pairs. Until recently only 19 species diagnoses (some of these very brief) and a few observations on their biology were available, and only two of the species described were known from central Europe. From southern Germany alone, Regenfuss has described 24 new species.

The majority of species have only been found in beetles. A few occur also or only in cockroaches, grasshoppers and bees (Table 18). The 18 known species of the genus *Eutarsopolipus* and the seven of the genus *Dorsipes* live exclusively on carabids, on which they are monohospital (confined to one host species) or, in a few cases, bihospital, i.e. occurring on two closely related host species. On the other hand, several podapolipid species can occur on one host, occupying strictly demarcated niches. Regenfuss reports the following examples: "*Eutarsopolipus pterostichi* lives beneath the elytra of the carabid *Pterostichus vulgaris* whilst *E. stammeri* is found as an endoparasite in its body cavity. As many as three ectoparasitic *Eutarsopolipus* species parasitize on *Pterostichus cupreus: E. abdominis* on the intersegmental skin of the abdomen, *E. thoracis* on the

Table 18. Distribution of the genera of the Podapolipidae in various hosts. The figures indicate the number of parasitic species. (According to Regenfuss, 1968)

Host	Coleoptera							Blattodea	Orthoptera	Hymenoptera
Parasite	Carabidae	Coccinellidae	Tenebrionidae	Chrysomelidae	Cerambycidae	Curculionidae	Scarabaeidae	Blattidae	Acrididae	Apidae
Chrysomelobia	—	—	—	1	—	—	—	—	—	—
Eutarsopolipus	18	—	—	—	—	—	—	—	—	—
Dorsipes	7	—	—	—	—	—	—	—	—	—
Tarsopolipus	—	—	—	—	—	—	1	—	—	—
Tetrapolipus	—	1	—	—	1	1	—	1	—	—
Podapolipus	—	—	2	—	—	1	—	—	2	—
Podapolipoides	—	—	—	—	—	—	—	—	3	—
Locustacarus	—	—	—	—	—	—	—	—	3	1

thorax, between the episternum II and episternum III, and *E. squamarum* on the cutaneous squames of the elytra... The strict spatial separation of synhospital parasitic species on their host is well demonstrated by the species *Eutarsopolipus vernalis* and *Dorsipes cryptobius* on the carabid *Pterostichus nigrita:*

1. After reaching their host the larvae of *E. vernalis* immediately seek out the space below the elytra at the posterior end of the animal, move anteriorly across the abdominal tergites and climb into a fold of the cutaneous wings (morphologically its ventral aspect). In this one location they grow as parasites, finally moulting and even remaining in the same place as adult females. After hatching from the eggs the young larvae wander back over the abdominal tergites. They can be found in large aggregations on the last covered tergites, apparently awaiting an opportunity to leave the host (this is presumably provided, as in other species, by copulation of the host).

2. The larvae of *D. cryptobius*, on the other hand, move forwards along the lower side of the elytra. They alight on a certain small area near the back of the upper surface of the cutaneous wings. At this point they insert their beaks and commence to grow and develop. Usually only one, but occasionally two of these larvae are found on each wing. The larvae finally become rigid and adult females emerge. The latter climb into the thoracic intersegmental fold between episternum II and episternum III, where again usually only one female parasite is found on each side. When hatched from the eggs the young larvae move to the posterior of the animal along the underside of the elytra.

Each of the two species of mite is so strictly confined to its own sharply delineated area on the host that this suffices for the recognition of which of the two synhospital species is involved."

Their peculiar method of dispersal provides an explanation for the host specificity of the mites. "In a number of cases the larvae have been observed to change to a new host during copulation, so that they invariably reach another individual of the same species. This appears to be the sole method of distribution since the number of parasites on a host is only observed to rise steeply during the short reproductive period of the latter" (Figs. 35, 36). Further, young beetles are free of parasites up to the time of copulation. Regenfuss concluded from

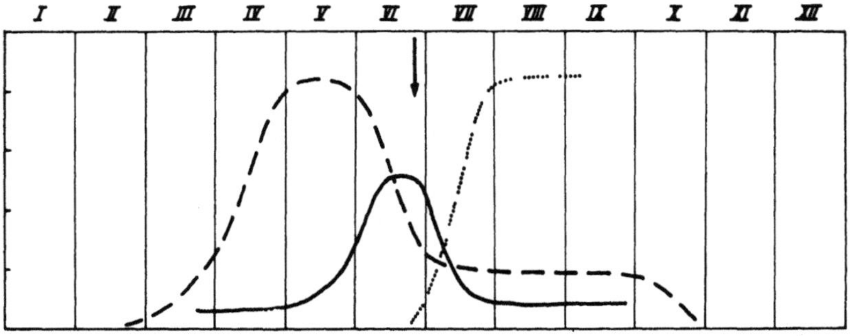

Fig. 35. Schematic representation of seasonal variations in infestation of a spring breeder *(Pterostichus nigrita)* with the mite *Eutarsopolipus vernalis*. *Broken line:* seasonal occurrence of *P. nigrita* (from Thiele, 1961); *arrow:* emergence of young beetles (dotted line); *continuous line:* infestation = parasitized beetles in percent. From Regenfuss, 1968 (slightly modified)

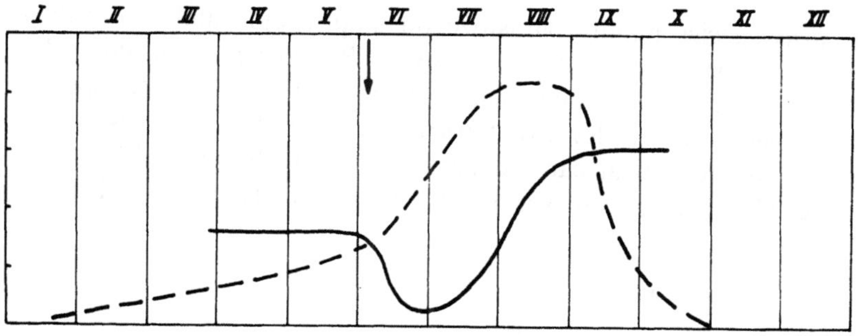

Fig. 36. Schematic representation of seasonal fluctuations in infestation of an autumn breeder *(Pterostichus vulgaris)* with the mite *Eutarsopolipus pterostichi*. *Broken line:* seasonal occurrence of *P. vulgaris* (from Thiele, 1961); *arrow:* emergence of young beetles; *continuous line:* Parasitic infestation = parasitized beetles in percent. From Regenfuss, 1968

his discovery of their mode of dispersal that "we have to assume from these results that many species of carabids survive for several years and reproduce repeatedly during the successive years". This conclusion has found increasing confirmation (see Chap. 6.B.I.).

The extent to which parasitism occurs is dealt with at the end of this chapter. The exact degree of damage caused to carabids by Podapolipidae seems to be unknown.

Acari: Sarcoptiformes. Typical parasites such as the itch mite belong to the order of the sarcoptiformes. Members of the Canestriniidae family have often been cited as parasites of carabids although only Sturani (1962) has provided any exact data concerning the hosts of the following species:

Canestrinia procrusti Berlese	on *Carabus coriaceus*
Canestrinia carabicola	on various *Carabus* species
Canestrinia procera	on *Procerus gigas*
Canestria sardica Vitzthum	on *Carabus morbillosus*

Acaridiae (= Tyroglyphidae) also occur in masses on carabids, although apparently only in over-moist cultures. Delkeskamp (1930) wrote of *Carabus nemoralis* that the beetles looked as if dusted with white powder. However, this is a further case of phoresia and not of parasitism. Like Delkeskamp, I myself observed (in *Pterostichus nigrita* and *P. angustatus*) that the mites rapidly left the beetles if the substrate dried out. Beetles afflicted with extremely large numbers of parasites suffered damage.

Hymenoptera: Proctotrupoidea. Proctotrupoids are the most important parasites of carabids among the Hymenoptera. Usually only the males of Proctotrupoidea are winged, and the wingless females dwell characteristically on the ground. Proctotrupoidea are therefore typical parasites of soil arthropods. The females lay their eggs on the carabid larvae. According to Weidemann (1965) Proctotrupoidea are relatively nonspecific as regards host, but his data reveal that *Codrus* species mainly parasitize on staphylinids and *Paracodrus* on elaterids. *Amara* species are listed as hosts of *Proctotrupes gravidator* L., whilst Briggs (1965) reports *Harpalus aeneus* as being a host for this species and *Harpalus rufipes* as a host for *Proctotrupes gladiator* Hal. Species of the genus *Phaenoserphus* are typical carabid parasites although some also attack staphylinids (Fig. 37). Proof of host specificity based on cultures has only been

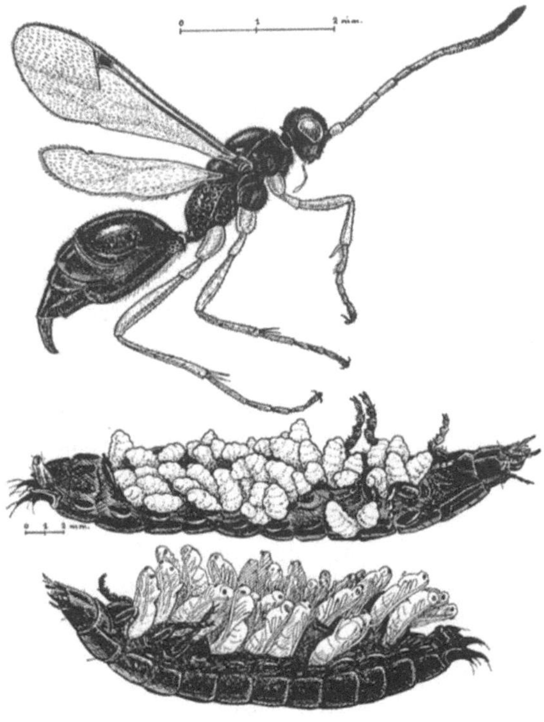

Fig. 37. Female of the Proctotrupoid genus *Phaenoserphus (above). Centre:* a larva of *Carabus glabratus*, from which the parasite larvae are emerging. *Below:* the pupae of the parasite on the dried-up larva. From Sturani, 1962

obtained in a few cases. The most common carabid parasite seems to be *Phaenoserphus viator* Haliday, which, according to Weidemann, who has summed up the older literature, has been shown to parasitize on *Carabus granulatus*, *C. violaceus*, *Pterostichus niger*, *P. vulgaris* and *Nebria brevicollis*. Davies (1967) observed *P. vexator* Nixon in *Notiophilus biguttatus* and I, personally, have isolated *P. dubiosus* Nixon from *Nebria brevicollis*.

In England *Phaenoserphus viator* was found in *Nebria brevicollis*, *Pterostichus vulgaris* and *P. madidus*, *Phaenoserphus pallipes* (Latreille) in *Notiophilus* spp., *Nebria brevicollis*, *Agonum dorsale*, *Pterostichus vulgaris* and *Calathus fuscipes*. "*P. viator* is gregarious and occurs mostly in winter-active larvae, *P. pallipes* is solitary and usually parasitizes summer-active larvae" (Critchley, 1973).

Sturani (1962) obtained no fewer than 40 adult specimens of *Phaenoserphus* from one larva of *Carabus glabratus*. According to the little information available carabid populations do not appear to be very heavily infested. I personally found that only four of 55 individuals of *Nebria brevicollis* caught at the third instar harboured parasites. In soil samples taken from different fields over the course of four years Scherney (1959a) found only two *Carabus* larvae infested with parasites. Only Davies (1959) expresses the opinion that the Proctotrupoidea exert a considerable influence on carabid populations in some places (see Chap. 3.C.II.). Critchley (1973) found that the level of parasitism by *Phaenoserphus* varied from year to year and from field to field. In 3rd instar larvae of *N. brevicollis*, death was caused in four cases by these parasites at rates of 25% (n=294), 16% (n=346), 4% (n=182), and 9% (n=144).

Hymenoptera: Braconidae. These Hymenoptera appear to parasitize on carabids far less frequently than the previous group. The larvae of an unidentified species of braconid were in rare cases found in Canada as parasites on *Pterostichus vulgaris* (Rivard, 1964b). A single case of the occurrence of a braconid species, identified as *Microtonus carabivorus* Mues., in a *Galerita* species was reported by Thompson (1943, quoted from Rivard, 1964b). Luff (personal communication) was so kind as to inform me of his finding a species of *Microtonus* "in large numbers" in *Harpalus rufipes*.

Hymenoptera: Mutillidae. The velvet ant *Methoca ichneumonides* Latr. is a parasite specialized on Cicindelinae. Its behaviour was graphically described by Burmeister (1939, see Figs. 38, 39): "The wingless female penetrates the hideout of the *Cicindela* larva, is grabbed and lifted up by the mandibles of the larva, but remains unharmed on account of her slender build. Only in this position can the *Methoca* female apply its paralysing sting to the throat or between the thighs of the beetle larva. When the *Methoca* is freed due to

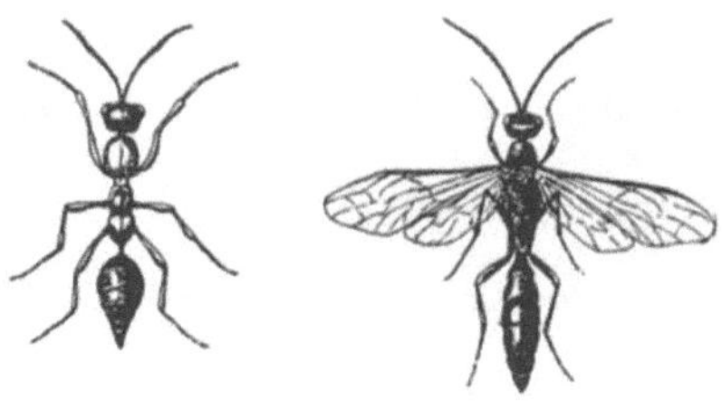

Fig. 38. The mutillid wasp *Methoca ichneumonides*, a specialized parasite of Cicindelinae. From Burmeister, 1939

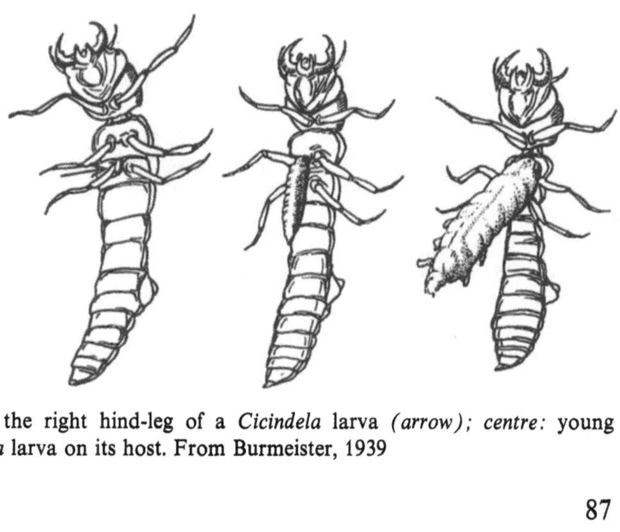

Fig. 39. *Left: Methoca* egg on the right hind-leg of a *Cicindela* larva *(arrow)*; *centre:* young larva; *right:* fully grown *Methoca* larva on its host. From Burmeister, 1939

the waning of its victim's strength it invariably lays its egg on the larva, near the hip joint of one of the hind legs. When this is done, it seals the sandy hideout. The larva soon hatches but remains on the spot where the egg has been layed, fastening itself onto the intersegmental skin of the *Cicindela* larva, which it slowly proceeds to drain. After about four weeks, when the *Methoca* larva is fully grown, it releases its host, which so far has only been paralysed but now dies. The parasite proceeds to spin itself a cocoon in a convenient spot in the cicindelinid hole and pupates".

Diptera: Larvaevoridae (=Tachinidae). After the Proctotrupoidea, these are the next most frequently encountered insects parasitizing carabids. The following species have been mentioned in the literature:

Parasite	Host	Reference
Viviana cinerea FALL.	*Carabus solieri*	Sturani (1962)
	Carabus nemoralis	Sturani (1962)
	Carabus problematicus	Sturani (1962)
	Carabus hortensis	Sturani (1962)
	Carabus cancellatus	Sturani (1962)
	Carabus clathratus	Sturani (1962)
	Carabus glabratus	Sturani (1962)
	Carabus coriaceus	Sturani (1962)
	Carabus violaceus	Sturani (1962)
	Carabus purpurascens	Sturani (1962)
	Carabus hispanus	Sturani (1962)
	Carabus rutilans	Sturani (1962)
	Carabus auronitens	Sturani (1962)
	Carabus pyrenaeus	Sturani (1962)
Biomyia georgiae B. & B.	*Calosoma sycophanta*	Rivard (1964b)
Eubiomyia calosomae COQ.	*Calosoma* spec.	Rivard (1964b)
Freraea denudata ZRTT.	*Carabus scheidleri*	Sturani (1962)
Freraea gagathea R. D.	*Carabus monilis*	Lindroth (1949)
	Harpalus rufipes	Lindroth (1949)
Frontina austera MEIG.	*Carabus cancellatus*	Sturani (1962)
	Carabus violaceus	Sturani (1962)
	Carabus hortensis	Sturani (1962)
	Amara spec.	Nielsen (1909)
	Broscus cephalotes	Herting (1960)
	Calathus fuscipes	Herting (1960)
	Harpalus rufipes	Herting (1960)
	Pterostichus spec.	Nielsen (1909)
	Zabrus tenebrioides	Herting (1960)
Weberia pseudofunesta VILLEN	*Amara aulica*	Lindroth (1949)
	Harpalus rufipes	Lindroth (1949)
Dinera grisescens FALL.	*Harpalus aeneus*	Briggs (1965)
Larvaevoridae gen. sp.	*Pterostichus lucublandus*	Rivard (1964b)
Larvaevoridae gen. sp.	*Pterostichus vulgaris*	Rivard (1964b)

II. The Role of Parasites in Distribution of Carabids and in the Regulation of Their Population Density

Most authors judge the role of parasites in governing the population density and distribution of carabids to be small.

The only information we have as to the part played by parasitic fungi in the field is a little concerning Laboulbeniales. In southern Germany, Scheloske (1969) found 145 species of arthropods to be infested with 80 different species of Laboulbeniales. Of these parasitic species, 23 were found in carabids. Of 133 species of ground beetles investigated, 61 proved to be more or less seriously infested. The percentage of infested individuals within a single species ranged from 0.6 to 67.9. Species in which more than 50% of the individuals were infested included *Clivina fossor* (one species of *Laboulbenia*), *Patrobus atrorufus* (six species), *Harpalus puncticollis*, *Pterostichus minor*, *Agonum obscurum* and *A. dorsale*. Many more Laboulbeniales are found in moist than in dry habitats. Scheloske pointed out that host damage due to Laboulbeniales is very small. Nevertheless, of 12,596 individuals of infested and non-infested species of carabids 33.2% harboured Laboulbeniales. In a polder region in the Netherlands, recently colonized by carabids, 33 out of 89 species were reported to be infested with 11 different species of Laboulbeniales (Meijer, 1975).

Despite their being of widespread occurrence in nature in some places, gregarina seem to cause no damage to the carabid populations. Delkeskamp (1930) cites Wellmer's finding that 24 individuals out of 28 of the species *Broscus cephalotes*, and 24 out of 25 specimens of *Carabus coriaceus* were infested with intestinal gregarina parasites. Delkeskamp writes: "If gregarina did any great damage to their hosts the latter would long ago have been exterminated at this high rate of infestation", although he judges the harmful effect of coelom gregarinae to be larger (see Chap. 3.C.I.).

Rivard (1964a) dissected more than 5500 carabids (over 40 species) and found only 19 individuals, or 0.3%, to harbour parasites. Of these, seven cases involved mermithids, 10 braconids and two larvaevorids. The present author found that less than 10% of the *Nebria brevicollis* larvae investigated were infested with Proctotrupoidea.

Nevertheless, it cannot be claimed that our knowledge of the degree of parasitization in carabid populations is in any way representative. It may turn out that mites of the Podapolipidae family are of considerable importance. In the tables contained in his monograph, Regenfuss (1968) lists 6918 carabids belonging to 78 species, of which 837 (or 12.1%) had parasites. He comments: "The list reveals the extent of the material investigated but not the intensity of the parasitic infestation, which is known to vary greatly according to season as well as to show considerable differences within the collecting area." Of some carabid species only a few specimens were investigated, but of other species from several hundred up to more than a thousand individuals were studied. Nevertheless, parasites were found in 24 out of 78 species investigated. In some genera *(Harpalus, Abax, Molops, Brachinus)* no parasites were found despite ample material. Some species were severely attacked: of 1324 specimens of

Pterostichus vulgaris almost 40%, and of 608 individuals of *Pterostichus nigrita* 22% were parasitized by two species. Nevertheless, the epidemiology of carabid parasites has not so far been seriously studied. Regenfuss made his observations in southern Germany. Carabid material from western and northern Germany, including the two species of *Pterostichus* mentioned above, was found to be completely free of Podapolipidae. Regenfuss made the further observation (see Chap. 3.C.I.) that the mites only attack carabids during the reproductive period of the latter.

On the whole, the evidence so far at our disposal leads to the cautious verdict that parasites play no decisive role in the regulation of carabid populations.

Davies (1959) recorded a case in which varying degrees of parasitization might have been responsible for the differences in frequency of carabids in two habitats. Of two forest habitats, A and B, B showed considerably higher figures for freshly moulted adults of *Notiophilus biguttatus* and *N. rufipes* in two succeeding years. Larvae collected in A and bred further in the laboratory only produced adult beetles in six cases, and in 24 cases the presence of the proctotrupoid parasite *Phaenoserphus vexator* could be shown. From the larvae collected in B, however, 34 adult beetles were obtained and only five cases of parasitization were established. Davies considers it possible "that a smaller influx of immature adult beetles in the oak-sycamore 'mull' conditions of plot A may be ascribed to higher mortality as a result of parasitism among the larval population... the idea may at least be suggested here, that the 'mor' condition of the beech area, plot B, with a thicker and more complex litter layer, by providing protection from attack by the *Phaenoserphus* parasite, perhaps itself imposes a *Notiophilus* population structure of the kind indicated". This interpretation appears plausible, but requires confirmation from laboratory experiments.

Recently, Critchley (1973) found that species of *Phaenoserphus* were responsible for 20%–25% mortality in the larvae of several carabid species investigated, so that at least in some places, and in certain years, these parasites may be an important factor in the regulation of the population density of ground beetles.

D. The Predators of Carabids

I. A Survey of the Most Common Species Preying on Carabids

1. Insectivores: Hedgehogs and Shrews

Although little exact information is available, these mammals probably eat large numbers of carabids. Krumbiegel's (1932) is the only report concerning the hedgehog *(Erinaceus europaeus)*. He wrote: "...*Procrustes* and *Carabus* species are also found in its excrements and are as a rule devoured whole or at the most after removal of only the elytra. It is remarkable that *Erinaceus* and the Soricidae are not deterred by the acrid digestive secretions disgorged by carabids when attacked...".

Some data concerning shrews as enemies of carabids have been given by Murdoch (1966). He cites Lavrov (1943), who states that beetles form the main food source of shrews (*Sorex araneus* L.) in summer, and an unpublished thesis by Cowcroft (1954), according to which shrews of this species voraciously devour large numbers of carabids. "*Agonum fuliginosum* were very quickly seized and eaten by a captive water shrew. It is interesting that the elytra were spat out by this shrew and that elytra were fairly common in the marshes in spring". The author considers it likely that predaceous shrews account for his observation of a substantial reduction in numbers of *Agonum* and *Pterostichus* species in spring. Greene (1975) stated that the vagrant shrew (*Sorex vagrans*) in the steppe region of southeastern Washington was "perhaps the most important mammalian predator" on cychrines.

2. *Insectivores: The Mole (Talpa europaea)*

On account of the long-standing controversy as to whether the insectivorous mole is beneficial or detrimental, investigations on its nutrition have long been carried out.

We are indebted to Schaerfenberg (1940) for a particularly thorough study: his article includes a review of the investigations of earlier authors. Of these, Tauber found adult carabids (*Pterostichus* and *Amara*) in only two out of 50 stomachs investigated, and carabid larvae in seven. Sachtleben found adult beetles, including a few carabids, in 22 out of 140 stomachs. In the course of a more extensive study Hauchecorne, on the other hand, reported carabids as being "frequent" in 200 mole stomachs, and out of 130 cases in which beetle larvae were found 22 included carabid larvae.

Schaerfenberg personally analysed 300 stomachs. In evaluating the data he lists separately the number of stomachs in which a particular prey occurred and how many individuals of any one group of prey were found in the entire material. Adult beetles were found in 123 cases. Taking this figure as 100%, the carabids, with 26%, occupy the second place in the list, which is headed by Melolonthinae with 27.6%, the Elateridae occupying the third place with 20.3%.

Beetle larvae were found in 247 stomachs. Of these, carabids were present in 25.3%, Curculionidae in 33.2%, Melolonthinae in 34.4% and Elateridae in 62.8% of the stomachs. In all, 180 adult beetles had been consumed, of which 39 were carabids (27.6%), followed by Melolonthinae (22.5%) and Elateridae (15.5%). Among the much larger number of beetle larvae consumed (1256) carabids occupied only the fifth place with 90 (7.3%), the first four places being taken by Elateridae (24%), Melolonthinae (20.8%), Curculionidae (16.4%) and Coprophaginae (12.7%). More than half of the carabid larvae belonged to the genus *Pterostichus*.

This confirms that carabids represent a habitual prey for moles, but, in common with beetles as a group, they do not constitute a very large proportion of its food. Moles cannot therefore be expected to play a significant role in regulating the numbers in carabid populations. The moles capture more larvae than adults because most of the larvae lead a subterranean existence.

However, the above data originate from moles caught on cultivated land. Oppermann (1968), in a recent investigation, has shown that the proportion of earthworms found in the prey decreases from meadow via deciduous forest to pine forest, whereas the proportion of beetle larvae rises, and particularly the carabid larvae play a large role in deciduous forests.

3. Bats

Bats were long assumed to feed exclusively on flying insects caught whilst themselves in flight. The comprehensive year-round analyses of their excrements made by Kolb (1958) have revealed that at least some species of bats consume large quantities of insects from the ground. Having located them by their scrambling noises, the bats alight next to the insects, stalk them if necessary and pick them up, as Kolb showed experimentally and documented in films.

So far, such evidence has mainly been obtained for *Myotis myotis*. At least 14 classifiable species of carabids have been found in the food of this species of bat, at least 12 of them incapable of flight. "The remarkable fact has emerged that not only do carabids account for a substantial portion of their food, but that they are in fact the basic diet of mouse-eared bats altogether. In spring, immediately after the bats return from their winter quarters, the excrements consist almost entirely of carabid remains. Their share in the total food matter fluctuates considerably over the year and sometimes sinks as low as 10%, but it is never entirely absent." The following table summarizes the detailed information obtained by Kolb in 1956 concerning the composition of the diet of mouse-eared bats in Bamberg:

Season	Percentage of carabids in food		
Spring	almost	100%	
Beginning of May		60%:	occurrence of cockchafers
2nd half of June		10–20%:	masses of *Tortrix viridana*
July		40%	
August		60%	
September/October	about	30%:	occurrence of dung beetles
November: Carabids again account for the larger portion of the nutrition			

"No other group of insects can be followed so well throughout the entire year, so that it can safely be said that ground beetles form the staple diet of mouse-eared bats. Their percentage share in the composition of the food decreases in proportion to the increasing availability of other food species, but never completely disappears.

"This is not to say, of course, that carabids form the larger part of their diet, which would be absolutely incorrect, but since they are found in the mouse-eared bat at all times of year, even if sometimes in small numbers, they can be regarded as the iron rations to which the animals can resort in time of need" (Kolb, 1958).

4. *Rodents—Mice*

A considerable proportion of animal matter was found in the yellow-necked field mouse, *Apodemus flavicollis* Melchior, in a forested region in Czechoslovakia (Obrtel, 1973). Remains of prey were found in 78% of the 256 stomachs investigated: 97% of these were arthropods of the forest floor, 13.8% of them carabid larvae (together with only 0.3% staphylinid larvae) and 7.8% adult Coleoptera (of these 2.9% carabids). These results suggest that the mice could exert a considerable influence on the carabid larvae. Greene (1975) reported that deer mice *(Peromyscus maniculatus)* fed on many cychrines caught in pitfall traps.

5. *Birds*

It far exceeds the scope of this book to attempt to examine the enormous quantity of material concerning the nutrition of birds. For our purpose it suffices to pick out a few studies which illustrate the finding that nearly every kind of bird can prey upon ground beetles.

Kovačević and Danon (1952, 1959) published stomach analyses on 136 species of birds belonging to 34 families in Yugoslavia. They recorded whether or not carabids were among the prey without mentioning species or genus of the beetles. Although usually only a single individual of any one species of bird was investigated (occasionally a few of one species) carabids were found in 31 species belonging to 17 families, or half of those investigated. No doubt if larger numbers of any one of the species mentioned were to be investigated carabids would be found in far more species. The families are listed below (in brackets the number of species in which carabids were found):

Podicipidae (2)	Picidae (2)
Anatidae (1)	Corvidae (2)
Falconidae (2)	Certhiidae (1)
Rallidae (1)	Turdidae (4)
Otididae (1)	Sylviidae (4)
Charadriidae (2)	Motacillidae (1)
Scolopacidae (1)	Sturnidae (1)
Strigidae (1)	Fringillidae (4)
Upupidae (1)	

Even the low number of 47 individuals of the genus *Lebia* that were found as prey in birds in North America is, according to Lindroth (1971b), spread over 22 species of birds from 16 families (including, apart from those already mentioned, Icteridae, Tyrranidae, Vireonidae, Laridae, Phasianidae, Tetraonidae and, as proof that birds which catch their prey only during flight also eat carabids: Apodidae, Hirundinidae and Caprimulgidae). The data suggest that birds are among the most important species preying upon carabids.

It is a very interesting fact that not only do carabids provide food for adult birds but can also form an important part of the diet of the nestlings of smaller song-birds. Employing the neck-ringing method, Pfeifer and Keil (1958), in Germany, found the following percentages of beetles in nine species of birds investigated, only one of which had received no beetles at all (Table 19). Only qualitative information is given regarding carabids found in the food.

Table 19. Mean percentages (from 4–5 years) of beetles in the food of nestlings of nine species of song-bird. (According to Pfeifer and Keil, 1958)

	% Beetles (mean)	Carabids
Starling	10.1	Yes (no details)
Field sparrow	2.7	Not mentioned
Great tit	2.9	Not mentioned
Blue tit	0.7	Not mentioned
Marsh tit	0.0	—
Nuthatch	11.3	Apart from *Xylodrepa quadripunctata* especially the ground beetles *Calosoma inquisitor, Pterostichus niger, P. vulgaris, Carabus nemoralis, Abax ater.*
Pied flycatcher	9.3	Not mentioned, preferably Cantharidae
Garden redstart	13.8	Not mentioned
Spotted flycatcher	16.6	Not mentioned

Even where carabids are not specifically mentioned it can be assumed with certainty that they are among the prey. A remarkable fact is that the nuthatch feeds its young a high proportion of very large carabid species.

Some birds are definitely specialized to a certain extent on carabids, at least locally. This has been established for the shrikes (Laniidae). Mansfeld (1958) recorded 175 carabids in the food of the red-backed shrike (plus 12 larvae), together with only 59 Silphidae, 53 Elateridae, 26 Curculionidae, 11 Staphylinidae, 20 Cerambycidae and seven *Cassida* (Chrysomelidae). Here, too, nestling diet was analysed.

It is relatively easy to establish that every possible kind of bird eats carabids, but it is a much more difficult matter to assess the extent of the influence exerted on carabid populations. Stein (1960) investigated the insect population of two oak-hornbeam stands in Münsterland (Westphalia, Germany), which he considered, on the basis of his studies, to be identical as to vegetation and microclimate. Stand R, however, as a result of conservation measures, harboured three to four times as many birds as stand A. Stein found that the insect population in the former area was reduced, and interpreted this as being due to the birds. Nearly all of the more common carabids were found in smaller numbers in this area. A total of 2595 carabids was found in the area with fewer birds, whereas only 1806 were recorded in the area with more birds, which means a reduction in carabid numbers of about 30% in the second area.

Since their prey has been especially well analysed quantitatively a separate chapter has been devoted to birds of prey and owls.

6. *Birds of Prey and Owls*

Thanks to the extremely comprehensive investigations of Uttendörfer (1939) on the food of these birds we are well informed regarding their importance as carabid predators.

Birds of prey have apparently little influence on carabids. The latter have only occasionally been identified in the following species: Hobby (*Falco subbuteo* L.), sparrow hawk (*Accipiter nisus* L.), buzzard (*Buteo buteo* L.), honey buzzard (*Pernis apivorus* L.), short-toed eagle (*Circaetus gallicus* Gm.) and lesser spotted eagle (*Aquila pomarina* Brehm). The kestrel (*Falco tinnunculus* L.) plays a somewhat greater role, its prey consisting of up to 10% insects. According to Uttendörfer, Csiki investigated the contents of 94 stomachs in Hungary and found 520 insects in 50 of them, including one *Cicindela*, one *Calosoma*, two *Carabus* and 10 small carabids. Madon reported on 306 "pellets" from France and Switzerland, containing 333 vertebrates and 1073 insects, 56 of the latter being carabids. But of 151 insects found in the contents of 30 stomachs in France, only one was a carabid.

The red-footed falcon (*Falco vespertinus* L.) represents a much greater danger to carabids than the kestrel. In 88 stomachs containing numerous insects Csiki found four *Cicindela campestris*, one *Calosoma*, 28 *Carabus* and 205 small carabids, including *Bembidion*. The largest share of the prey, however, consists of ants.

Sufficient knowledge has been gained from thorough and extensive investigations on "pellets" for the most important owls to be listed according to the extent of the danger they present to carabids.

The results of Uttendörfer's investigations on "pellets" in Germany are summarized below:

Species of owl	Number of pellets investigated	Number of these pellets containing carabids
Long-eared owl (*Asio otus*)	85	3 = 3.5%
Barn owl (*Tyto alba*)	69	4 = 6 %
Tawny owl (*Strix aluco*)	138	40 = 29 %
Little owl (*Athene noctua*)	32	28 = 88 %

These results suggest that the little owl is one of the more dangerous enemies of carabids. One sample revealed as many as 98 carabids plus only seven vertebrates. "Ground beetles, large and small, account for the majority of the insect species identified in the little owl..." (Uttendörfer, 1939). The same author cites comparable results obtained by Madon in France: 2555 insects were found in stomach contents and "pellets", of which 58% were beetles, including three *Calosoma*, 82 *Carabus* and 365 smaller carabids. The carabid family was only exceeded in numbers by the 449 grasshoppers.

Similar results were reported from England (Hibbert-Ware, cited by Uttendörfer, 1939), where the following insects were identified in the prey of the little owl:

10,200 Earwigs
7000 Carabidae
3500 Staphylinidae
1600 Curculionidae
600 Elateridae
1700 Geotrupes

Especially frequent prey of the little owl were the following species:

Species of carabid	Percentage nightactivity according to Thiele and Weber (1968)
Pterostichus madidus	37–100%
Carabus violaceus	100%
Carabus nemoralis	20–100%
Nebria brevicollis	90– 99%
Harpalus aeneus	76%
Pterostichus vulgaris	67–100%
Abax ater	83–100%

It thus appears that mainly night-active carabids fall a prey to the little owl: the fact that it also occasionally hunts by day accounts for the presence of *Cicindela campestris* on one occasion.

A recent paper by Smeenk (1972) recorded his thorough investigations on the prey of the tawny owl in the Netherlands. By far the most numerous insects among its victims were the beetles. Of the 367 individuals found, carabids were, with 27%, second only to the Scarabaeidae (43%) in frequency and were distributed over 18 large and small species: the eurytopic forest carabids *Carabus nemoralis* and *Pterostichus oblongopunctatus* were the most common.

7. *Frogs and Toads*

An investigation carried out by Zimka (1966) in Poland revealed that carabids constitute the most important prey of *Rana arvalis* in the forest ecosystem (Fig. 40).

Highly active and nimble animals are more frequently caught by frogs than slow, sluggish species. Zimka established a close positive relationship between what he terms the "consumption coefficient" and the "mobility coefficient" (Table 20).

The consumption coefficient is given by

$$P_i = \frac{\text{Number of prey individuals in stomach contents}}{\text{Abundance of prey individuals}}$$

The mobility coefficient is given by

$$M_i = \frac{\text{Number of prey individuals in traps}}{\text{Abundance of prey individuals}}$$

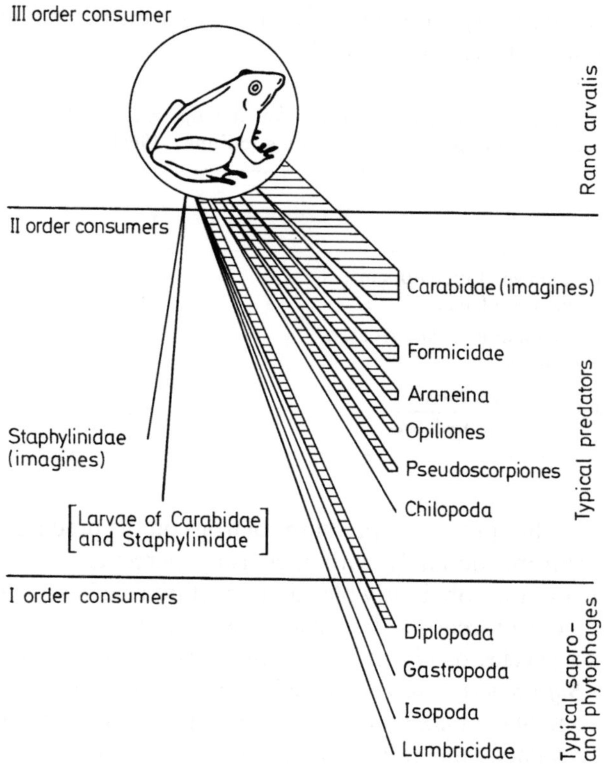

Fig. 40. Diagram illustrating consumption by *Rana arvalis* of different groups of litter fauna in forest habitats in Poland. Width of arrows indicates value P_i (see text), that is the number of animals contained in a stomach in relation to their density in the habitat. From Zimka, 1966

Table 20. Relationship between consumption- and mobility coefficients for animals of the forest floor caught by *Rana arvalis*. (According to Zimka, 1966)

	P_i	M_i
Araneina	20	41
Chilopoda	2	2
Pseudoscorpiones	9	6
Formicidae	33	75
Opiliones	12	149
Carabidae	100	533
Predators (mean)	*29*	*161*
Non-predators (Staphylinidae, Lumbricidae, Diplopoda; mean)	*5*	*20*

These figures show that, on account of its preference for quickly moving prey, the frog catches far larger numbers of predatory arthropods, including carabids, than saprophages.

Comparative studies involving various stations showed that the consumption coefficients of the frogs for predators, as compared with those for non-predators, increase disproportionately with increasing percentage of predators in the macrofauna of the soil (Table 21).

Table 21. Numbers of predacious soil arthropods caught by frogs at various stations with different percentages of predators. (According to Zimka, 1966)

	Station investigated		
	1	2	3
Percentage of predators in the macrofauna	51	75	81
Consumption of the predators by frogs	0.04	0.05	0.24
Ratio $\frac{P_i \text{ predators}}{P_i \text{ non-predators}}$	1.30	2.80	5.30

The "predatory pressure" of frogs on the predaceous soil arthropods increased disproportionately with increasing percentage of the latter in the soil fauna. Frogs thus obviously play an important role in regulating the position of carabids in the biocenosis. Larochelle listed (1974d) 62 species of carabids as prey of six species of North American frogs. In 342 individuals of the American toad *(Bufo americanus)* Larochelle (1974c) found 98 carabid species in the stomach contents (a maximum of 17 individuals in one stomach), and 19 more in excreta. Larochelle therefore describes this toad as the "champion beetle collector".

8. *Ants (Formicidae)*

In review articles on the subject (Burmeister, 1939; Scherney, 1959a) no particularly large role has been attributed to ants as enemies of carabids, mainly on the basis of the publications of Wellenstein (1929, 1954). He observed (1954) that carabids placed in ant heaps "...as a result of their speed, and their protective armour and sharp mandibular pincers, were successfully able to ward off ants".

More recent studies by Kolbe, however, reveal that ants can, in fact, exert a very considerable influence on carabid populations. Kolbe (1968a) caught soil-dwelling Coleoptera in an oak-birch forest in a series of five traps, of which trap one was set up at 190 m and trap five only 10 m from a row of seven nests of *Formica polyctena*. The following carabid catches were made:

Trap	1	2	3	4	5
Carabus coriaceus	1	—	—	—	—
Carabus violaceus	1	—	—	—	—
Carabus problematicus	1	5	2	—	—
Carabus glabratus	1	—	—	—	—
Trechus quadristriatus	1	2	1	—	1
Trichotichnus laevicollis	1	1	1	5	—
Pterostichus oblongopunctatus	7	6	23	20	—
Pterostichus niger	4	—	—	—	—

Trap	1	2	3	4	5
Pterostichus strenuus	—	—	—	1	—
Abax ater	12	17	12	28	1
Abax parallelus	—	1	1	1	—
Molops elatus	3	5	12	9	—
Sum of individuals	32	37	52	64	2
Sum of species	10	7	7	6	2
Number of *Formica polyctena* individuals caught	—	1	327	332	1010

These results indicate that in the vicinity of the *Formica* nests a sharp decrease in numbers of both species and individuals of carabids, and according to Kolbe of all other soil-dwelling *Coleoptera*, can be observed.

Further investigations by Kolbe (1969) show that carabids are attacked and severely injured by ants, which means that the negative correlation between frequency of ants and carabids has a real, causal connection. *Abax ater* individuals set loose on highly frequented ant paths were attacked. "I consider an attack to be successful if a worker ant takes a biting hold on a beetle. In most cases the ants clung onto extremities and antennae with their mandibles. As a rule, the experimental animals were able to shake off the attack in less than one minute and seek out a nearby hiding place. It cannot be said to what extent ant secretions entered any possible wounds" (Kolbe, 1969). Carabids that had been attacked by ants were subsequently observed in a terrarium under the optimum conditions also provided for the untreated control animals. Mortality rates were as follows:

Date	Percentage deaths I	II	III
17. 7.	0	0	0
23. 7.	0	0	0
30. 7.	0	5	0
6. 8.	0	10	0
13. 8.	10	20	0
20. 8.	10	20	0
27. 8.	30	30	0
30. 8.	50	50	0

I = 10 individuals of *Abax ater*, of which each animal had suffered 2–4 successful attacks by ants
II = 20 individuals of *Abax ater*, 6–10 successful attacks each
III = 20 control animals

Thus in 45 days mortality amounted to 50% in the attacked animals, whereas all control animals survived. Kolbe obtained comparable results using *Pterostichus oblongopunctatus*.

This seems to indicate that ants can considerably influence the carabids' chances of occupying a habitat. On this point Lindroth informs me: "Ants also play a significant negative role due to their occurrence in large masses, for example under stones, disturbing and putting other insects to flight. A reduction in number of species and individuals near ant nests is not necessarily due to predation alone."

Darlington (1971) places great emphasis on the role played by ants in tropical lowlands as competitors of ground-dwelling, wingless carabids and, on the basis of his investigations in New Guinea, claims, "that the tropical fauna, which is larger than the temperate one in any case, would be still larger if it were not for the presence of ants ...".

Similar conclusions were reached by van der Drift (1963) from his studies in Suriname, South America: "Carabids were notably scarce in our samples, and in the pitfall traps only a few species occurred frequently. It seems that the niche occupied by the Carabidae in the temperature zone is filled by Formicidae in the tropics". This claim seems to be somewhat exaggerated, in contrast to that of Darlington, and it should be borne in mind that the studies of van der Drift were carried out almost entirely on cultivated land, where the presence of man is probably very favourable to ants (the numbers caught by van der Drift would suggest this).

9. Predacious Flies—Asilidae

Lavigne's (1972) thorough scanning of the pertinent literature revealed that in 24 species of Asilidae a total of 25 species of Cicindelinae were identified as prey. The observations were distributed over 12 different countries in North, Central and South America, Europe, Africa and Asia. No shadow of doubt remains as to the not insignificant role played by Asilidae as predators on Cicindelinae.

10. Araneae (Spiders)

The vague comment is often encountered in the older literature that "ground spiders" are hostile to carabids. More explicit information on this point is provided by Tretzel (1961). The Agelenid species *Coelotes terrestris* (Wider) preys upon certain species of carabids. Its living quarters are self-constructed galleries in the ground, connected by a corridor with a web-lined, funnel-like trap. In a pine forest such webs were encountered both at the foot of tree trunks (tree web) or in the moss, and frequently contained two carabid species, *Synuchus nivalis* and *Pterostichus oblongopunctatus*. The following table shows, however, that compared with other insects (particularly ants and Tenebrionidae) carabids do not account for a large part of the prey in the case in question (Table 22).

Tretzel warns against placing too much value on the above figures since the fragments in the web are difficult to identify and count. In both cases carabids accounted for rather more than 10% of the total. Greene (1975), on the other hand, cites older reports according to which in Washington carabids constituted 40% of the prey of the black widow spider *Latrodectus*.

Table 22. Prey found in the webs of *Coelotes terrestris*. (According to Tretzel, 1961, summarized)

	Tree web	Moss web
Tenebrionidae	47	3
Carabidae	11	19
Other Coleoptera	5	3
Coleoptera: larvae	14	18
Lepidoptera	2	5
Dermapteroidea	—	2
Heteroptera	1	2
Diptera	—	2
Formicoidea	2	73
Other Hymenoptera	1	—
Undetermined insects	17	26
Araneae	1	3
Total	101	156

II. What Part is Played by Predators in Regulating the Distribution and Population Density of Carabids?

Predacious enemies are undoubtedly an important factor in the regulation of the population density of carabids. They are preyed upon by innumerable animals, including almost every kind of bird (insectivorous, omnivorous and birds of prey), and in particular by insectivorous mammals. Bats are capable of playing an important role in reducing carabid numbers, and frogs can be reckoned among the most dangerous enemies. In addition, other hitherto unmentioned insectivorous vertebrates probably play a not inconsiderable role, as, for example, Carnivora[6] among the mammals, as well as lizards.

Arthropods known to be important enemies of carabids (ants, predacious flies and ground spiders) have been studied in some detail. The combined effect of so many enemies cannot fail to have a regulatory influence on the population density of carabids, although no quantitative estimations seem to be available. In woodland investigated by Stein (1960, see above), he found that an 350% increase in the bird population led to a reduction of about 30% in carabids. However, the only evidence that a predator *(Formica polyctena)* can keep a habitat, otherwise suitable for carabids, almost completely free of the species was brought by Kolbe (1969). Nevertheless, this, too, only applies to the immediate vicinity of ant nests. The significance of predators in determining the presence or absence of carabids or certain species thereof in a particular habitat is thus probably to be regarded as slight.

---

[6] Neal (1948) reports *Carabus* and *Pterostichus* as prey of badgers in England and Maser (1973) records skunks *(Mephitis mephitis)* as catching Cicindelinae in North America. Greene (1975) reports that these skunks as well as raccoons *(Procyon lotor)* raided his pitfall traps and thus preyed on cychrines. Larochelle published lists of mammals (1975a), amphibia and reptiles (1975b) preying upon carabids.

E. Defence Mechanisms of Carabids

I. Mimicry in Carabids

Only a few cases of possible mimicry have been reported in carabids. Balsbaugh (1967) and Hemenway and Whitcomb (cited by Lindroth, 1971) made some contributions on the subject which Lindroth mentioned in connection with reports of his own investigations. It appears that mimicry is confined to members of the Lebiini tribe. Their model is presumably provided by certain species of Chrysomelidae of the subfamily Alticinae (flea beetles). A close connection exists between Lebiini and the latter since Lebiini (in addition to *Brachinus*) are the only parasitic carabids. They probably live in the flea beetles (see Chap. 3.F.VIII.), which ensures their occurrence in one and the same habitat.

By no means all *Lebia* species resemble the Alticinae species that have been suspected or proved to be their hosts. Further, it is far from being the invariable rule that a similarity must exist between all *Lebia* species and any Alticinae. In some cases, however, as for example between *Lebia vittata* F. and *Disonycha alternata* Ill. and between *Lebia viridis* Say. and several arboreal species of the genus *Altica* in North America, the external likeness is very striking. In both cases Lindroth observed isolated specimens of *Lebia* among the masses of the corresponding Alticinae (Lebiini do not usually occur in high population densities).

Lindroth suspects that this is a hitherto unobserved form of Batesian mimicry. Despite their very obvious colouring, flea beetles can easily escape from birds on account of their ability to jump. It therefore seems plausible that insect-eating birds learn to associate the distinctive colouring of the Alticinae with the fact that they are rarely to be caught, thus establishing a protective mechanism that also benefits the mimetic species. The following observations seem to offer support to such a hypothesis:

1. Of 1783 Alticinae identified in the course of analyses of stomach contents in North American birds, 95% were found in species that pick up their prey on or very near to the ground. Alticinae are thus seldom caught by birds that catch insects on plants.

2. Lebiini apparently possess no defence mechanism in the form of an unpleasant taste. In analyses of stomachs of birds, 47 specimens of 10 different species of Lebiini were found, including one suspected mimetic form. In one case, six specimens of *Lebia grandis* were found in the stomach of a grey plover *(Squatarola squatarola)*.

The larvae of an African genus, *Lebistina*, related to *Lebia*, parasitize on flea beetles of the genera *Diamphidia* and *Polyclada*. These are poisonous Alticinae and are used by Bushmen in the preparation of an arrow poison highly toxic to warm-blooded animals. A very convincing similarity exists between *Lebistina subcruciata* Frm. and *Diamphidia nigroornata* Stål and between *L. holubi* Pér. and *D. vittatipennis* Baly (Fig. 41). In these cases the model is protected by chemical means.

These few examples of mimicry in carabids have been put to discussion by Lindroth and require further investigation.

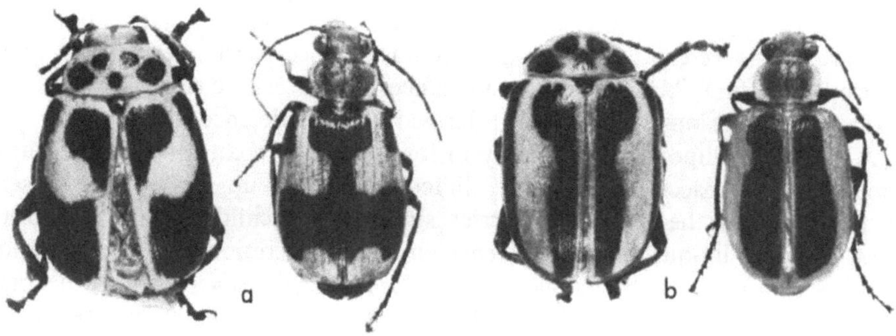

Fig. 41 a and b. Two South African species of the carabid genus *Lebistina* with their respective host of genus *Diamphidia*. These are flea-beetles famous for their production of arrow poison. 3 × actual size. (a) *Diamphidia nigroornata (left)* and *Lebistina subcruciata (right)*; (b) *D. vittatipennis (left)* and *L. holubi (right)*. From Lindroth, 1971 b

II. The Biochemical Defence Weapons of Carabids

Carabids, in common with all members of the suborder Adephaga, possess pygidial glands which produce defensive secretions. The gland exhibits great variability in structure and even greater variety in the nature of the secretions produced. Elucidation of the chemical structure of the latter is largely due to the work of Schildknecht and his co-workers (reviewed in Schildknecht *et al.*, 1968). These defence substances form a highly interesting chapter in biochemical ecology, from which deductions concerning the evolution of the family can be made. We shall return to this later (see Chap. 9.E.).

A Comparative Survey of the Defence Substances. 81 central European species were investigated. Carabids can be sharply divided into groups according to the class of substance produced:

Isovaleric acid and isobutyric acid
Methacrylic acid, tiglinic acid
Formic acid
Quinones (benzoquinone, toluoquinone; 2-methyl-3-methoxy-p-benzoquinone)[7]
m-Cresol
Hydroperoxide and hydroquinone

These groups of substances are as a rule characteristic for fairly narrowly defined taxonomic units. Further substances, sometimes in addition to the abovementioned, occur in carabids that belong to quite different taxonomic units, i.e.

Alkanes (paraffins): frequent;
Aliphatic ketones
Salicylaldehyde
Salicylic acid methyl ester: the last three are rare.

[7] Only in *Clivina fossor*.

Since isovaleric acid and isobutyric acid are weak acids and hardly poisonous, they are considered to have been acquired as defence substances early in phylogenesis (see Chap. 9.E.). All other substances, however, are more or less highly toxic cellular poisons and represent derived types of defence substances.

"The lack of lipophily, especially in formic acid, which, for example, makes penetration of insect cuticle more difficult, is made up for by the presence of an alkane in the role of a carrier substance" (Schildknecht *et al.*, 1968).

A hitherto unique form of defence involving secretions from the pygidial gland is exhibited by the female of *Pterostichus lucublandus* towards the males of the species (Kirk and Dupraz, 1972). "When a female is receptive, she will usually mate with a newly introduced male within 5 min. However, an unreceptive female approached by a male will run from him. Then, if pursued, she discharges a blast of liquid substance from the tip of her abdomen toward him. A male receiving the discharge immediately stops running and ... within 10 s, his movements become so uncoordinated that he may roll over on his back and be unable to regain his footing, his legs become stiff, and movement ceases. He remains in his deathlike coma for 1–3 h, but ca. 15–30 min after the first signs of revival, recovery appears to be complete". The females are immune to their own secretions. In the author's opinion the females also make use of their weapon to prevent the males from consuming the eggs at the time of laying.

The "Explosive Chemistry" of the Bombardier Beetles

The defensive behaviour of the bombardier beetles (Brachinini) has been very graphically described by Schildknecht, Maschwitz and Maschwitz (1968): "If one chances to turn up a stone beneath which a community of bombardier beetles has taken up quarters, they at once give a demonstration of their unique explosive weapon. At the same time as trying to escape in all directions or to dig themselves into the ground they fire off an evil-smelling discharge which is accompanied by a distinctly audible sound. The strong heat produced by the discharge can also lead to a flash which helps to scare off an attacker in the dark".

The composition and mode of action of the secretion was explained by Schildknecht *et al.* (1968). The paired anal glands of the Brachinini each consist of the actual lobe of the gland, a collecting bladder and a strong, highly sclerotic "firing chamber". The glands produce two initially inactive precursors for the explosive reaction, hydroquinone and hydrogen peroxide. The solution in the collecting bladder contains about 10% hydroquinone and 25% hydrogen peroxide. By means of the orifice muscle the solution is propelled into the firing chamber, into which a 40–60% protein solution consisting of catalases and peroxidases is poured by glands in its walls. The enzymes split the peroxide into water and oxygen and oxidise hydroquinone to p-quinone which is then expelled under the propelling action of the liberated oxygen.

Eisner (1972) and Aneshanley *et al.* (1969) calculated that the heat content of this mixture upon extrusion is 0.19 cal/mg, which is sufficient to raise the temperature of the solution to 100° C. Using microcalorimeters and thermistors an energy content of 0.22 cal/mg and a temperature of 100°C were, in fact, measured. On films and photographs it could be seen that the substance extruded consisted of "vapour as well as droplets". Eisner attempted to "prey" on *Brachinus* himself. "The moment I took the beetle between the lips it discharged and the sensation of heat was immediate and disagreeable".

III. Sound Production by Carabids

The literature contains next to nothing concerning the ability of ground beetles to produce sounds. Not until 1972 was the structure connected with sound production in *Cicindela tranquebarica* subjected to closer investigation by Freitag and Lee. Tiger beetles possess a pars stridens on the dorsal aspect of the costae and subcostae of both wings. By vibration of the alae this "file" is drawn across the "scraper" (plectrum) on the elytra and a soft but clearly audible sound is produced. In the same year, some older papers concerning the stridulation apparatus of carabids were listed by Larochelle (1972c).

Freitag and Lee have found the same type of "alary-elytral type" of stridulation apparatus in 62 Cicindelinae and 13 other carabids of very widely differing systematic status. The present author has personally observed that *Cychrus caraboides* is also capable of sound production. (Note at correction: Stridulation in this species had been reported as early as 1906 by Bagnall, cited by Greene 1975, who reports on this behaviour in North American cychrines of the genus *Scaphinotus*.) Stridulation and chirping apparatus in *Carabus irregularis* have been described in detail by Bauer (1975a).

Concerning the significance of stridulation, the above authors write: "Although we are not certain of the function of sound produced by species of the genus *Cicindela*, our personal field and laboratory observations suggest that the sound produced in tiger beetles is used for warning. That is, tiger beetles stridulate when they are disturbed by other organisms." This is confirmed by my own observations on *Cychrus*. The animals produce the sound as soon as they are picked up between two fingers.

According to Freitag and Lee there is no evidence so far that stridulation of Cicindelinae is in any way connected with their sexual behaviour.

Elaphrus cupreus and *E. riparius* stridulate by rubbing the two dorsal rows of stridulatory bristles over two areas of parallel ridges on the inner surface of the elytra. Although stridulation occurred in all stress situations it was never observed in connection with normal copulation. The chirping of *Elaphrus* therefore serves rather to scare off enemies than for intraspecific communication (Bauer, 1973), although the stridulation of *E. riparius* has scarcely any effect on the common sandpiper, *Tringa hypoleucos;* if at all, the bird was only irritated at the very beginning and if not hungry (Bauer, 1975a). Concerning stridulation in the cychrines Greene cites a remark of Wheeler (1970) that *Scaphinotus* spp "can quickly intimidate a field mouse, *Peromyscus leucopus* Raf., for some time after a single encounter...".

In a more recent paper, Bauer (1976) tested the behaviour of common sand pipers with individuals of *E. cupreus* which either were normal or unable to chirp after removal of the stridulation apparatus. "Stridulating beetles were released more often and swallowed less spontaneously. In approximately 20% of the captures, the beetles discharge their pygidial defense glands. The stridulation behaviour, therefore, is interpreted as a warning signal—comparable to warning colours—which is associated with defense secretion".

F. Nutrition of Carabids

The digestive apparatus of a large number of carabids was described in detail by Reichenbach-Klinke in 1938, his article also including a review of what was at that time known concerning their nutritional habits.

Carabids appear to be primarily carnivorous; exceptions to this rule will be considered later (see Chap. 3.F.VI.). Three types can be distinguished, according to their manner of food uptake. This is regarded as being particularly primitive in the case of the Omophroninae, in which no trace of extraintestinal digestion is detectable. This is also by no means the general rule for other carabids. The majority of Harpalinae, as well as certain Carabinae (e.g. the genus *Nebria*), constitute a second group in which all transitional stages from the ingestion of chewed but largely undigested fragments of food, to a potent extraintestinal digestion can be found. In *Harpalus*, *Amara* and *Clivina* no true extraintestinal digestion is present: although *Harpalus* and *Amara* pour a secretion onto their prey, preoral digestion is not considered to play a large role since undigested musculature of the prey can be found in the crop. Some Harpalinae such as *Broscus cephalotes* and *Pterostichus niger*, on the other hand, normally produce secretions which can be seen to cause discoloration and decomposition of the flesh of their prey.

Only in the third group is digestion exclusively extraintestinal. To it belong mainly species of the genus *Carabus*, as well as *Calosoma*, the Cychrini and apparently all Cicindelinae.

In what follows we shall examine first of all the more fundamental data concerning the nutrition of carabids. The numerous investigations that have been carried out with the definite aim of estimating the role carabids might play if applied to combatting insect pests form the subject of a separate chapter (see Chap. 4.A.I.1.).

I. Laboratory Observations on Choice of Food

Laboratory observations form the bulk of our knowledge concerning the nutrition of species with extraintestinal digestion, in which crop analysis is thus impossible. Investigations of this kind have mainly been carried out on members of the genus *Carabus*.

Numerous observations were contributed by Jung (1940), who also evaluated the work of earlier authors, of whom von Lengerken (1921) deserves special mention. Scherney later carried out extensive quantitative investigations on numerous species of *Carabus* which he summarized in 1959a.

All of these authors consider earthworms to be a significant source of food for *Carabus* species. Jung only observed attacks on larger slugs (*Arion* species) if they had first been decapitated. According to Scherney (1955, 1959a), on the other hand, *Agriolimax agrestis* is readily attacked and ingested by *Carabus auratus*, *C. cancellatus* and *C. granulatus* even when intact. The two latter species even climbed potato plants in pursuit of the slugs.

Almost every kind of insect within reach, adult or larva, is consumed by *Carabus* species. Although von Lengerken found that very hairy caterpillars were not touched by *Carabus* (not even when they were hungry), Jung reported the reverse: "The spiny, prickly caterpillars of butterflies are just as little immune to attack as the thickly furred caterpillars of Arctidiidae or those covered with poison hairs, such as *Euproctis chrysorrhoea* L., or *Macrothylacia rubi* L." An exception is formed by the small ermine moths (*Hyponomeuta padellus* L. and *H. malinellus* Z.), their larvae being rejected even if offered in small pieces. If a *Carabus* bit into such a morsel by chance it immediately spat it out and cleaned its mandibles thoroughly on a clump of earth.

Jung states that carrion originating from vertebrates is accepted by almost all species of *Carabus* but, according to Scherney (1959a), it should not have decayed too far, nor, on the other hand, be too dried out. I personally have kept (and successfully bred) many species of Carabinae and Harpalinae for months and even years on a diet of pure beef. Jung describes how *Carabus auratus* greedily sucks out "the foul, inky fluid from the inside of decaying prepupae and pupae of *Timarcha coriaria* Laich".

II. Nutrition in the Field, as Revealed by Analyses of the Contents of the Digestive Tract

This method of investigation is only applicable to carabids that tear up and ingest their prey with little or no extraintestinal digestion.

Although few investigations of this kind are so far available, those of Smit (1957), Skuhravý (1959), Dawson (1965), Penney (1966) and Hengeveld (unpublished) have been carried out with great care[8].

In general, a broad spectrum of prey species was revealed in the carabids under investigation, and widely differing preferences with regard to food were observable. The most extensive material was studied by Skuhravý (1959), and consisted of 2382 individuals from 12 field-dwelling species. In 409 cases the crop (and where also investigated, the intestine, too) contained identifiable fragments of food, in 540 cases a thick, brown soup, and in 1433 cases the crop was empty.

The results were evaluated semiquantitatively and are summarized in Table 23. Large species differences in composition of diet are revealed. Many species

[8] Larochelle has started to collect world lists of prey for various carabids on the basis of his own investigations and from the literature (Cicindelinae: 1974a; *Chlaenius:* 1974b; Cychrini: 1972a).

Table 23. Arthropods as prey of field-dwelling carabids. (According to Skuhravý, 1959)

	Arachnida	Acari	Formicoidea	Lepidoptera larvae	Aphidoidea	other Homoptera	Curculionidae	Staphylinidae	small Carabidae	Cantharidae larvae	Coccinellidae larvae	Silphidae larvae	other Coleoptera (incl. larvae)	Tenthredinidae larvae	Thysanoptera	Chrysopa larvae	Heteroptera	Dermaptera	Collembola
Pterostichus cupreus (summer/autumn)	2	2	—	5	5	4	—	1	—	1	1	—	—	—	1	1	—	—	—
Pterostichus cupreus (spring)	3	1	2	2	—	2	1	—	—	1	—	—	1	1	—	—	—	—	—
Pterostichus lepidus	4	1	4	5	5	2	—	1	1	—	—	—	—	—	—	—	1	1	1
Pterostichus vulgaris	4	1	3	5	2	3	3	3	—	1	1	—	—	—	1	—	—	—	—
Pterostichus macer	—	—	3	2	5	5	—	—	—	—	—	1	—	—	—	—	1	—	—
Harpalus rufipes	1	—	3	3	5	2	2	—	—	—	—	—	—	—	—	—	1	—	—
Harpalus aeneus	1	—	1	—	2	—	—	—	—	—	—	—	1	—	—	—	—	—	—
Agonum dorsale	—	1	1	3	5	1	2	2	—	3	—	—	1	—	1	—	1	—	—
Calathus fuscipes	—	1	5	5	5	2	4	—	—	—	1	1	—	3	1	—	4	—	—
Index total (for *P. cupreus* highest value only)	13	6	22	28	34	19	12	7	1	5	3	2	3	4	4	1	8	1	1

The numbers indicate (each referring to 50 dissected individuals per species):
5 = in 8–15 ind.; 4 = in 6–7 ind.; 3 = in 3–5 ind.; 2 = in 2 ind.; 1 = in 1 ind

also consume varying quantities of plant matter, and in some cases the composition of the food varies according to season.

Skuhravý calls attention to the following observation: after leaving its winter quarters in spring, *Pterostichus cupreus* ingests up to two-thirds plant matter and up to one-third animal matter, of which spiders and ants make up the larger part. In the second period of activity of the year, on the other hand, animals such as butterfly larvae, aphids and cicadas account for up to four-fifths of its nutrition.

Pterostichus lepidus is almost completely carnivorous (butterfly larvae, aphids, ants). *Pterostichus vulgaris* ingests up to nine-tenths animal matter, preferably caterpillars, spiders and ants. The diet of *P. macer* consists of up to three-fourths animal matter and is mainly made up of aphids and ants. Again, three-fourths of the food of *Agonum dorsale* consists of animal matter, largely aphids, caterpillars and larvae of Cantharidae. *Harpalus rufipes* consumes only 50% animal matter (aphids and ants), whereas *Harpalus aeneus* proves to be almost 100% phytophagous.

The reverse is true of *Calathus fuscipes*, a purely carnivorous species, in whose food aphids, caterpillars and ants predominate. The thorough investigations of Smit (1957), who performed quantitative analyses on the nutrition of *Calathus erratus* and *C. ambiguus*, provide the basis for drawing comparisons. The two species also mainly eat ants and aphids (Table 24). Although the animals were

Table 24. The percentage of different prey in the food of *Calathus erratus* and *C. ambiguus*. (According to Smit, 1957)

Prey	Percentage in the food of	
	C. erratus	*C. ambiguus*
Ants	28	20
Aphids	27	33
Heteroptera	15	15
Hymenoptera	11	14
Larvae	10	6
Araneida	3	5
Rest	6	7
Total	100	100

taken from the same habitat (a dune region near the Hague), certain differences were found in the composition of their food. These were manifest less in the overall distribution of prey in their diet than in the seasonal differences between the composition of the diet of the two species. *C. erratus* achieves its maximum activity 1/2–1 month earlier than *C. ambiguus*, i.e. from August to the beginning of September, whereas the latter species is not fully active before mid-September. Smit separates these two maxima by a line dividing "pre-" from "post-season". It appears that the proportion of aphids consumed rises for both species (steeply in *C. erratus* and only slightly in *C. ambiguus*) up to September, which is the time when aphids are most abundant (Table 25). At the same time, the proportion

Table 25. Ants and aphids as food of *Calathus erratus* and *C. ambiguus*. (From Smit, 1957)

	Calathus erratus			*Calathus ambiguus*		
Season	n	aphids eaten per beetle	ants eaten per beetle	n	aphids eaten per beetle	ants eaten per beetle
Pre-season	87	0.24	0.37	31	0.32	0.16
Post-season	19	0.63	0.05	47	0.46	0.32

of ants sinks drastically in *C. erratus* whereas it continues to rise in *C. ambiguus*. The latter species thus seems to exhibit a strong preference for ants whilst these are eaten by *C. erratus* only as a substitute for aphids, which it prefers.

Penney (1966) reports finding quite different preferences in *Nebria brevicollis* from deciduous forests in Scotland. Analyses of 305 crops revealed the following distribution of prey:

Small Diptera	38%	
of which:		
Culicidae		25%
small flies (mostly Sciaridae)		13%
Collembola	32%	
Mites	23%	
Spiders	4%	
Small earthworms	3%	

In this case, seasonal fluctuations in composition of prey were small.

We are indebted to Dawson (1965) for her thorough analysis of the prey of three moor-dwelling species of carabids. The animals most commonly found as prey in this investigation are shown in the following table (Table 26).

Table 26. Percent individuals of three species of carabids with remains of following prey in their crops. (According to Dawson, 1965)

	Agonum obscurum	*Agonum fuliginosum*	*Pterostichus diligens*
Number of carabids analysed	399	176	326
Remains of the following groups found in:			
Thysanoptera	5%	6%	6%
Diptera: *Forcipomya* sp. larvae	2%	9%	2%
Other Diptera	13%	11%	5%
Aphids	6%	5%	3%
Other Homoptera	11%	15%	15%
Collembola: *Podura* sp.	55%	58%	47%
Mites	18%	14%	19%
Opiliones	6%	7%	5%
Spiders	14%	13%	14%

The percentages are means taken from investigations made between April and October. Collembola and mites head the menu in each case. The figures for other prey animals vary greatly according to the numbers available from season to season. Thus in spring and summer Diptera play a large role whereas spiders attain a maximum in September.

The distribution of prey resembles that found by Penney in forest-dwelling *Nebria brevicollis*. The prey of the three moor-dwelling carabids investigated by Dawson show great similarities and seem to be mainly dependent upon supplies. No specialization on particular groups of prey animals is detectable in the different carabid species.

Following completion of this manuscript, food lists for 75 Cicindelinae from all over the world, and for 45 species of Chlaenius, another genus of world-wide distribution, have appeared (Larochelle, 1974a, b). These are the first lists of a comprehensive nature and confirm the wide variety of food consumed by carabids.

III. Oligophagous Predators

In the majority of species so far investigated a broad spectrum of prey has been found. A high degree of specialization in the adults, as far as is known, occurs in only a few genera.

1. Cychrus

That this genus specializes in hunting snails is reflected in its body structure (see Chap. 1.A.). The author's own investigations have shown that *C. caraboides* will not accept beef in captivity and can only be kept alive if fed with snails. The numerous North American Cychrini live on snails, too (Lindroth, personal communication, and Larochelle, 1972a). Larochelle records only a few cases in which insect larvae or plant matter were consumed.

Greene (1975), who described the prey catching behaviour of North American Cychrini in great detail, found that *Scaphinotus* adults were more successful in feeding on cracked snails as compared with intact ones whereas *Cychrus hemphilli* as the European species of the genus readily feeds on undamaged snails. Species of both genera attacked slugs as well, but while *Scaphinotus* also fed on earthworms, *Cychrus hemphilli* attacked the worms, but "all individuals then hesitated, broke off the attack, and backed away".

2. Notiophilus

In the vicinity of Erlangen in the Franconian Jura (Germany) Schaller (1950) observed a massive occurrence of a Collembola, *Hypogastrura bengtsoni* in 1945 and a concurrent increase in numbers of *Notiophilus biguttatus*. Laboratory observations extending over several months showed that this species of carabid can only be kept on a diet of Collembola, particularly *Onychiurus*, *Isotoma*, and *Hypogastrura* species. Mites belonging to the groups Oribatei and Gamasiformes could not be coped with by *Notiophilus* on account of their hard cuticle. But not even small fly maggots or other small ground animals were caught. Schaller therefore assumed that *Notiophilus* is specialized on Collembola.

According to Schaller, *Notiophilus* only catches moving individuals of Collembola, which it can perceive at a distance of several centimetres. He concludes that *Notiophilus* is well equipped to perceive movements. "This is readily understandable in view of the remarkable size of the eyes of this species". Schaller's observations indicate that in catching its prey the species relies entirely upon optical impressions, which agrees well with my own observations that *Notiophilus* is only active during the daytime. *Notiophilus biguttatus* provides a good illustration of the fact that laboratory observations do not always provide sufficient evidence of specialization on one particular type of prey. Crop content analyses in England revealed that only 18% of the prey consisted of Collembola whereas mites accounted for 67%. The remainder was made up of Diptera and other arthropods. Nevertheless, there were seasonal variations in the prey spectrum: in autumn the mites were especially predominant, but even in spring the Collembola hardly equalled them in numbers (Anderson, 1972). Larvae of *Notiophilus biguttatus*, according to recent observations, appear to be specialized on Collembola to a greater extent than the adults, as is indicated by the long sickle-shaped mandibles and the slender transition from head to thorax (Bauer, 1975b).

Notiophilus is much more successful in its attacks on Collembola than other carabids which hunt their prey during the day, e.g., *Asaphidion flavipes*. The reason (as deduced from cinematographic analysis) is that *Notiophilus* can better calculate the direction and distance of its attack, not because it is faster. In *Notiophilus*, 60% of the catching attempts were successful as opposed to a maximum of 4% in four other species of carabids (Bauer and Völlenkle, 1976).

3. *Calosoma*

In their natural surroundings the species of this climbing genus appear to feed mainly on butterfly caterpillars (Burgess, 1911; Burmeister, 1939). This will be dealt with in detail in Chapter 4. More recent laboratory experiments made by Dusaussoy (1963) show, however, that *Melolontha* larvae are also accepted (see also Chap. 4.A.4.).

4. *Dyschirius*

Many species of this genus live in large masses on sea coasts and the margins of inland waters, feeding for the larger part on staphylinids of the genus *Bledius*, and occasionally *Heterocerus* (Bro Larsen, 1936; Burmeister, 1939). They catch the staphylinids either on the ground or in their characteristic subterranean lairs. Since the legs of *Dyschirius* species are specialized for digging, they can easily extract *Bledius* from its quarters. *D. globosus*, widespread in the Palaearctic, also occurs in drier places, where it hunts staphylinids of the genus *Trogophloeus* (Burmeister, 1939).

Lindroth (1949), whose investigations should be consulted for details, performed comprehensive experiments on the relationships existing between *Dyschirius* and *Bledius*. He arrived at the conclusion that *Dyschirius* species consume many kinds of insects. "In the places where they live the fauna is extremely poor in species, due to a sparsity or complete lack of taller vegetation, and consists mainly of animals that can feed on the sand algae, i.e. mainly *Bledius*. The species of *Dyschirius* to be found is primarily determined by the soil conditions... The almost complete dependence of *D. obscurus*—and of *D. impunctipennis*—on *Bledius arenarius* is undoubtedly connected with the ... largely similar demands of these three species with regard to type of sand". Apparently, therefore, no direct obligatory interdependence exists between *Bledius* and *Dyschirius*.

5. *Nebria complanata*

This is a south and west European species exhibiting an extremely one-sided, but by no means obligatory, choice of food (Rudolph, 1970). It is found in coastal regions where, as larva and adult, it feeds almost entirely upon the amphipods *Talitrus saltator* and *Talorchestia brito*. In its upper intestine Davies (1953) found a small portion of fragments that might have been from isopods *(Oniscus, Amphidiscus, Armadillidium)*.

In 1956 Green reported the specialization of *N. complanata* on *Talitrus saltator*. In the same habitat he observed that *Bembidion laterale* chose the amphipods *Corophium volutator* (Pallas) and *Talitrus saltator* besides the larvae of Dolichopodidae.

The extreme specialization of *N. complanata* is not based on a genetically anchored adaptation, but is rather due to the overwhelming dominance of the above-mentioned amphipods in its environment. According to Rudolph, the animals accept many insects and meat in the laboratory.

6. *Consumption of Insect Eggs by Carabids*

Van Dinther and Mensink (1971) succeeded in a most elegant way in testing the extent to which carabids prey upon insect eggs in the field. The investigations were performed with the aim of estimating the importance of carabids in combatting the cabbage root-fly (*Erioschia brassicae* Bché.) in cabbage fields (see Chap. 4.A.I.1.).

Eggs of the house-fly *(Musca domestica)*, labelled with P^{32} in the laboratory, were stuck onto pieces of cardboard and exposed in a cabbage field. Carabids were caught in pitfall traps. Widely differing percentages of radioactive individuals of the various species were found, from which their importance as egg predators could be judged (Table 27).

The experiment reveals a very sharp distinction between *Bembidion* species and all others. Only the three species of *Bembidion*, particularly *B. lampros* and *B. tetracolum*, prey to a larger extent upon the eggs. Other species seem to consume eggs only occasionally. Larger species of carabids *(Broscus, Pterostichus, Harpalus)* show little interest in the eggs.

Table 27. Percentages of radioactive carabids after feeding with marked eggs of *Musca domestica* in a field experiment. (From van Dinther and Mensink, 1971)

	Number tested		Percentage radioactive	
	1968	1969	1968	1969
Bembidion lampros	364	308	7.7	6.8
Bembidion tetracolum	137	125	9.5	8.0
Bembidion femoratum	39	52	5.1	5.8
Nebria brevicollis	91	75	2.2	4.0
Broscus cephalotes	319	224	0.6	0.4
Calathus melanocephalus	241	118	1.2	3.4
Pterostichus lepidus	45	27	2.2	0
Amara spreta	99	126	2.0	2.4
Harpalus rufipes	73	185	1.4	0.5
Harpalus aeneus	31	35	3.2	0

IV. Quantities of Food Consumed by Adult Carabids

Scherney (1959a, 1961) carried out extensive feeding experiments on adults of various genera (Table 28) using earthworms, butterfly caterpillars, potato beetle larvae and pupae, saw-fly larvae and field snails as food.

Table 28. Average daily food consumption of different carabid species. (From Scherney, 1959a)

Species	Own weight	Average daily food uptake	x-times own body weight
Carabus auratus	0.640 g	0.875 g	1.36
Carabus cancellatus	0.560 g	0.775 g	1.38
Carabus ullrichi	0.680 g	1.051 g	1.55
Pterostichus vulgaris	0.150 g	0.507 g	3.38
Harpalus rufipes	0.120 g	0.278 g	2.31
Pterostichus cupreus	0.055 g	0.112 g	2.03
Calathus fuscipes	0.061 g	0.062 g	1.01
Nebria brevicollis	0.060 g	0.051 g	0.84

Most species take up at least their own body weight in food daily (the only exception is *Nebria brevicollis*, with slightly lower values). Furthermore, the calculated ratio of amount of food to body weight varies greatly from species to species, the highest value being that for *Pterostichus vulgaris*, which consumes more than three times its own weight daily.

For the sake of comparison, it is of interest to note that saprophagous soil animals take up much smaller quantities of food. Van der Drift (1951) collected data for diplopods and found a maximum uptake in *Glomeris marginata*, with 70% of its body weight, and a value of 51% for *Cylindroiulus silvarum*. The differences between species showing very different types of organization

Table 29. The consumption of house-fly eggs and larvae by different carabid species. (According to data of van Dinther, 1966)

Carabid species	Average wt (mg)	Average no. of eggs consumed/beetle/day	% of days on which eggs were consumed	Average wt of larvae consumed/beetle/day (mg)	The same in % of the wt of the predator	% of days on which larvae were consumed
Bembidion lampros	2.7	11.0±0.5	96	2.2±0.3	82	92
Bembidion tetracolum	8.4	27.8±1.2	98	5.4±0.5	64	100
Calathus melanocephalus	20	28.6±2.7	71	9.9±0.4	50	100
Calathus erratus	41	62.6±5.5	76	28.4±1.4	69	100
Amara spreta	34	102.5±10.3	85	10.5±0.7	31	73
Harpalus aeneus	55	182.2±12.2	92	13.7±1.2	25	86
Harpalus rufipes	134	54.7±18.3	56	37.1±3.6	28	98
Pterostichus lepidus	97	19.2±8.4	31	72.5±5.5	74	98

also seem to be smaller among the saprophagous diplopods than among the carnivorous carabids.

In order to obtain a basis for judging the importance of carabids as entomophages in field cultures van Dinther (1966) recorded the quantities of eggs and larvae of the house-fly consumed by a number of species (Table 29). Related to the body weight of the predator the quantities consumed were far below the values found by Scherney. The differences may be due not only to the fact that the beetles received a uniform diet in van Dinther's experiments, but also that different species of carabids were involved. It is interesting that the partially phytophagous *Amara* and *Harpalus* species ate considerably less than the other genera.

V. How do Carabids Recognize Their Prey? An Analysis of the Releasing Stimuli

A thorough analysis of the releasing stimuli by means of which carabids recognize their prey has only been made by Faasch (1968) on *Cicindela hybrida*. Since Cicindelinae are extreme day-active animals (see Chap. 6.A.III.), the results are not necessarily applicable to all carabids.

Observations were made on 150 individuals. The natural behaviour of the animals in their wild environment as well as in the laboratory indicates that prey of the size of a *Cantharis* can be perceived at a distance of 20–30 cm, and those of the size of an ant (3 mm) at about 10 cm. Cicindelinae react mainly, but not exclusively, to moving objects. They run up to such objects, take a short bite and withdraw briefly before renewing their attack. Faasch carried out a number of experiments using dummies. It could be proved statistically that dark objects on a light background have a high releasing value, whereas light objects on a dark background are much less effective. For this purpose discs of 6 mm diameter were used. The depth of the object was immaterial since halved beads of 6 mm diameter were in no way preferred to discs of the same size. In choice experiments an optimum releasing reaction was finally achieved by using discs 3 mm in diameter. Elongated objects (cylindrical beads) were no more effective than round ones in releasing the reaction.

The shape of the dummy on the other hand, does play a role in releasing the prey-catching response. It could be proved that a homogeneous black disc of 6 mm diameter was preferred to a double cross of the same total area (Fig. 42). Similarly, a black square was chosen in preference to one with a pattern. Similar results were obtained from further experiments of this nature, so that the possibility that Cicindelinae are capable of form perception cannot be excluded.

A moving bead of 3 mm diameter (on a rotating white disc) was chosen in preference to a stationary one of equal size (Fig. 43). No significant difference in choice was recorded, however, between moving and stationary beads of 5 and 9 mm diameter. A stationary bead of the optimum diameter of 3 mm was preferred to larger, moving ones.

Fig. 42. Mean of positive reactions *(height of columns)* of *Cicindela hybrida* to a homogeneously black decoy next to a patterned one of the same size. From Faasch, 1968

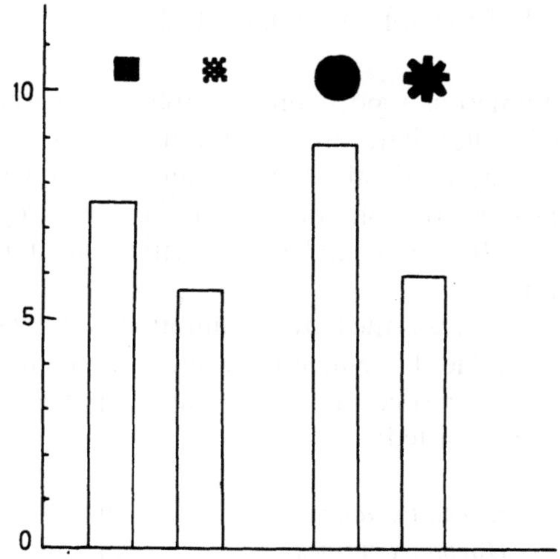

Hence it appears that the movement of the object of prey is only of limited importance as a releasing stimulus. This is understandable if it is borne in mind that *Cicindela* also finds food in the course of wandering aimlessly, by picking up in its mandibles objects such as insects or pieces of wood that chance to lie in its path. A chemical sense is involved in distinguishing between food and other objects: if black discs of 6 mm diameter were smeared with the juice of squashed flies, an average of 12.3 reactions were elicited as compared with only 3.1 using clean discs. However, the chemical sense is not involved in perception of distant objects as is shown by the fact that *Cincindela* even try to catch moving objects if they are separated from them by a glass plate. In nocturnal species a chemical sense might be of considerable use in localizing prey.

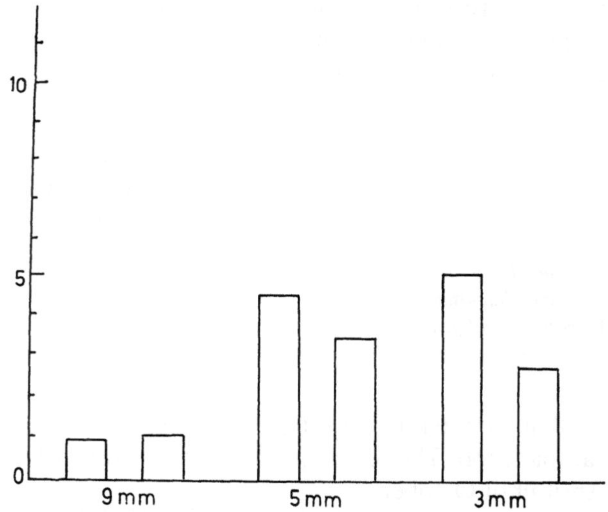

Fig. 43. Mean of positive reactions *(height of columns)* of *Cicindela hybrida* to moving decoys *(left)* next to stationary ones *(right)*. Three different sizes of decoy were employed. From Faasch, 1968

VI. Phytophagous Carabids

Numerous reports on phytophagous carabids exist in the literature (for reviews of earlier literature see, among others, Burmeister, 1939; Jung, 1940; Scherney, 1959a; Lindroth, 1949; Johnson and Cameron, 1969). Many earlier authors held partial phytophagy to be an exception in carabids, apart from *Amara* and *Harpalus*, and specialization on a pure plant diet was considered to be rare.

Whereas the latter assumption is still valid, the former requires some modification. The above-mentioned analyses of intestinal contents carried out by Skuhravý 1959 revealed the following proportion of plant matter in the diet of the species listed:

Harpalus aeneus	in 100% of the individuals
Harpalus rufipes	in 50% of the individuals
Agonum dorsale	in 25% of the individuals
Pterostichus macer	in 25% of the individuals
Pterostichus cupreus	
in spring	in 67% of the individuals
in summer	in 20% of the individuals
Pterostichus vulgaris	in 10% of the individuals
Pterostichus lepidus	in 0% of the individuals
Calathus fuscipes	in 0% of the individuals

For a whole series of genera in their natural environment, therefore, plant matter plays a considerable role in their nutrition, at least at certain seasons. The minority of species in Skuhravý's list are exclusively carnivorous.

Dawson (1965) determined the percentage of individuals of the species studied in whose crops remains of plant and animal matter were found. Since both were frequently encountered in one and the same crop the sum of the percentages given below exceeds 100.

	n	Percentage of guts containing animal remains	Percentage of guts containing plant remains
Agonum obscurum	438	94	65
Agonum fuliginosum	193	93	70
Pterostichus diligens	348	94	75

Plant matter is thus regularly encountered in these moorland species of carabids, too. The results given are mean values from investigations made from April until October.

The plant remains included unidentified fragments of higher plants; pollen from Leguminosae, Cruciferae, Caryophyllaceae, Umbelliferae, Compositae and the genera *Filipendula* and *Thalictrum*; fungal hyphae and spores, particularly of rust fungi, including *Triphragmium ulmariae* (Dc.) Link, a species that parasitizes on *Filipendula*; diatoms. Some of these, pollen and spores for example, might have entered the carabids in the intestinal tract of prey such as Collembola and mites.

These findings throw fresh light on the results of earlier laboratory feeding experiments. Jung (1940) reported that *Carabus* species, but especially *Carabus auratus*, eagerly consumed large quantities of squashed fruit, including pears, apples, strawberries, plums, greengages and bananas, although sour fruit was less readily accepted. Concerning the attitude of *C. auratus* to fruit Jung writes: "One receives the impression that the beetles almost overeat themselves in their greediness. Within an unbelievably short time the insects had filled themselves up and bored enormous holes in the fruit. Only pitiful remains of the meal were left over. The animals even picked up the dirty earth-coated remnants in their mandibles and tried to suck them dry."

Jung also reports on similar observations made by earlier authors on other *Carabus* species. Von Lengerken (1921) had seen *C. auratus* feeding on mushrooms and the poisonous bulbous agaric. Ankel (1916) found a *Carabus* individual eating dandelion flowers.

Scherney (1959a) reported from his comprehensive feeding experiments that *Carabus* species invariably chose animal matter in preference to plant matter (apple slices, decaying apples, unripe corn cobs and unripe ears of cereal, strawberries and ripe or decayed cherries) offered at the same time, and only ate the vegetable fragments in the complete absence of water.

On the other hand, Jung's observations tend to contradict the idea that *Carabus* only eats fruit in an attempt to satisfy its need for water, since he found that soaked wheat bread was also acceptable to *Carabus*. The animals poured out their gastric juices over the substrate. Eighteen individuals of three species of *Carabus* consumed an entire soaked bread roll within six days. Jung kept *Carabus granulatus* and *C. cancellatus* for six weeks on a diet consisting entirely of bread, without noticing any detrimental effects.

All of these results indicate that phytophagy is possible for many carabids and plays a substantial role in their nutrition. The recent data from analyses of the intestinal contents of wild beetles, cited above, confirm earlier statements made by Lindroth (1949) based on a comprehensive survey of the existing literature and on his own observations. His summary, shown below, is based upon extensive tables:

Animal matter is consumed by:	99 species, of which 31 only observed in captivity
Vegetable matter is consumed by:	85 species, of which 40 only observed in captivity
Thereof exclusively animal matter:	53 species
Thereof exclusively plant matter:	37 species
Mixed nutrition:	48 species

Lindroth concludes: "These figures clearly indicate that carabids consume large quantities of vegetable matter. In addition, no less than 48 of 138 species investigated here, i.e. 35%, were able to utilize both animal and vegetable matter. I am convinced that this is a normal state of affairs for carabids and that in the future careful feeding experiments will reveal the highly omnivorous character of these animals in general".

Johnson and Cameron (1969) collected data on the uptake of plant matter by more than 150 species of ground beetles. Their own laboratory investigations on 14 field species (three *Amara*, three *Anisodactylus*, three *Agonum*, three *Harpalus*, *Agonoderus pallipes* and *Pterostichus vulgaris*) showed that they all ate grass seeds in larger or smaller quantities, even if animal matter was offered at the same time (larvae, pupae and young, still soft beetles of *Hyperodes*-Curculionidae).

Adults of *Amara* species are reported to feed almost exclusively on ripe plant seeds, for which purpose they climb up into the vegetation, but they consume animal matter as well. The latter plays a greater role at the larval stage, although plant roots are also eaten underground (Burmeister, 1939).

An adult *Amara cupreolata* can consume 10–20 seeds of *Poa annua* in one night, which means an uptake of one-eighth to one-fourth of its body weight in the form of seeds, if ten seeds weigh 3.3 mg and the beetle 25 mg (Johnson and Cameron, 1968). The degree of specialization of phytophages can be large or small. *Amara ingenua* is highly polyphagous but also consumes animal substance. In choice experiments the species left untouched only three of the 13 types of plant seeds offered (*Anthriscus*, *Vicia*, *Viola*). The consumption of the other ten varied from 3% to 94%. The latter figure indicates a clear preference for *Polygonum aviculare*, a species that is never absent from the natural habitat of *A. ingenua*. "Its low growth form is particularly advantageous since the animal can readily reach the fruits" (Lindroth, 1949).

In general, the occurrence of Cruciferae seeds favours the presence of *Amara* and the seeds of Umbelliferae that of *Harpalus* species (Lindroth, written communication).

Zabrus tenebrioides, the cereal ground beetle, is a highly specialized phytophage. It climbs the stalks and completely devours the milky, ripe grains, leaving the outer protective layer of the harder ones. The beetles choose rye, wheat, barley and maize in preference to oats, which are rarely accepted. In autumn *Zabrus* eats the young leaves of the newly sprouting winter crop.

Zabrus larvae construct vertical shafts 20–40 cm deep, in the ground, into which they drag leaves and stalks of the cereal plants (Burmeister, 1939). According to Gersdorf (1937) and reports from Burmeister, *Zabrus* does not reject meat but in fact rather prefers fresh meat to lettuce seeds if both are offered.

An extreme case of specialized plant nutrition is exhibited by *Ditomus clypeatus* (Schremmer, 1960). The Ditomini constitute a separate tribe distributed in Mediterranean regions. Its mode of nutrition has been described in great detail by Schremmer. The animals live near the sea coast where he observed them climbing over the fruits of *Plantago lanceolata* and gnawing out single seed capsules which they then carried off to their underground lairs (Fig. 44). Three of the lairs investigated were built on the same basic plan. The beetles had a seed depot in a special storage chamber. First of all the *Plantago* fruits were carried into a depot where they underwent a fermentation process, following which the individual seeds were stuck onto the ceiling of the main storage chamber.

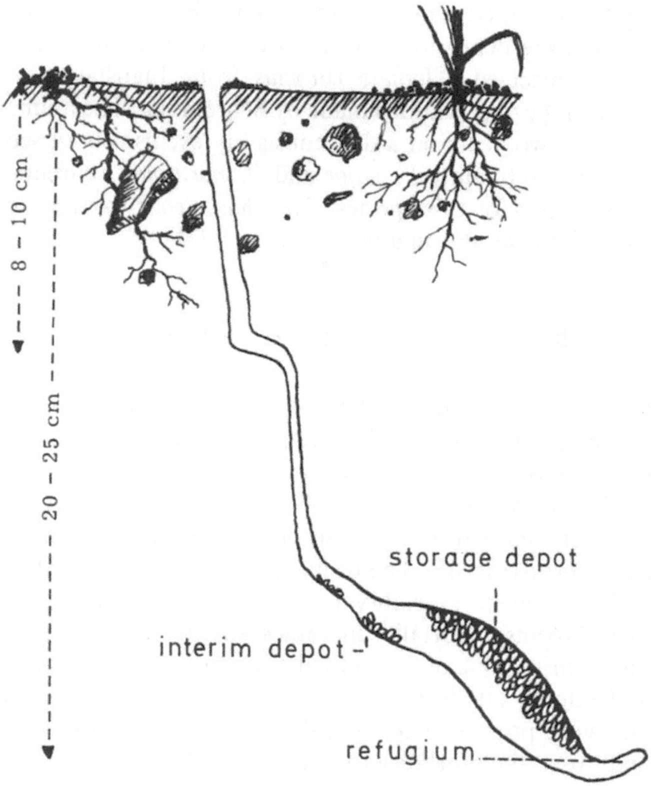

Fig. 44. *Above:* Section through the lair of *Ditomus clypeatus* with its store of plantain seeds. *Below: Ditomus* collecting seeds. From Schremmer, 1960

Up to 180 seed capsules were found in such a depot. The structure of the lair and the depository can be observed in small plexiglass containers. In June 1969 I received a female *Ditomus* from Jugoslavia, on which I was able to confirm the observations made by Schremmer. The animal survived in my laboratory for two years on a diet consisting exclusively of seeds from various species of *Plantago* (mainly *P. major* and *P. maritima*). Schremmer assumes monophagy on *Plantago* for this species. *Synuchus nivalis* carries seeds of *Melampyrum* into its lair in a similar manner (Manley, 1971).

VII. The Nutrition of Carabid Larvae

No successful analyses of the intestinal contents of larvae have been reported in the literature. Luff (1974) found a few annelid chaetae and unidentified bristles in larvae of *Pterostichus madidus*.

The development of carabid larvae on a variety of diets was the subject of a particularly thorough quantitative analysis made by Ferenz (1973) on *Pterostichus nigrita* (Fig. 45). The results clearly showed that the most rapid development and the lowest mortality (<10%) were obtained on a diet of insect larvae (meal worms). Mortality increased sharply if beef was fed alone, although even then almost 50% of the individuals achieved metamorphosis, whereas according to Rudolph (1970) the larvae of *Nebria complanata* could only be kept alive on living prey. Protein-containing artificial foodstuffs were practically useless and raised the mortality to as much as 100% in *P. nigrita*. A diet of earthworms raised the death rate even more than beef.

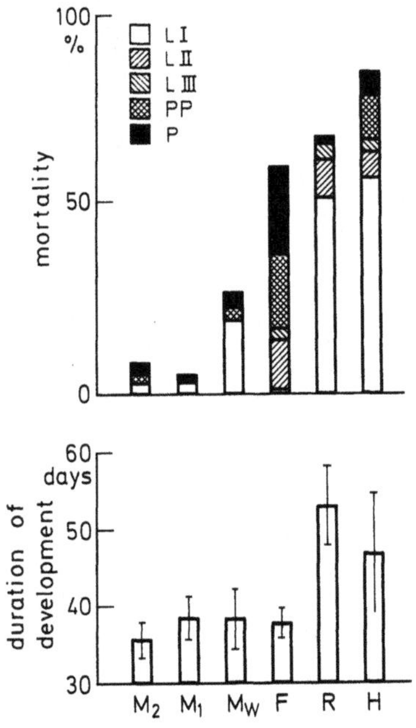

Fig. 45. Effect of various types of nutrition on the mortality of the individual developmental stages and total duration of development (mean ± standard deviation) of *Pterostichus nigrita*. M_2: 2×2 meal worm pieces per week; M_1: 2×1 meal worm piece per week; M_w: 1×2 meal worm pieces per week; *F*: chopped beef; *R*: earthworms; *H*: protein-containing artificial foodstuff (dog biscuits). From Ferenz, 1973

The larvae of other carabids, on the other hand, and particularly species of the genus *Abax*, even appear to be specialized on earthworms. Löser (1972) only succeeded in breeding *Abax ater* and *A. parallelus* in the laboratory on a diet of earthworms, and Lampe (1975) found the same to apply to *A. ovalis*, after Thiele (unpublished) had tried unsuccessfully to rear the species on meat and freshly killed insects. The findings of Neumann (1971) suggest that the same type of specialization may even exist in the field. *Abax ater* was found to have penetrated reforested areas of the lignite workings near Cologne only if recolonization by earthworms had already occurred.

Examples of this kind suggest that the larvae may be far more specialized with respect to diet than the adults. *Abax ater*, *parallelus* and *ovalis*, for example,

Table 30. Nutrition of *Pterostichus madidus* larvae as revealed by choice experiments. (According to Luff, 1974)

Invariably consumed	Percentage consumed	Number of experiments
Nematoda	100	1
Annelida	100	15
Isopoda	100	8
Thysanoptera	100	8
Hemiptera: Jassidae	100	3
Lepidoptera larvae	100	2
Diptera: larval Tipulidae, Bibionidae	100	20
Coleoptera: carabid eggs	100	10
Coleoptera: larval Hydrophilidae, Staphylinidae, Circulionidae	100	15
Sometimes consumed		
Aphids	86	7
Diptera adults	83	18
Collembola	79	14
Diptera: larval Drosophilidae, Stratiomyidae	75	12
Acari	70	10
Araneae	67	3
Diplura	67	6
Mollusca: Limax	50	8
Coleoptera: larval Carabidae, Cantharidae	40	10
Coleoptera adults	36	14
Phalangida	33	6
Diptera: puparia of *Drosophila*	25	4
Never consumed		
Mollusca: eggs of Limax	0	6
Diptera: larval Syrphidae	0	15
Coleoptera: larval Halticinae	0	5
Plant materials: seeds; leaves of both grasses and moss; fruit of strawberry; apple	0	17

can be kept for years, and even lay eggs, in the laboratory on a purely meat diet (Thiele, 1961, 1964a, 1968c). On the other hand, *Pterostichus nigrita, P. oblongopunctatus, P. angustatus, P. coerulescens, P. cupreus* and *Agonum assimile* could be bred over several generations in the laboratory on a pure beef diet (Thiele, 1968c) although the example provided by *P. nigrita* shows that improvement of the diet can bring a much higher degree of success in breeding. The results nevertheless show that extreme specialization with respect to diet is also rather the exception than the rule for the larvae of carabids.

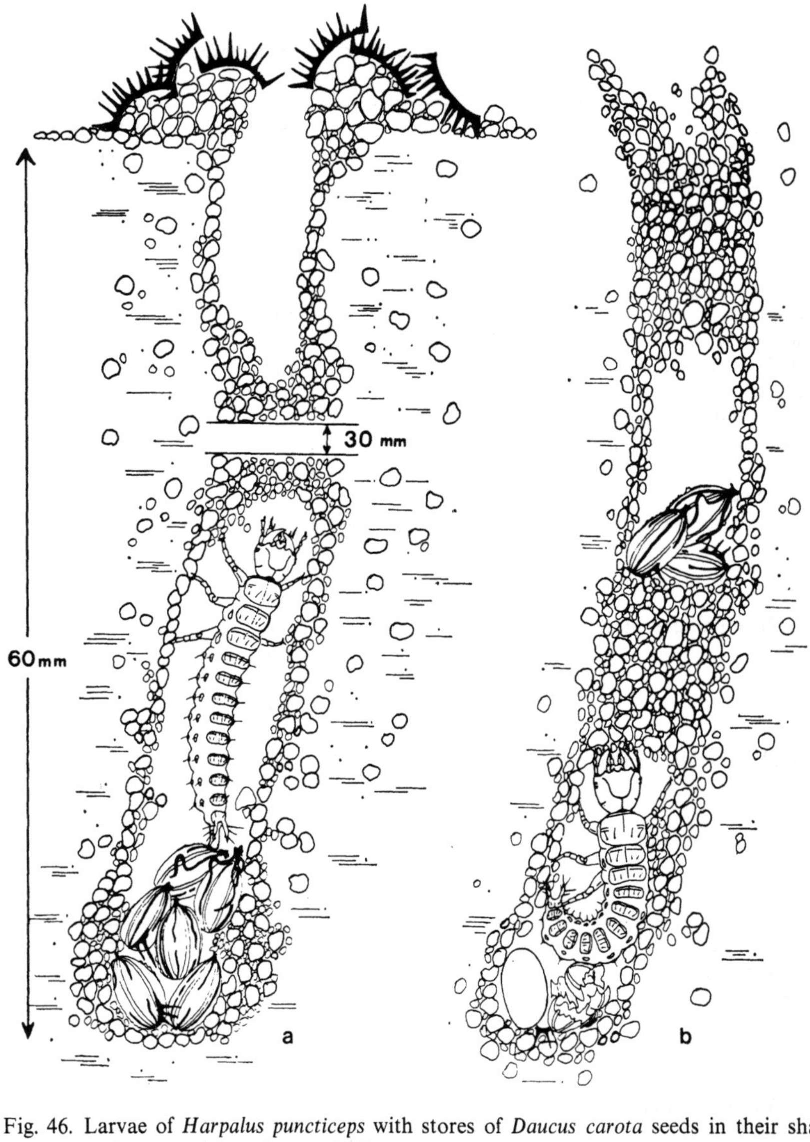

Fig. 46. Larvae of *Harpalus puncticeps* with stores of *Daucus carota* seeds in their shaft-like lairs. From Brandmayr and Brandmayr, 1975

This was further illustrated by the results of feeding experiments on larvae of *Pterostichus madidus* (Luff, 1974; Table 30). They accepted a wide variety of food, but it was remarkable that they totally rejected vegetable matter although the adults are partial phytophages. *Amara communis* is also carnivorous at the larval stage although the adults eat seeds (Burakowski, 1967). "In the Adephagid beetles, the adults tend to have more conservative feeding habits than the larvae, in which fluid-feeding and partial external digestion are apparently common. This is, of course, partly due to the simplified mouth parts of the larvae..." (Evans, 1965).

In the course of a campaign to combat the termite *Reticulitermes flavipes* (Kollar) in Hamburg, Weidner (1957) observed a larva of *Pterostichus (vulgaris?)* consuming a termite larva, and assumed, with good reason, that the larvae of *Harpalus aeneus*, too, prey upon these termites. Unfortunately, nothing is known of prey-predator connections between carabids and termites in their true native areas.

The consumption of vegetable matter by carabid larvae has seldom been reported. The soil-inhabiting larvae of *Harpalus rufipes* probably consume seeds of *Chenopodium album* in earthworm channels (Briggs, 1965). Larvae in the laboratory consumed ripening seeds of a variety of plants but grew most rapidly on those of *C. album*. An enormous concentration of *Harpalus rufipes* larvae was recorded in places where the plant was found (29–44/yd^2 as compared with a normal value of 2–3/yd^2). The larvae of *Harpalus pennsylvanicus* and *H. erraticus* in North America collect grass seeds in their living quarters (Kirk, 1972), which is reminiscent of the behaviour of *Ditomus* adults (see also *Carterus*, Chap. 3.B.IV.). The larvae of *Harpalus puncticeps* are purely phytophagous. They prefer seeds of *Daucus carota*, which they store in the small cells in which they undergo metamorphosis (Fig. 46; Brandmayr and Brandmayr, 1975).

Quantities of Food Consumed by Carabid Larvae

Quantitative data on the nutrition of carabid larvae are almost completely lacking. The most exact investigations are still those of Burgess (1911) on *Calosoma sycophanta*. If fed on brown-tail moth caterpillars of the fourth instar and gipsy-moth caterpillars of the fourth to sixth instar the larvae required 67–95 individuals in order to complete their development in an average of 28 days. Most effective was a diet of gipsy-moth caterpillars of the sixth instar alone, the larvae requiring an average of 41 (37–52) caterpillars and completing their development up to metamorphosis in only 14 days. Sturani (1962) observed that if fed only on snails of the species *Helicigona arbustorum* the larvae of *Carabus olympiae* required on an average the following quantities: during the first stage three snails of 12 mm diameter, during the second stage three snails of 18 mm diameter, and in the third stage three fully grown snails of approx. 24 mm diameter.

The Prey-Capture Behaviour of Carabid Larvae

Sturani (1962) also observed that what he termed the longimandibular larvae feed almost exclusively upon snails in contrast to the larvae of the brevimandibular

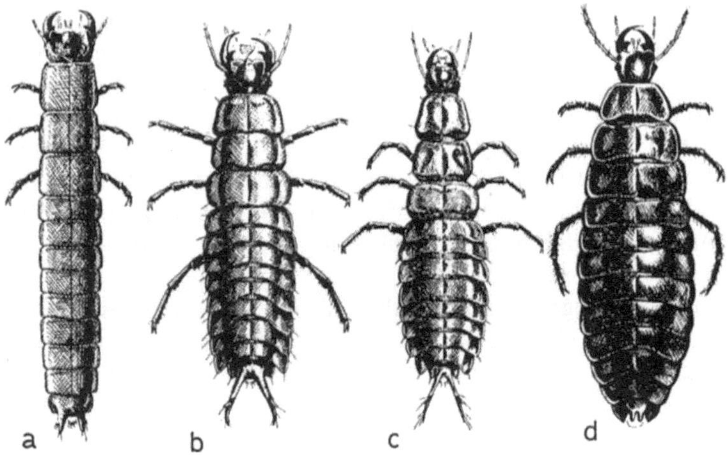

Fig. 47a–d. A brevimandibular carabid larva: (a) *Carabus cancellatus;* and three longimandibular carabid larvae; (b) *Carabus depressus;* (c) *Carabus cychroides;* (d) *Cychrus italicus.* From Sturani, 1962

form, which are polyphagous. The longimandibular differ from the brevimandibular larvae in having a broad, flat body (Fig. 47), most pronounced in *Cychrus* larvae which depend entirely upon snails as prey, and whose adults are also specialized snail hunters. The larvae enter the snail shell in a most characteristic manner. They crawl along between the shell and the soft body of the snail with their ventral side towards the shell. In addition to their shape, this enables

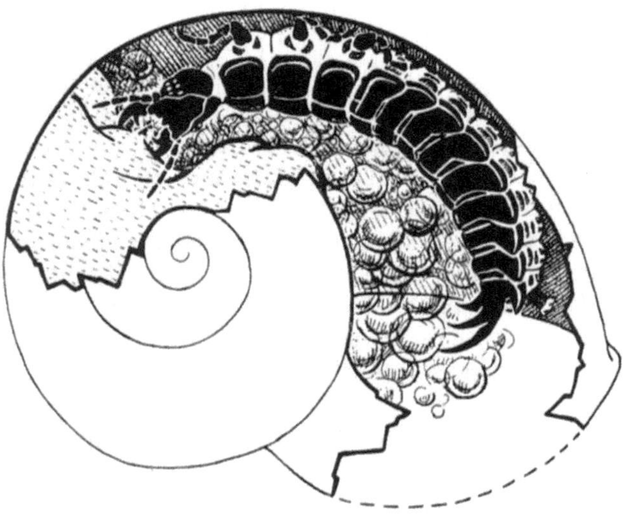

Fig. 48. Posture assumed by a longimandibular larva of *Carabus* in penetrating the shell of a snail. From Sturani, 1962

them to avoid getting their stigmae blocked with the slime exuded by the snail when it is attacked (Fig. 48). A very similar behaviour of the larvae of North American Cychrini of the genus *Scaphinotus* was reported by Greene (1975).

A most peculiar type of prey-catching tactic is exhibited by the larvae of Cicindelinae (Faasch, 1968, who also lists older literature). They dig more or less vertical shafts in the ground, 15–20 cm in depth at the first larval instar of *Cicindela hybrida* and *campestris*, and about 40 cm in depth at the third instar. The larvae lie in wait near the surface, their head capsule sealing the shaft. Figure 49 a–d shows the sequence of events in capturing their prey. Experi-

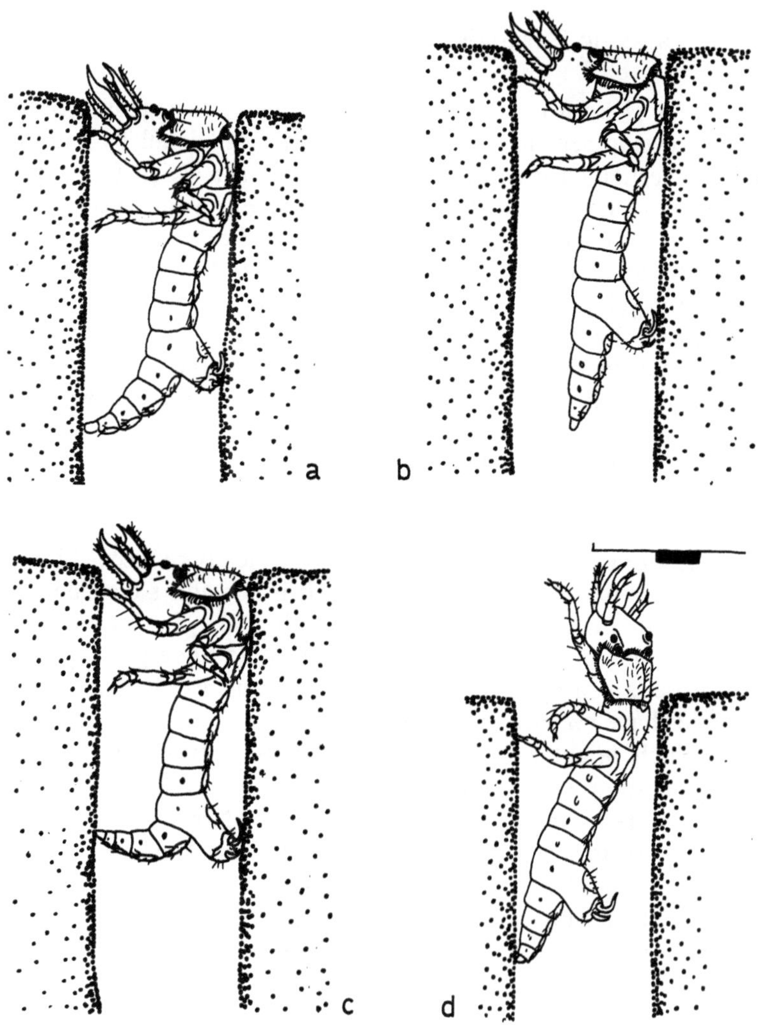

Fig. 49 a–d. Larvae of *Cicindela hybrida* catching prey. (a) The larva lying in wait at the entrance to its hole. (b) The larva sees an object approaching. The abdomen is straight. (c) Abdomen curved and pressed against the wall, the animal ready to spring at the approaching object. (d) The larva springs at the object (a decoy in this case) in order to seize it. From Faasch, 1968

ments involving dummies showed that the releasing stimulus is the same as in the adults. A dark object on a light background releases the prey-catching manoeuvre in the larvae just as in the adults, whereas light objects on a dark background have no effect. Three-dimensional dummies are not preferred to flat ones, and black discs of 5 mm diameter have an optimum releasing effect at a distance of 8 mm. As a rule the larvae do not react to objects beyond their reach, so that they have been held capable of estimating distances. Prey perception is entirely visual and, in contrast to the adults, there is no evidence of form perception. A further difference between larvae and adults is that, within their prey-catching radius, the former only react to moving objects.

VIII. Parasitic Carabids

As far as is known, the larvae of the species-rich genus *Lebia* are parasites. Their development has been described in most detail by Lindroth (1954). The cases investigated involved the larvae of *Lebia* which parasitize on the larvae or pupae of Chrysomelidae. The southern European species *L. scapularis* was observed on *Galerucella luteola* Müll., and the North American *L. grandis* on the Colorado beetle, *Leptinotarsa decemlineata* Say.

The development of the larvae of *L. chlorocephala* on *Chrysomela varians* Schall. was closely observed by Lindroth. The first larval instar (Ia) of *L. chlorocephala* is a normal carabid larva, which penetrates the pupal cavity of a *Chrysomela* pupa and feeds upon it. Stage Ib arises from stage Ia without a moult and is characterized by an enormous stretching of the intersegmental skin which permits an increase of body length from 2.7 mm to 11 mm. Stage Ib moults to produce the II and, in contrast to other carabid species, last larval stage. This stage takes up no food. All tergites and sternites have disappeared, and the mouth parts, legs and cerci are drastically reduced. Lindroth was unable to carry his broods beyond this stage. It is nowadays considered to be doubtful whether, in *Lebia*, a prepupal stage with buds of wings and compound eyes precedes the pupal stage (i.e. that *Lebia* would undergo a hypermetamorphosis) as was described by Silvestri (1904) for *L. scapularis*. Prothetelic stages in various carabids have been described as malformations by Paarmann (1967), an explanation which might also have applied to Silvestri's observations. In any case it is a question which requires further investigation (see also Chap. 3.E.I.).

Erwin (1967) showed that the larvae of the North American species *Brachinus pallidus* parasitize on the pupae of Hydrophilidae. The first larval instar penetrates the pupal chamber of the host and although the ensuing stages occasionally leave the host pupa for a brief period, they remain within the pupal chamber. Whereas most carabids have only three larval instars[9], *B. pallidus* has five: the first four instars are active and take up food, but the fifth remains inactive

[9] According to Bílý (personal communication) three *Amara* species of the subgenus *Celia* (*A. ingenua*, *cursitans* and *municipalis*) were consistently found to have only two larval instars. In laboratory bred *Pterostichus angustatus* and *P. nigrita* very occasionally "dwarf beetles" emerged from pupae that had been preceded by only two larval instars. They had a reduced number of antennal segments or exhibited even greater abnormalities (Paarmann, 1967).

Fig. 50. *Brachinus pallidus*, fifth larval instar. Natural length 6–10 mm. Description in text. From Erwin, 1967

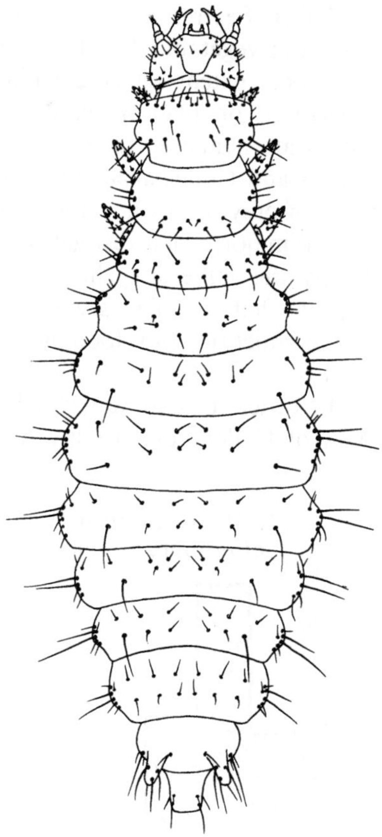

and moults to a pupa after four to five days. Both the fourth and fifth instars are very different from the usual type of carabid larvae (Fig. 50): the legs are much shortened, the head capsule is unpigmented and the abdominal segments are soft and greatly expanded. The fifth instar has no stemmata. Erwin also suspects other Brachinini of being parasitic, and he lists the other genera for which a parasitic mode of life is assumed, although in most cases further studies have not been carried out. They include *Pheropsophus* (with degenerated later larval instars that live on mole cricket eggs), *Pelecium* (on beetle pupae and millipedes), *Arsinoe* (on Tenebriodae larvae), *Sphallomorpha* (familiar from ant nests) and *Orthogonius* (lives among termites).

IX. Concluding Remarks on Carabid Nutrition

As a result of the investigations so far carried out we know that the majority of carabid species enjoy an extremely varied menu. It is surprising that at least some species of field-dwelling carabids consume a not inconsiderable proportion of plant matter. However, apart from the consumption of carrion, the most widespread form of nutrition is predatory.

Cases of extreme specialization are rare among predators, as is an exclusively animal or vegetable diet. Examples of the latter are *Zabrus*, which is specialized to a large extent on cereal plants, and *Ditomus*, which is even more highly specialized on plantain seeds. It appears that a large number of nutritional niches are available to most carabids, and as is seen in the case of *Calathus erratus* and *C. ambiguus*, two closely related species can favour quite different types of prey in one and the same habitat.

The majority of investigations, however, reveal that carabid species of approximately the same size show very similar preferences with respect to prey in any one habitat. The great variety of prey demonstrated indicates as a rule that competition for food need not necessarily play a large role (see Chap. 3.A.I.).

In order to ascertain whether three carabid species whose diet consists to 50% of Collembola occupy different nutritional niches due to their preying

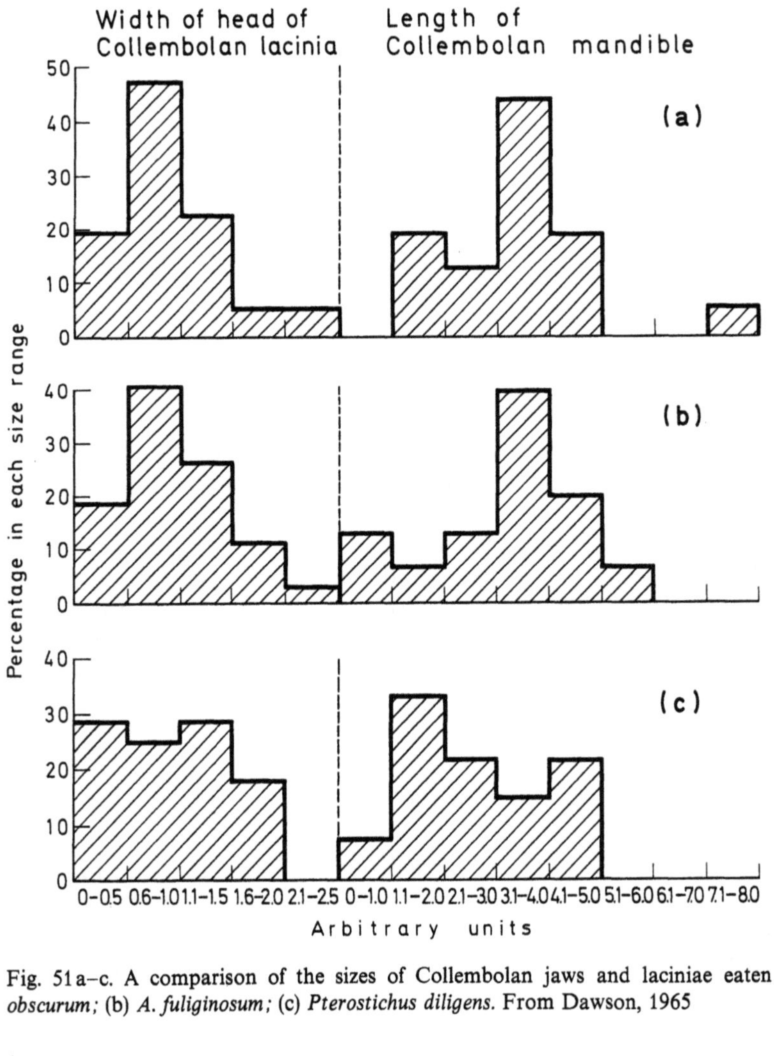

Fig. 51a–c. A comparison of the sizes of Collembolan jaws and laciniae eaten by (a) *Agonum obscurum;* (b) *A. fuliginosum;* (c) *Pterostichus diligens.* From Dawson, 1965

upon animals of different size Dawson (1965) measured the size of the prey. *Agonum obscurum* and *Pterostichus diligens* are 5 mm long, *Agonum fuliginosum* 6.5 mm. As far as this was possible, the length of the mandibles and the width of the laciniae of the Collembola from the crop contents of the beetles were measured. Almost identical curves for distribution of prey size were obtained for the two *Agonum* species, whereas *Pterostichus diligens* obviously preferred smaller Collembola (Fig. 51). Dawson writes: "But again this difference in diet between *P. diligens* and the two *Agonum* species probably does not mean that *P. diligens* prefers smaller Collembola, but merely that its feeding cycle and habitat made smaller ones more available to it. The proportion of young Collembola in the population is highest in September, October and November (Weis-Fogh, 1947–48). *P. diligens* had a peak of feeding activity in September when 72% had full guts, compared with 35.5% of *Agonum fuliginosum* and 11.5% of *A. obscurum*. It was during this month and October (when 36.5% *P. diligens* were still feeding) that all but two of its very small springtails were eaten. Also *P. diligens* lives in the soil more than do *Agonum obscurum* and *A. fuliginosum*, and the smaller species and instars of springtails are found there because larger ones cannot get into the spaces between the soil particles".

A thoroughly investigated example such as this shows that there is almost no actual nutritional competition at all between *Agonum* and *Pterostichus diligens* in their common habitat.

G. Parameters in Reproduction and Development Which are of Importance for the Biology of Populations

In the preceding sections the relations between carabids and biotic factors were considered. Such factors exert an influence on the numbers of carabid populations. In what follows, some aspects of their reproductive biology will be dealt with, including fecundity and life span, these being important parameters in the development of carabid populations.

I. Fecundity

The number of eggs laid within one reproductive period varies greatly from species to species. Furthermore, the numbers laid by different females of one and the same species under the same conditions in laboratory-reared stock tend to vary (e.g. between one and 105 larvae were recorded per female for *Pterostichus nigrita*). It is not proposed to give an exhaustive account of the scattered literature on this topic but to touch upon some of the ecological aspects of fecundity.

Females of the cave-inhabiting genus *Aphaenops* and its relative *Aepopsis robinii* bring only one large egg to maturity at a time, and this at relatively long intervals (Richoux, 1972), in the case of *Aphaenops cerberus* about 30 days (Deleurance, 1964a). The total number of eggs laid by one female is therefore presumably very small. But the number of eggs is usually much higher in other carabids.

The first reports on this subject were those of Burgess (1911) on *Calosoma sycophanta*, a species that lives as a rule for two to three years. The females usually only lay larger numbers of eggs after their second overwintering. The mean number of eggs for the total of all laying periods (two to three) amounted to 100 in the field and 139 in the laboratory, the highest number on record being 653.

A very high figure has been reported for *Pterostichus chalcites* (Kirk, 1975). Two ♀♀ (kept together with four ♂♂) laid 209 eggs altogether during one egg-laying period, were then unproductive for 34 days, were kept one month at +5°C and afterwards laid another 493 eggs. These eggs totalled 702, in other words, a mean of 351 eggs per female.

Extensive records are available for *Pterostichus angustatus* and *P. oblongopunctatus* (Paarmann, 1966). The former lays about twice as many eggs as the latter, closely related species. The number of eggs laid was:

	P. angustatus		*P. oblongopunctatus*	
	mean	maximum	mean	maximum
bred ♀♀	105	320	48	129
trapped ♀♀	136	261	65	105

The mean figures for trapped wild animals are higher than those for laboratory-reared animals. The numbers of eggs from all laying periods were added together (usually one period, sometimes two, rarely—in the case of *P. angustatus*—three).

The fecundity of *P. angustatus* is therefore double that of *P. oblongopunctatus*. It is tempting to regard this high fecundity as an adaptation to the instability and scattered nature of the habitats of *P. angustatus* (Chap. 8.F.).

Apparently the number of eggs does not depend on the size of the carabid, since the small *P. angustatus* lays as many eggs as the large *Calosoma sycophanta*, and the small *P. oblongopunctatus* as many as the larger *Carabus* species (according to Scherney, 1959a, a mean of 56 for *C. auratus*, 45 for *C. cancellatus*, 41 for *C. granulatus* and 22 for *C. ullrichi*).

By comparison, the brood-tending carabids lay fewer eggs, the batches consisting of only about 15 in *Abax ovalis* (Lampe, 1975) and an average of 16 per batch for *A. parallelus* (Löser, 1970). The latter species, however, is peculiar in that the females lay twice, at intervals of one to two months, and occasionally even three times.

It has recently been shown for larval overwinterers that a considerable part of a population can overwinter after reproducing and can even lay a second batch of eggs. In spring breeders, however, this type of behaviour is an exception, one example already mentioned being *Calosoma sycophanta*. For *Poecilus* species, Krehan (1970) observed in the field that the proportion of overwintered females in the reproducing spring population is extremely small. In the laboratory, nevertheless, it was possible to attain repeated ripening and deposition of eggs by treatment with a photoperiod appropriate for triggering maturation. In this way females of *P. angustatus* were induced to lay up to three batches, and those of *P. oblongopunctatus* often two batches, at intervals of about six months to one year (Paarmann, 1966).

Agonum assimile is very long-lived in the laboratory (no attempts have been made to ascertain whether this is also the case in nature). If animals of this species were exposed to three-month periods of short-day following egg deposition they could be induced to reproduce successfully up to five times in the ensuing long-day, finally achieving an adult age of three years (Neudecker and Thiele, 1974, Table 31). The average number of progeny (counted as larvae), however, decreased progressively from 27.6 to 5.7.

Recently the first attempt to estimate the total number and biomass of eggs produced by carabids in a particular habitat, per year, was reported by Grüm (1973). In a deciduous forest (Tilio-Carpinetum) the total number of eggs per 100 m^2 amounted to 4221 (=1837 mg dry weight) and in a mixed coniferous-deciduous forest (Pino-Quercetum) 3547 (=1956 mg dry weight). The estimation involved the most common carabids (four species of *Carabus* and four of *Pterostichus*) and was based on a calculation of the density of females from trap catches and a determination of the number of eggs produced by one female as revealed by dissection of animals taken from field catches throughout the year. The biomass of eggs produced is as high as 50% of the biomass of the females of the total carabid population (Grüm, 1975).

II. Life Span

It was previously assumed that most carabids do not live much longer than one year in the open (from hatching of the larvae from the egg to completion of the first reproductive period as adults). This assumption was made for larval and adult overwinterers. In the meantime it has been shown for many larval overwinterers that a large per cent of the population whose larvae hatched in late summer, in addition to their period of reproduction in the ensuing summer, go through a second reproductive phase subsequent to overwintering as adults (see Chap. 6.B.III.). This implies a total minimum life span of two years. In some species development normally takes two years from hatching to the first egg deposition.

In the laboratory, adults may live much longer. The record is still held by *Carabus auratus* (Nickerl, 1889), that survived in captivity for five years (Kern, 1912). I myself have kept a female specimen of *Abax ovalis* in captivity for three years. It was caught wild, and must have achieved metamorphosis at least one year prior to this, which makes a total adult age of four years, preceded by almost one year for earlier development. However, this was the longest-lived of a group of 45 experimental animals caught wild, 33% of which reached an age of two years in the laboratory. In one case I was successful in maintaining *Abax ater* in the laboratory for three years and nine months.

Table 31. Number of progeny of pairs of *Agonum assimile* with more than one reproductive period. (According to Neudecker and Thiele, 1974)

Reproductive period	Treatment			Number of pairs	A/B (%)	C	D	Age at end of reproduction
1. Period	4 wk LD	+20 wk SD	+12 wk LD	20	100/100	28	3–53	36 wk
2. Period	12 wk LD	+16 wk SD	+12 wk LD	16	100/100	22	9–38	64 wk
3. Period	12 wk LD	+ 8 wk SD	+20 wk LD	14	100/72	24	0–81	92 wk
3. Period	12 wk LD	+25 wk SD	+14 wk LD	18	89/89	15	0–34	88 wk
4. Period	14 wk LD	+16 wk SD	+12 wk LD	15	93/87	11	0–24	116 wk
5. Period	12 wk LD	+ 8 wk SD	+12 wk LD	12	58/42	6	0–25	136 wk

Column A: Percent females laying eggs. Column B: Percent females with larvae. Column C: Average number of larval progeny per pair. Column D: Extremes of larvae per pair.—LD: long day; SD: short day; wk: weeks. One group was observed from the first through the third reproductive period, another from the third through the fifth

Agonum assimile was reported as living for three years under laboratory conditions (Neudecker, 1974), and a female of *Ditomus* survived for two years in my laboratory following its capture. A pair of *Omophron americanum* was held in captivity for the same length of time by Frank (1971a). Including preimaginal stages, this implies a total life span of three years. He emphasized, however, that the large majority of species caught did not survive longer than one year in captivity.

Of six females of *Calosoma sycophanta* one died after overwintering four times and reproducing three times, at an age of about four years (Dusaussoy, 1963). *Brachinus* species can also be long-lived, a male specimen of *B. sclopeta* surviving in the laboratory for five years (Wautier and Viala, 1969).

The life span of the species *Harpalus smaragdinus* was measured in the field using marked animals. Of the 109 animals retrapped one or more years after marking, 72 had overwintered once, 19 twice, 14 three times and 4 four times: the oldest individual achieved an age of 1469 days, or almost four years. The longest life span recorded for a specimen of *Amara infima* was 1099 days, 1470 for *Harpalus anxius*, 675 for *H. aeneus* and 352 days for *Pterostichus lepidus* (Schjøtz-Christensen, 1965).

Mortality during development in the field seems to be different in spring and autumn breeders. In three species of spring breeders (*Carabus arcensis, C. nemoralis, Pterostichus oblongopunctatus*) density from the egg stage until the 3rd larval stage (a span of 50 days) was reduced to figures ranging from 13.2% to 32.9%; in three autumn breeders (*C. glabratus, C. hortensis* and *P. niger*) down to figures ranging from 7.4% to 26.5% (during a span of 280 days). The mean total life span of the spring breeders was smaller than that of the autumn breeders, if one considers the duration of one whole generation, but the life span of the adults is longer in the spring breeders (247–327 days) than in the autumn breeders (103–141 days). The active stages suffer a higher mortality; this holds true for larvae as compared with eggs, prepupae, and pupae, as well as for active adults as compared with hibernating ones (Grüm, 1975a).

H. The Importance of Carabids for Production in Ecosystems

Studies on the productivity of terrestrial ecosystems, especially secondary production, have only recently been intensified. Thus only a few studies in this field have paid special attention to carabids. Nevertheless, they are such that certain conclusions as to the importance of our particular group of animals in the ecosystem can be drawn from them.

A rough estimate of the role played by an animal group in the productivity of the ecosystem is provided by an estimate of the average standing crop. Kaczmarek (1967) collected data of this kind for forests of three different vegetational types in Poland.

Excerpts from his results are shown in Table 32. The figures for predatory invertebrates of the litter and soil layers are much higher than those for frogs

Table 32. Average standing crop (in kcal/m^2/year) of several groups of animals in three types of forests in Poland: Vaccinio myrtilli-Pinetum (VmP), Pineto-Quercetum (PQ), and ecotone Pineto-Quercetum/Alnetum (PQ-A). (According to Kaczmarek, 1967)

Habitat	Phytophagous insects	Saprophagous animals	Predatory invertebrates	Frogs	Birds
VmP	2.47	3.79	1.20	0.06	0.03
PQ	2.14	9.00	2.60	0.19	not estimated
PQ-A	5.75	22.80	5.75	0.85	0.14

and birds, which are further up the nutritional chain. Unfortunately, no figures are available for carabids, but the author's quantitative diagrams show (Figs. 52, 53) that the *Carabus* species alone almost equal the remaining predacious soil invertebrates.

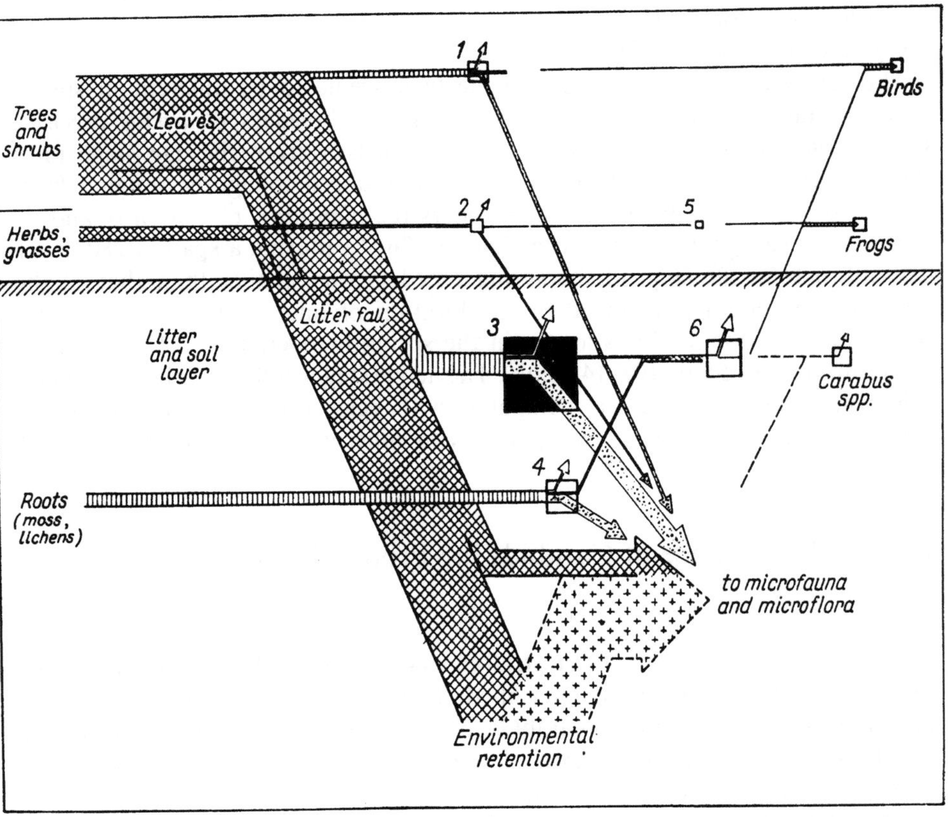

Fig. 52. Diagram of the energy flow in a pine-oak forest (Pino-Quercetum) in Poland. The squares represent quantitatively the average standing crop of *(1)* leaf-feeders of tree and shrub canopies, *(2)* phytophagous invertebrates of herb layer, *(3)* saprophagous meso- and macrofauna, *(4)* phytophagous meso- and macrofauna of litter and soil layer, *(5)* predatory invertebrates of herb layer, *(6)* predatory meso- and macrofauna of litter and soil layer. Notice the outstanding importance of the genus *Carabus* as compared with other predators. Width of bars between the different trophic levels represents quantitatively the energy turnover per year. From Kaczmarek, 1967

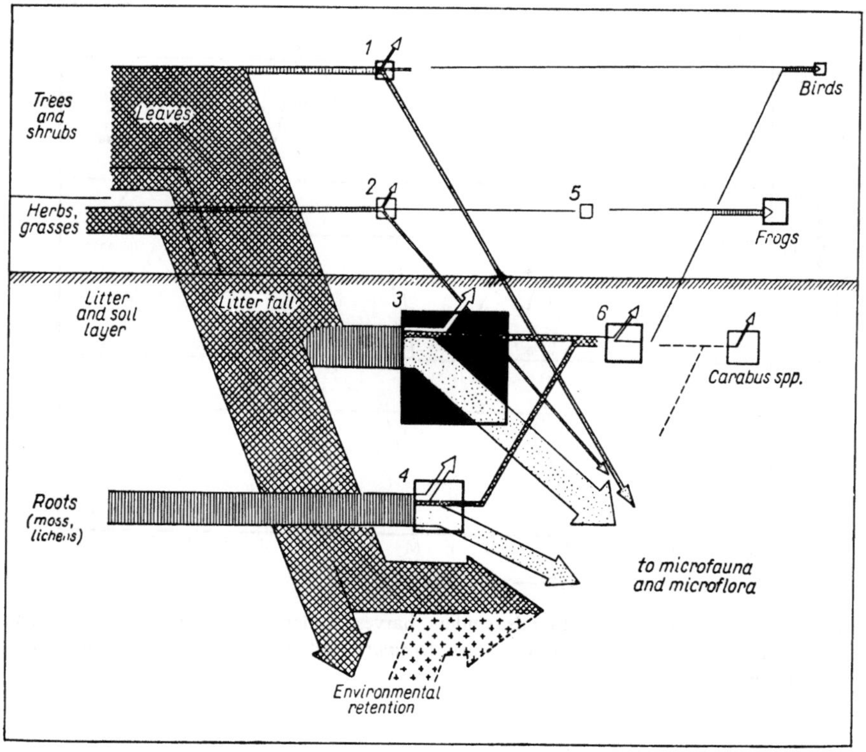

Fig. 53. Diagram of the energy flow in the ecotone pine-oak forest/alder forest (Pino-Quercetum/Alnetum). Explanation see Figure 52. Here, too, the role of *Carabus* is striking. From Kaczmarek, 1967

Within the framework of the International Biological Programme Weidemann (1971a, 1972) undertook detailed studies on the position of carabids in a forest ecosystem in the German hills (Solling near Göttingen), at an altitude of 500 m. A variety of parameters was used in estimating the role of carabids in the ecosystem. As compared with that of Chilopoda, spiders and harvest spiders the activity density (see p. 13) of carabids occupies an intermediate position (Fig. 54). Their abundance (number of individuals per unit of area) is very low compared with that of Chilopoda and spiders (Fig. 55), but their biomass, on the other hand, is comparable with that of other predatory arthropods (Fig. 56), in agreement with the findings of Kaczmarek.

A more exact determination of the energy flow through a population has to be calculated according to the formula

$$A = P + R,$$

where A indicates the total energy assimilated by the population in the period of time under consideration (gross production), P is the net production of potential energy stored in the population in the form of body substance, and R is the quantity of energy assimilated but reutilized for life processes, of which, in

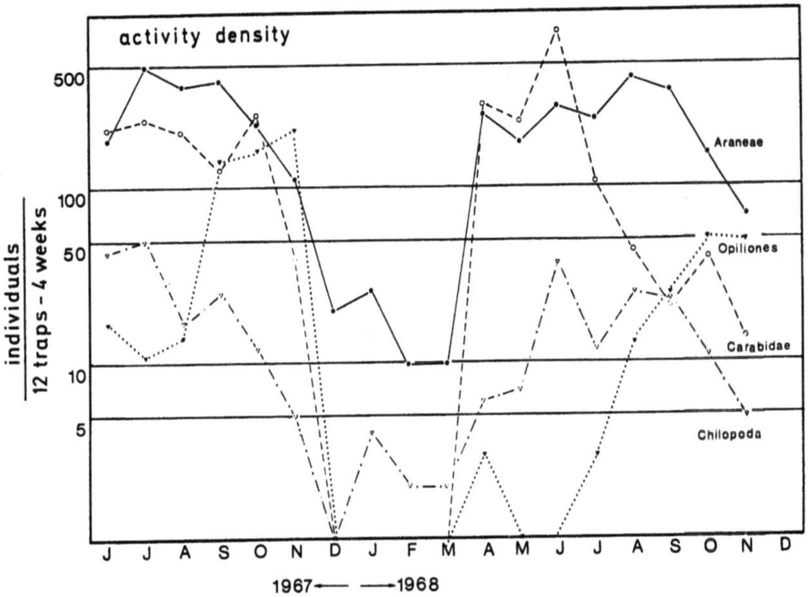

Fig. 54. Activity density of chilopods, spiders, harvest spiders and carabids in a tall beech forest in the Solling (German hills) in two years. *Ordinate:* individuals/12 traps in four weeks. From Weidemann, 1972

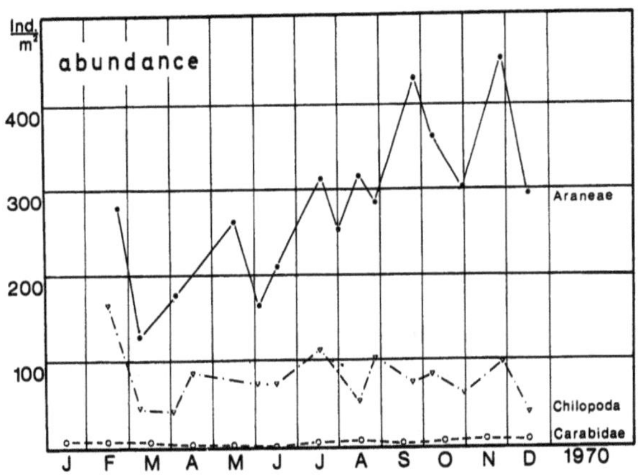

Fig. 55. Abundance (individuals/m²) of chilopods, spiders and carabids in the Solling over the course of the year 1970. From Weidemann, 1972

this case, only the most important part, that used up in respiration, was determined. Waste products and excrements were thus neglected, which seems permissible in view of the fact that only a preliminary idea of the size of production was required.

A calculation of production necessitates:

1. construction of an energy-balance sheet for an average individual, which in turn involves (a) measurement of production due to growth (P_g), whereby this has to be followed through all stages of development under field conditions

Fig. 56. Biomass (mg/m²) of chilopods, spiders and carabids in the Solling in the course of 1970. From Weidemann, 1972

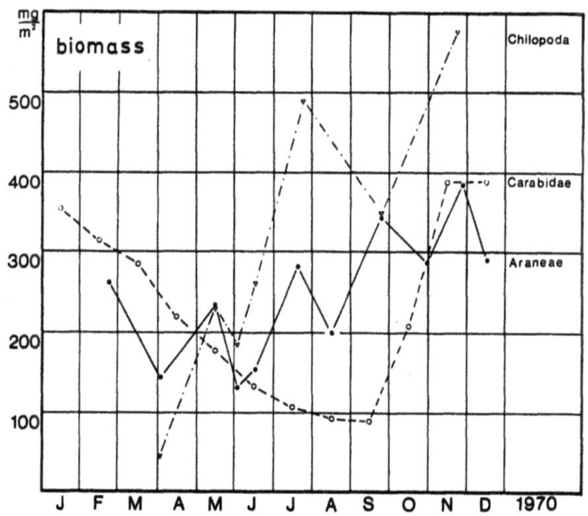

(an example is given in Fig. 57 for *P. oblongopunctatus*), (b) the measurement of production due to reproduction (P_r), which for univoltine species in the given period of one year usually equals the average number of eggs laid by one female, (c) measurement of respiration at all stages of development under field conditions over the course of the year.

2. From the energy balance thus obtained for an average individual, that of an entire population can be calculated, on condition that a precise analysis of the population movements during this period of time is available (one year in this case).

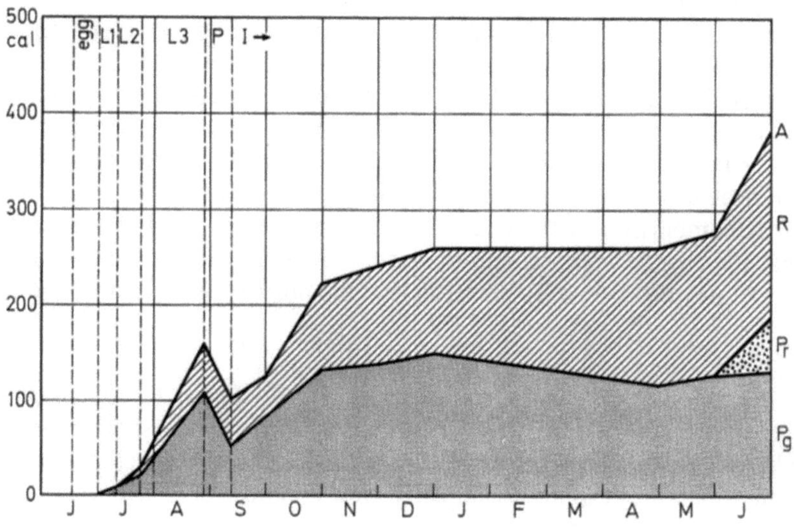

Fig. 57. Cumulative energy balance of an average individual of *Pterostichus oblongopunctatus*. Explanation and abbreviations in text. From Weidemann, 1972

From such measurements Weidemann was able to construct an energy balance sheet for one year for *Pterostichus oblongopunctatus* and *P. metallicus* in a beech forest in Solling. This is a species-poor ecosystem, in which the two species studied accounted for almost all of the carabids found, so that the data can be considered as representative of the entire carabid population.

Table 33. A year's energy balance for populations of two carabid species in Solling (German Mittelgebirge). Beech forest, 120 years old. The numbers indicate kcal/ha. For further explanations see text. (According to Weidemann, 1972)

	Pterostichus oblongopunctatus May 69–April 70		*Pterostichus metallicus* July 69–June 70	
P_r	2578		563	
Least-P_g	2115		716	
Least-P		4693		1279
Least-R (egg-adult)	2628		5181	
R (adults)	3555		1929	
Least-R		6183		7110
Least-A		10876		8389
P/R		0.7590		0.1799
$\frac{P \times 100}{A}$		43.15%		15.25%

All values were calculated in kcal/ha (Table 33). Since the abundance of larvae could not be determined exactly it was taken to equal the number of beetles in the ensuing generation, which is necessarily a minimum value. This means that the energy balance for the whole population is also a minimum value. Since the largest energy turnover is in the third larval instar, whereas losses in larval populations take place rather in the preceding stages, the values are probably only very little below the true figures.

A comparison of the energy balance of the two *Pterostichus* species shows that the energy flow through the populations is very similar (*P. oblongopunctatus*:*P. metallicus* = 1.3:1). Nevertheless, the portion of the gross production made up by the net production $\left(\frac{P \times 100}{A}\right)$ differs greatly, being about three times as large for *P. oblongopunctatus* as for *P. metallicus*. *P. oblongopunctatus* can store much more energy and develops within a much shorter period of time, whereas *P. metallicus* stores less energy and its larvae develop slowly and overwinter. This may explain the unusually low P/R ratio of *P. metallicus*. Furthermore, the adults of the species live for several years (see Chap. 6.B.III.).

A careful analysis of the energy turnover of *Nebria brevicollis* was made by Manga (1972). His values, which are expressed in Joules, have been converted into kcal for the sake of comparison with those of Weidemann. Manga's studies include accurate measurements of population density of the larvae. The total turnover of the population (8060 kcal/ha/year) lies in the same region as that of the species investigated by Weidemann. The highest rate of oxygen consumption was shown by the larvae I, with 0.9202 mm^3/mg/h, which is about double that of the larvae III or of active adults, and four times that of prepupae, pupae or diapausing adults. Respiration accounted for about 51.58% of the total turnover of the population (for comparison: *Lithobius* species—Chilopoda: 95%). 65% of the total turnover is due to the larvae.

Weidemann found evidence that *P. metallicus* preys upon Diptera, Diptera larvae and larvae of Curculionidae (*Phyllobius argentatus* and *Polydrosus undatus* among others), besides Collembola. Studies on the biological production of weevils in the forests investigated by Weidemann show that the production of *Phyllobius argentatus* populations amounted to 2300 kcal/ha/year. Theoretically, therefore, *Pterostichus metallicus* should have no difficulty in exterminating the *Phyllobius* population if it were to consume these animals alone. In fact, the number of *Phyllobius* adults sinks by about 30% shortly after emerging in June, a time at which *P. metallicus* shows a high nutritional uptake and accumulation of reserve substances. Since *P. metallicus*, like other carabids, is highly polyphagous there is no danger of its extinguishing the prey population, especially in view of the fact that the composition of its prey varies with availability. This was also found by Smit (1957) for *Calathus species* (see Chap. 3.F.II.). The majority of carabid species move in a very adaptable nutritional network that can even vary considerably for different species living side by side in one and the same habitat. This was investigated and vividly described by Skuhravý (1959) for field carabids (see Chap. 3.F.II.). The data of Weidemann show "that carabids can exert a considerable influence on other populations".

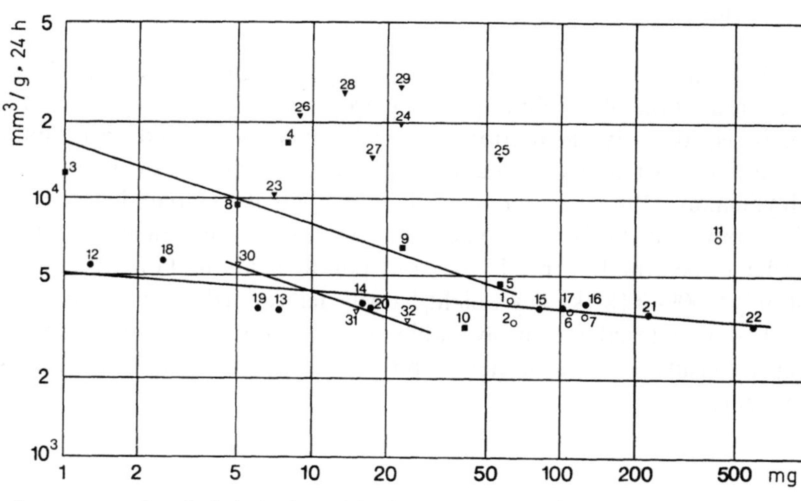

Fig. 58. Respiratory rates (mm^3 O_2/g body weight in 24 h) of epigaeic predatory arthropods at 15°C in relation to body weight. *Open circles* and *squares (1–11)*: carabids; *solid circles (12–22)*: spiders; *solid triangles (23–29)*: harvest spiders; *open triangles (30–32)*: chilopods. From Weidemann, 1972

The question arises as to the relative significance of carabids in the ecosystem as compared with other predatory arthropods of the soil surface. To this end Weidemann (1972) collected data on the respiration rates of carabids, spiders, harvest spiders and Chilopoda (Fig. 58), and came to the following conclusions: "The respiration rates of spider and carabid adults of the same weight class are very similar to one another. The oxygen consumption of carabid larvae, in contrast, is markedly higher than that of spiders or chilopods of comparable weight. The respiratory rate of lighter chilopods (<10 mg) is higher than that of spiders and that of heavier specimens (>10 mg) is lower. The respiratory rate of harvest spiders is the highest of all. Bearing in mind the above-mentioned order and the mean biomass it can be tentatively postulated that the energy flow through the population of spiders and chilopods in the beech woods investigated is of a similar order of magnitude to that of the carabids". Among predatory soil arthropods it is only considerably higher in ants. Horstmann (1974) found the highest values in *Formica polyctena* Förster, with 81 kcal/m^2/year. He cites values of 35–65 for other species of ants. This means that the energy turnover for certain ant species is one to two orders of magnitude greater than that of carabid species, which according to the data cited above, amounts to about 1 kcal/m^2/year.

The annual production of all the carabids in different types of woodland in Poland was lowest in a pine forest (Vaccinio myrtilli-Pinetum: 0.765 kg/ha/yr dry weight) and highest in an alder forest (Circaeo-Alnetum: 2.010 kg/ha/yr dry weight), "and seemed to be related to the amount of nitrogen in the leaf-fall" (Grüm, 1976). The production for two single species during one generation (dry weight) amounted to:

Carabus nemoralis: 1017 mg/100 m^2 = 0.102 kg/ha
Carabus glabratus: 1779 mg/100 m^2 = 0.178 kg/ha
(Grüm, 1975b)

The data of Kaczmarek (1963, see p. 135 and Figs. 52, 53) reveal that the production of carabids per unit area, for example, is higher than that of birds. Hardly any calculations of the biomass of carabids per unit area have so far been attempted. A figure of about 15 kg/ha (fresh weight) as given by Heydemann (1962a) in one case, is very high if compared, for example, with the biomass of birds in woods, which was found by Thiele (1959) to be 0.5–1 kg/ha in central Europe. In rye and potato fields in Poland, a total biomass of only 0.009–0.187 kg dry weight/ha was determined (Kabacik-Wasylik, 1975).

From material so far available we can conclude that carabids probably play a considerable role at their trophic level (as primary and, above all, secondary consumers).

Chapter 4

Man and the Ground Beetles

By no means the least important biotic factor confronting ground beetles in their various native ecosystems is Man himself. It therefore seems appropriate to follow the chapter on carabids and biotic factors with a consideration of the connections between this family and our own species. Nowhere has a comprehensive survey of the relevant literature yet appeared, which is one more reason for our devoting space to a topic that clearly warrants much closer attention in the future. Kulman (1974) has pointed out how little is so far known concerning the ecology of North American carabids, and gives a short survey of the investigations carried out in this region on biological control involving carabids.

The connections between the manifold species of the ground beetle family and man are of a twofold nature. On the one hand, the animals exert an influence, usually beneficial but sometimes detrimental, upon man, and on the other, the ground beetles are influenced by human interference within the ecosystem. We shall deal with both of these aspects in the following.

A. The Importance of Ground Beetles to Man

I. Ground Beetles as Entomophages and Potential Benefactors

1. Experiments on Pest Destruction Using Ground Beetles

The most extensive and thorough experiments on this topic were carried out by Scherney on *farmland* in Bavaria (Germany). He was mainly concerned with the effect of carabids on potato beetles. Of 262,000 carabids caught in various parts of Bavaria in the period 1957–1959, 79% belonged to 11 species which might be of possible value in a biological pest-control programme. Only five of these were consistently encountered in larger numbers on fields: *Carabus auratus*, *C. cancellatus*, *C. granulatus*, *Pterostichus vulgaris* and *P. niger* (Scherney, 1962a).

Experiments designed to clarify the efficacy of these species in destroying potato beetle larvae were carried out near Puch, Bavaria, on potato plants growing in concrete containers, 106 cm in diameter (Scherney, 1959b). Each of these containers harboured two *Carabus* individuals, a population density corresponding to that found under natural conditions in potato fields in the vicinity. In 1956 four repetitions were made with two *Carabus cancellatus* and four with two *C. granulatus*. 500 eggs of the potato beetle had also been introduced into each container. After 33 days the mean values for larvae and pupae from these eggs were as follows:

With *Carabus cancellatus*	With *Carabus granulatus*	Control
22.5	20.25	450

In 1957 the results, using an additional *Carabus* species, were:

With *Carabus auratus*	With *Carabus cancellatus*	With *Carabus granulatus*	Control
43	27	30.5	443

Taking various factors into consideration these figures point to a consumption of six to six and one-half larvae and pupae/*Carabus*/day.

In both years additional experiments were carried out on plots of 4 m^2 fenced off within potato fields, and each containing 16 plants. Into each of these enclosures six specimens of *Carabus* and 4000 eggs of the potato beetle were introduced. Here, too, the results indicate a sharp reduction in the potato beetle stages (Table 34).

Table 34. Developmental stages of the potato beetle found after 40 days in field experiments with and without *Carabus*. (According to Scherney, 1959)

	I *Carabus cancellatus*	II *Carabus granulatus*	III *Carabus cancellatus* (2nd exp)	IV *Carabus granulatus* (2nd exp)	V without *Carabus*
1956	274	393	339	326	2967
1957	826	627	873	583	3164

The practical importance of the destruction of potato beetles is shown in the increase in weight of potatoes harvested from the various plots (Table 35).

Table 35. Weight (in kg) of potatoes harvested from plots infested with potato beetles, with and without *Carabus*. (According to Scherney, 1959)

		1956	1957
Exp I	*Carabus cancellatus*	14.6	12.2
Exp II	*Carabus granulatus*	12.6	11.1
Exp III	*Carabus cancellatus* (2nd exp)	11.9	10.8
Exp IV	*Carabus granulatus* (2nd exp)	12.9	11.5
Exp V	Without *Carabus*	7.3	6.7
Exp VI	Control without potato beetle	15.9	14.5

Compared with the control without potato beetles (Exp. VI) the crop reduction in Experiment V amounts to 54% in both years, whereas the mean reduction on plots with *Carabus* amounts to 18% in the first year and 21% in the second. This means that the damage was reduced to one-third, solely due to the activity of *Carabus.*

In 1958 the experiments were extended to larger plots (Scherney, 1960b) of 12 m^2, each bearing 46 plants. In the controls without potato beetles 45.7 kg (=100%) per plot were harvested. The experiments were varied so as to simulate varying degrees of potato-beetle infestation. The number of *Carabus* individuals introduced into each plot corresponded to the natural population density of three Carabus/2 m^2. In these experiments, too, *Carabus* was responsible for a considerable decrease in damage (Table 36).

Table 36. Weight of potatoes (in kg) harvested in larger plots with *Carabus* and varying numbers of potato beetles (Percent reduction in crop weight in brackets). (According to Scherney, 1960b)

	Eggs or young larvae of potato beetle per plant		
	100	50	25
Without *Carabus*	28.7 (37.2 %)	35.5 (22.8 %)	42.7 (6.6 %)
With *Carabus*	39.1 (14.5 %)	41.9 (8.3 %)	44.7 (2.2 %)

In cases of low infestation it should be possible to use *Carabus* instead of chemical measures.

The experiments were repeated in 1959 with, in addition, some involving twice the density of *Carabus*. The following reductions in crop weight were recorded:

	Potato beetle eggs per plant	
	100	50
without *Carabus*	31.5%	18.6%
normal *Carabus* density	16.4%	10.7%
double *Carabus* density	6.6%	3.6%

In all experiments even a normal population density of *Carabus* reduced the damage to half or less (Fig. 59).

This very promising picture was impaired, however, by Scherney's (1962b) discovery that the density of *Carabus* found in the experimental area near Puch was by far the highest in the whole of Bavaria. Carabids obviously cannot be counted upon to be equally efficient in combating harmful insects in all areas.

The following results are also taken from Scherney's publication (1962a). Where 70–90 turnip-fly larvae (*Athalia rosae* L.) per plant were found in a mustard field, four individuals of *Pterostichus vulgaris* were caught per pitfall trap before pupation of the larvae as compared to 248 subsequently. This high

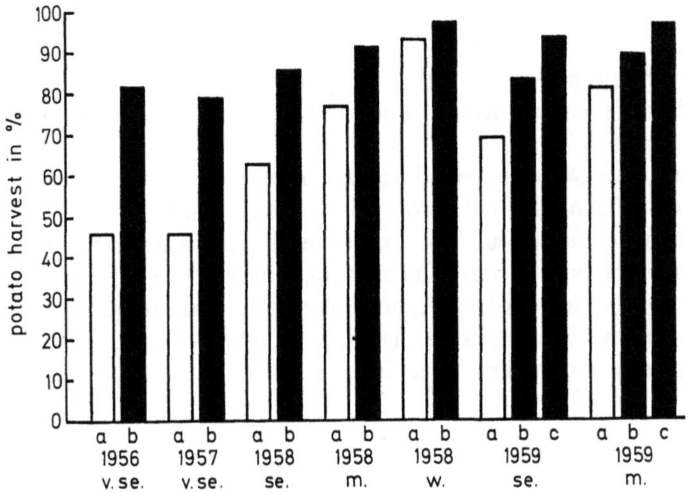

Fig. 59. Potato harvest from fields afflicted with potato beetles or not, with and without carabids. Degree of affliction: *v.se.*: very severe attack by potato beetles. *se.*: severe attack by potato beetles. *m.*: moderate attack by potato beetles. *w.*: weaker attack by potato beetles. *a*: without carabids. *b*: with carabids (natural population density). *c*: with carabids (twice the natural population density). From Scherney, 1960b

figure was put down to attraction of the predators by the large supplies of prey. In feeding experiments *P. vulgaris* consumed 15 *Athalia* larvae per day, a weight corresponding to four times its own.

The small *Clivina fossor*, a species equipped with burrowing feet, destroyed about 65% of the larvae and pupae of the blossom beetle (*Meligethes* spec.) in laboratory experiments in which the prey had been introduced 6–7 cm beneath the soil surface. "It can be concluded from this experiment that even the smallest pest larvae can be destroyed by predatory beetles in the soil".

An investigation involving 35 ha was carried out in order to estimate the losses of larvae of *Contarinia tritici*, the yellow wheat gall midge, during their migration from ears of grain to the soil where they pupate. Random samples were taken from 35 plants before migration and 35 subsequently from the soil. It was calculated that of 417,350,000 individuals 173,075,000 (42%) were lost during the migratory phase (24.7–4.8.1972 = 12 days). According to laboratory experiments predators could even have accounted for 290,000,000 larvae in 12 days. This indicates a high degree of predatory activity, the lion's share having been devoured by the carabids (e.g. *Pterostichus vulgaris*, *Agonum dorsale* and *Loricera pilicornis*), spiders and staphylinids (investigations made by Basedow, 1973, in West Germany).

The effect of carabids on one and the same pest appears to vary considerably according to the composition of the carabid population itself and the prevailing environmental conditions. This view is substantiated by the investigations of several authors on the predatory role sometimes assumed by small carabids towards the eggs of the cabbage root fly, *Erioischia brassicae* (Bché.). For a review see van Dinther and Mensink (1965), from which the following authors are cited: In England, Wright and Hughes (1959) found 90% destruction of the eggs by *Trechus quadristriatus* and *Bembidion lampros* as well as by a staphylinid, *Aleochara bipustulata*. From Canada, Wishart *et al.* (1956) reported 70% destruction of the eggs by predatory beetles, especially *Bembidion quadrimaculatum* and *B. nitidum*. Again in England, Coaker and Williams (1963) found 24% destruction of developmental stages by carabids (particularly *Bembidion lampros*) as compared to 34% by staphylinids. In Holland, however, Abu Yaman (1960) found that mortality due to predatory beetles (including *Bembidion*) was only 17%.

By means of the precipitin test (see below) Frank (1971b), in Canada, was able to demonstrate in an elegant way the effect of carabids on the red-backed cutworm (*Euxoa ochrogaster* Guenée), a noctuid caterpillar harmful to many field crops. In 1967, of 115 individuals belonging to 14 carabid species only three, i.e. 2.6%, devoured cutworms or pupae of *Euxoa*. In 1968, however, the test revealed that 123 (24.2%) of 507 individuals belonging to 14 species destroyed pests. These 123 individuals belonged to seven of the 14 species caught. Despite a sometimes even heavy attack by predators Frank concludes that: "Carabidae are important as predators of *Euxoa achrogaster* Guenée, but failed to eliminate the prey population in a study area in central Alberta". This conclusion seems to be reached in almost every case where the role of carabids as predators on harmful insects has been investigated.

The same holds true for carabids as predators of *forest insects*. Although *Pterostichus oblongopunctatus* and *P. metallicus* participate in the natural reduction of the population numbers of the bark beetle *Tryptodendron lineatum* Olv. in Czechoslovakia, their contribution, according to Novák (1960), should not be overestimated.

A valuable tool for determining the proportion of a population of predators that has consumed the pest in question is offered by the precipitin test. With its help, the effect of carabids on a population of the winter moth *Operophthera brumata* (L.) was studied by Frank (1967a) in England. An antigen was prepared from the pupae of the moth and injected into rabbits to produce a specific antibody, with the aid of which the crop contents of predatory arthropods could be tested. The following were revealed to have consumed pupae in the four days prior to their being caught:

21 out of 122 *Abax ater*
3 out of 18 *Pterostichus madidus*
1 out of 5 *Pterostichus vulgaris*, but only
6 out of 226 *Philonthus decorus* (staphylinids).

These results suggest that carabids in this case play a larger predatory role than staphylinids.

Further investigations on the changes in population density of coleoptera and winter moths provided the basis for judging the efficacy of the predators: in 1965 about 269 larvae achieved pupation per m^2 (Frank, 1967b). Of the 238 healthy individuals 45 were consumed by the three carabids mentioned (about 21%) and 51 by *P. decorus* (the latter with a much higher population density than the carabids). By way of comparison the shrew, *Sorex araneus*, ate 15 winter moth pupae. Added to this is the consumption of an unknown number of larvae prior to pupation and of moths following emergence. The effect of wire-worms, small mammals and the pupal parasite *Cratichneumon culex* (Muell.) (Hymenopt.) could not be determined quantitatively. According to these results, carabids play an important, but not necessarily decisive, role in controlling the winter moth. In regions like the western German hills, which are permanently afflicted with the pest, its annual large-scale occurrence indicates that under the conditions prevailing in the area it cannot be controlled by natural enemies alone.

Studies concerning the factors governing the population dynamics of the oak egger moth, *Tortrix viridana*, in western Germany also included investigations on the effect of carabids (Schütte, 1957). Only relative values for the part played by carabids were obtained: the number of carabids caught in the various habitats was multiplied by the number of caterpillars of the moth eaten per individual carabid per day in feeding experiments, to give the so-called "total aggression value". Since the values for the individual populations correlated well with the known values for the absolute numbers of caterpillars it can be assumed that, roughly speaking, the higher the population density of the caterpillar the greater the effectiveness of the carabids. Further observations showed that the maximum "total aggression value" of the carabids lagged behind the prey maximum to an increasing extent over the course of several years, a phenomenon also well-known for parasites. But since the observations were restricted to four years and only approximate values are available, the above statement cannot be considered as final. It seems, however, that carabids on their own would not be highly effective in controlling a permanent pest such as the oak egger moth. Investigations on the mortality factors of the Swaine jack-pine sawfly, *Neopridion swainei* Middleton (Hymenopt.) in Canada (Tostowaryk, 1972) also revealed a density-dependent effect for the predatory activity of carabids. Five of the 15 carabids caught attacked the cocoons of the sawfly (*Pterostichus punctatissimus, P. adstrictus, P. coracinus, P. pennsylvanicus* and *Sphaeroderus lecontei*, i.e. predominantly *Pterostichus* species). A carabid individual consumed on an average three cocoons per month, which corresponded to a population decrease of 1.0 to 4.4% of the prey species in a number of pine stands. Where the supply of prey was raised by the artificial introduction of cocoons into the forest floor the per cent of pupae attacked by carabids rose to 10.5%. In cases of a natural increase in population density of the sawfly, a higher percent of attack by carabids was recorded. No such effect was seen with wire worms.

A connection between prey density and the predatory action of carabids is, however, by no means the general rule. A comparison of pest-stricken forest habitats in Poland *(Lymantria dispar, Bupalus piniarius)* with healthy stands revealed no significant difference in either the numbers of species of carabids or in the structure of their associations and "no significant influence of the population density of injurious insects on the number of caught carabid beetles, both of all species and of species which can influence the gradation processes of phytophagous insects" (Leśniak, 1972).

2. *The Influence of Ground Beetles on the Invertebrate Fauna of Cultivated Ecosystems*

Clarification of this question can be expected from field experiments in which the development of the soil fauna on normal areas is compared with that on areas with a reduced carabid population. Studies of this nature have been made in Byelorussia (Dubrovskaya, 1970). Plots of 4 and 16 m^2 on a summer wheat field were enclosed by narrow ditches that were then sprayed with hexachlorane in order to prevent the entry of soil animals into the experimental areas. "It was assumed that carabids within the small plots, as very mobile insects ... would fall into the ditches and perish. The number of carabids in the isolated plots should therefore have been sharply reduced. Relatively immobile phytophages and saprophages within the plots would be considerably less affected".

In this way the carabids in the test plots were reduced by 34.8% as compared with the surrounding field, which led to a sharp rise in their potential prey, particularly earthworms (Table 37).

Table 37. Population density of soil animals on field plots with artificially reduced carabid populations, in % of the numbers on the untreated field. (According to Dubrovskaya, 1970)

Carabidae	65.2%
Earthworms	207.8%
Elateridae	175.5%
Staphylinidae	166.6%
Pea and bean weevils	136.3%
Others	86.6%

The author sums up her results as follows: "The methods used in this experiment were very crude and there could be many objections to the results. Nevertheless, in our view it does give some idea of the importance of carabids as a regulator of the abundance of the soil population".

3. *Do Ground Beetles from Hedges, Groups of Trees and Wind-Breaks Influence the Pest Fauna of Adjacent Cultivated Areas?*

This problem has been investigated over a broad belt stretching from England to the Ukraine, particularly in central Europe. Following the Second World War it was hoped that, in connection with replanting wind-breaks, the species-rich fauna of the hedges, including numerous entomophages and above all carabids, would penetrate into the surrounding cultivated areas and destroy the pests. The first thorough studies on the fauna of ridge hedges in Schleswig-Holstein in northern Germany (Tischler, 1948) led to the author's warning that the hedges could not be expected to resolve all the problems caused by harmful insects (Tischler, 1951).

Quantitative studies were carried out on this problem by Thiele (1964b) in the Bergisches Land in the west German hills (maximum 275 m above sea level). In three thoroughly investigated habitats 49%–94% of the individuals of the epigaeic fauna (mainly carabids) were found to be forest species (caught throughout the year). In another year, the proportion of forest individuals from one of the habitats was compared with that in neighbouring fields (based on catches made between the end of June and the beginning of August). The habitat under consideration was a bushy strip, about 10 m wide, flanked on both sides by fields. The proportion of forest individuals within the hedge amounted to 95%, 58% on its northern edge and 27% on its southern edge. On the adjacent grain field to the north only 2.5% forest individuals were found within 10 m of the hedge, and as few as 0.5% on the potato field to the south (Table 38; Fig. 13). None of the common species was equally distributed between hedge

Table 38. The occurrence of carabids and staphylinids in a hedge and the neighbouring fields. (From Thiele, 1964b.) Caught between 29.6. and 8.8.1957; in potato field only between 4.7. and 8.8.1957)

	Cereal field	Hedge N. side	Hedge interior	Hedge S. side	Potato field
			Number of traps		
	9	8	8	8	9
Species of hedge interior					
F *Pterostichus cristatus*	12	52	413	23	—
F *Patrobus atrorufus*	—	—	14	1	—
F *Agonum assimile*	—	—	13	—	—
F *Pterostichus oblongopunctatus*	—	2	8	1	—
Hedge species					
F *Abax ater*	4	29	40	22	2
F *Trichotichnus laevicollis*	—	2	2	9	—
F *Carabus problematicus*	1	4	3	2	1
Tachinus laticollis	—	1	—	1	—
F *Othius myrmecophilus*	—	1	—	1	—
Field species					
Pterostichus vulgaris	287	8	16	128	557
Harpalus rufipes	3	—	—	—	6
Stenus biguttatus	8	—	—	—	1
Oxytelus sculpturatus	2	—	—	—	3
Only or mainly in cereal fields					
Agonum dorsale	134	17	—	1	36
Agonum muelleri	136	1	—	—	11
Carabus granulatus	50	2	—	1	10
Loricera pilicornis	30	—	—	—	7
Amara spec. (mainly *plebeja*)	24	—	—	—	—
Philonthus fuscipennis	19	—	—	—	—
Agonum sexpunctatum	13	—	—	—	1
Carabus purpurascens	6	1	—	3	—
Lathrobium geminum	4	1	—	—	1
Stenus brunnipes	2	—	—	—	—

	Cereal field	Hedge N. side	Hedge interior	Hedge S. side	Potato field
			Number of traps		
	9	8	8	8	9
Only or mainly in potato fields					
Bembidion lampros	22	—	—	2	70
Bembidion quadrimaculatum	—	—	—	—	23
Trechus quadristriatus	—	2	4	3	8
Dyschirius globosus	—	—	—	—	10
Lathrobium longulum	—	—	—	—	3
Synuchus nivalis	—	—	—	—	2
Caught in this series with no detectable biotope affinity					
Tachinus rufipes	9	28	4	21	15
Stenus bimaculatus	3	5	1	—	1
Oxytelus rugosus	2	—	2	—	3
F *Carabus nemoralis*	2	—	—	—	1
Sum of more common species	773	156	520	219	772
Sum of forest species (F)	19	90	493	59	4
In percent	2.5	58	95	27	0.5
Only single individuals of the species caught	9	1	3	1	1
Overall total	782	157	523	220	773
Total in each trap	87.0	19.6	65.4	27.5	85.9

and field. In its species composition the hedge fauna gives the impression of an impoverished forest fauna.

Transfer and retrapping experiments were carried out on the field species *Pterostichus vulgaris:* 150 marked individuals were set free in both the northern and the southern field and on the northern and southern margins of the hedge. Of the 300 individuals set free in the hedge none was recovered in the hedge itself but 19 had returned to the fields. Of the 300 individuals set free in the fields 39 were recovered there, two at the margin of the hedge and none inside the hedge.

A similar procedure was followed with the forest species *Pterostichus cristatus,* by far the most common species within the hedge (Table 38). 150 marked individuals were set free inside the hedge. Of these, 18 were recaught in the same place and one had wandered to the margin of the hedge. 50 marked individuals were set free at each margin of the hedge, making a total of 100. This is a high concentration when compared with the normal population, but it did not seem to lead to any increase in migration to neighbouring fields. The experiments show that these species are strongly attached to the habitat of their choice and that artificial attempts at redistributing them are unsuccessful. The conclusion to be drawn is that: "the dominant carabids of the hedgerow, being forest species, scarcely penetrate at all into the fields, which are climatically unsuitable for them. The hedge fauna, therefore, cannot be expected to contribute to the maintenance of the biocenotic equilibrium of fields..." (Thiele, 1964b).

Results of this nature agree with qualitative investigations made by Tischler (1948) in northern Germany, although the hedges studied were on low dams and were drier than those in the Bergisches Land. Nevertheless, the fauna of forest and forest edge dominated in the oak-hornbeam hedges, accounting for 44% of all the insect species found. Eurytopic species, with their greater ecological flexibility, occupied the second place (38%), whereas species from grassy habitats accounted for only 18%.

Further investigations in central Europe have confirmed these findings. In the hills of Saxony in East Germany, at about 600 m above sea level, species characteristic of the neighbouring fields and pastures were encountered in young hedges on stony ridges that serve as wind-breaks (Hiebsch, 1964). Forest species were seldom found in these young hedges, but a ten-year hedge harboured more forest species than one in its fourth year. Hiebsch concluded that: "The mutual influence of the fauna of hedges and the adjoining fields is negligible and is of little importance in biological pest control".

In northern Germany Gersdorf (1965) studied the relationships between the carabid fauna of a 350 m × 400 m area of cultivated moorland used for grazing, and an 8 m deep and 10 m high hedge of mixed deciduous and coniferous trees and shrubs surrounding it. Forest animals accounted for more than half of the species found in the hedge and for 41% of the individuals. *Pterostichus niger* was a typical dominant species in the hedge, where 48 of its larvae were caught as compared with four up to 10 m from it and four more on the rest of the pasture. However, the species caught in the largest numbers, *Pterostichus coerulescens* (55% of all individuals), was only present by chance in the hedge itself, where 21 individuals were found as compared with 724 at a distance of 10 m. It "seems that the differences between fauna of hedge and pasture are not so striking as those between hedge and field...". Nevertheless, Gersdorf concludes: "Taken as a whole, the fauna of the pasture is not enriched by that of the hedge, even if a few individuals of seven out of eight species are found on it, of which five occurred only in the near vicinity of the hedge". The species spectrum of the pasture is thus scarcely influenced at all by the proximity of the hedge. Carabids are unable to control the large numbers of *Tipula paludosa* L. because the *Tipula* larvae hatch at a time when only a few carabids are present.

Further confirmation of the above results was obtained in England (Pollard, 1968c). *Trechus obtusus*, *Leistus ferrugineus* and *Abax ater* were confined to hedges, while *Nebria brevicollis*, *Agonum dorsale* and *Bembidion guttula*, although found in hedges, also occurred elsewhere. On the other hand, *Pterostichus vulgaris*, *P. madidus*, *P. macer*, *Harpalus rufipes*, *Carabus violaceus*, *Bembidion obtusum*, *B. lampros*, *Trechus quadristriatus*, *Loricera pilicornis* and others were almost exclusively found in fields. The situation is exactly the same as that prevailing on the continent, with the exception that in the British Isles, where forests are not so common, the forest element in both fields and hedges is less well developed.

In the Lugansk region of the eastern Ukraine the fauna of the extensive shelter belts (20–60 m deep) is ecologically closely related to that of the ravine forests, with 22 species in common as compared with only nine in common with the natural steppe. However, the fauna of the ravine forests, as would be expected from the very different type of climate prevailing in Russia, is a very different one from that to be encountered in the forests of central Europe. In addition to *Pterostichus oblongopunctatus*, *Calosoma sycophanta* and *C. inquisitor* the majority of species found are such as would be met with on fields and dry grasslands in central Europe, e.g. *Pterostichus coerulescens*, *P. cupreus*, *Amara bifrons*, *A. aulica*, *A. communis*, *Asaphidion flavipes*, *Brachinus explodens*, *Panagaeus cruxmajor* and others. In central Europe, on the other hand, the eastern elements of the open steppe are usually completely lacking (Ghilarov, 1961).

Only in Poland (Poznan province) do conditions differ from those so far described (Gorny, 1968b, 1971; Bonkowska, 1970). There, the wind-break strips investigated are quite different from the damp, dense, western and central European hedges in that they are very long (2000 m) and relatively broad (36 m). They are of loose growth and consist mainly of *Robinia pseudoacacia*, with a sprinkling of *Quercus rober*, *Larix decidua* and *Alnus glutinosa:* their microclimate is rather dry. Scarcely any specific species were found (one individual of *Pterostichus oblongopunctatus*, five *Pterostichus nigrita*), but surprisingly, in contrast to the situation in western central Europe, *Pterostichus vulgaris* plays a dominant role. Carabids from the shelter belts can penetrate into the field zone as far as the influence of the microclimate of the former extends. Gorny regards this as a "biocenologically beneficial fact, although all the forest species of Carabidae found in the marginal zone of the field belonged to the accidental group". In fact, of the six species which Gorny allots to this group (*Calosoma sycophanta*, *Carabus nemoralis*, *C. cancellatus*, *Amara familiaris*, *A. similata* and *Pterostichus vulgaris*) only 18 individuals were found in the shelter-belt and as few as ten in a neighbouring field. Just as in Russia, the species found in the shelter belt are mainly those known as field species in central Europe.

In Schleswig-Holstein in northern Germany the situation even in large wooded patches in fields corresponded to that in the hedges (Tischler, 1958). Comparison with fields revealed that relatively few animals are common to both. "The fauna of fields enclosed by hedges differs little whatever the distance from woodlands ... Only a few species such as *Nebria brevicollis* and *Carabus nemoralis* seem to settle in greater numbers in fields near forests than in those at a distance. Forests, and particularly their marginal regions, are of much greater interest to field animals in winter than in the vegetational season. On the other hand, no significant degree of interchange of small animals of the soil surface takes place between the interior of forests and the fields, either in summer or winter."

In spite of the generally few connections between the fauna of woodland and field some interplay is observable. At the southern foot of hedges on steep dams in Schleswig-Holstein the temperature rises and activity sets in earlier than elsewhere in the vicinity (Fuchs, 1969). In March innumerable individuals of *Bembidion lampros* were already active on the southeastern slope of such a hedge whereas those in the fields did not become active until May/June.

On a southern slope of 75° large numbers of *Agonum dorsale* were active in mid-April whereas on a slope of 30° only six individuals were found in 90 min. Large numbers of a moisture-loving field species, *Pterostichus vulgaris*, withdrew into the hedge in the hottest month. Concentrations of *Carabus granulatus* and *Pterostichus niger* were observed on the north side of the margin of a hedge. "Animals of this type are almost certainly incapable of existing in hedgeless agricultural country in our latitudes on fields far from woodland. This holds as well for *Asaphidion flavipes* L. (and) *Pterostichus nigrita* F. ...".

Fuchs also provides data indicating seasonal migrations of *Agonum dorsale* between field and hedge. The catches in different months for hedge and field margins are as follows:

	III	IV	V	VI	VII	VIII	IX	X
Hedge	70	490	445	95	45	125	205	310
Field margin	3	32	280	655	288	19	7	3

These figures show that *A. dorsale* leaves its winter quarters in March/April, migrates to the fields in May, returns in August and begins to withdraw to its winter quarters in October. In the Rhineland Thiele (1964b) observed that, following the grain harvest on August 5th, the catches of *A. dorsale* on the margins of hedges rose rapidly during August from the initial minium values in July (Fig. 60). In deciduous woods near Magdeburg Roth (1963) recorded the following catches:

I	II	III	IV	V	VI	VII	VIII	IX	X	XI	XII
0	0	6	44	32	2	2	5	2	33	0	1

The migratory movements of the species thus seem to be very similar all over central Europe.

It appears that hedges surrounding fields are of little interest to the natural enemies of agricultural pests, particularly the larger predatory carabids and staphylinids. The opportunity presented by hedges for avoiding the upheavals of field cultivation, and their advantages as winter quarters are exploited by only a few species. In fact, the carabids that do most to maintain the biocenotic equilibrium on cultivated fields have very little connection with the hedges (e. g. *Pterostichus vulgaris, P. madidus, P. cupreus, Carabus auratus, C. cancellatus, Harpalus rufipes, Calathus fuscipes;* see also Kirchner, 1960; Scherney, 1961).

This does not mean, however, that carabids are not, in a wider sense, useful in maintaining the equilibrium of a natural environment. "The amazing wealth of species in the hedges points to their being an important refuge for the original forest fauna of any area" (Thiele, 1964b). This is particularly true in sparsely forested countries such as England, of which Pollard (1968a) wrote: "that much of our fauna is of woodland origin, and that many species are now dependent to a large extent on hedges for their continued existence in agricultural areas."

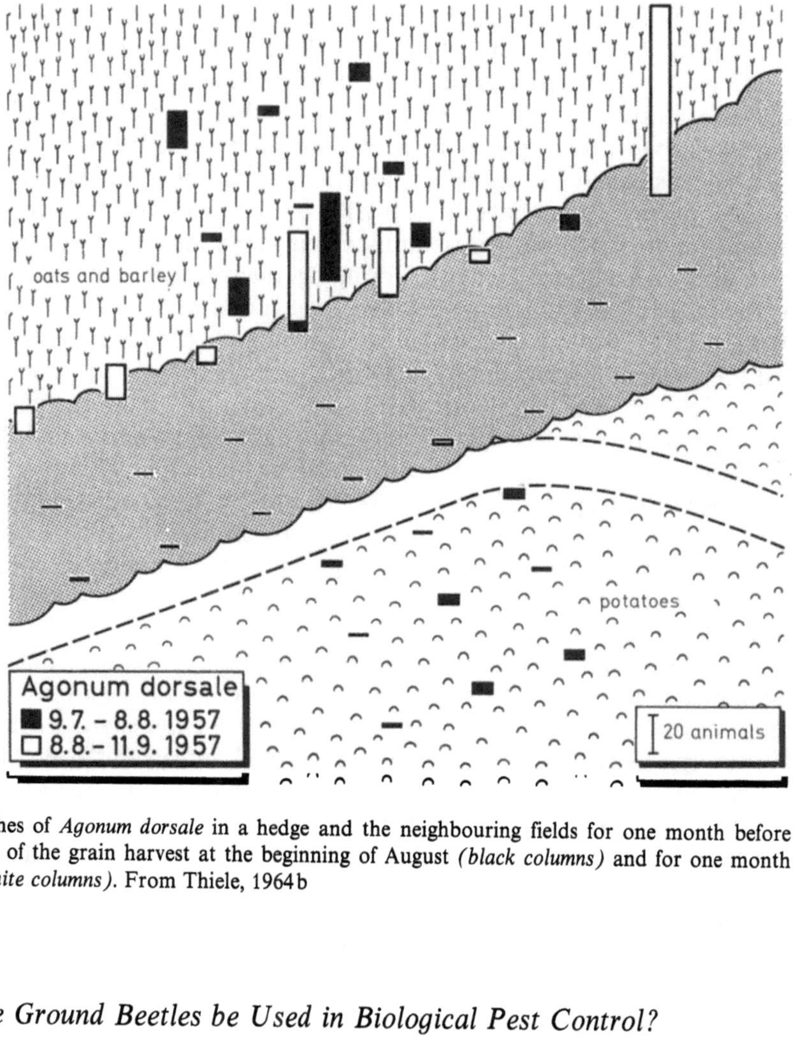

Fig. 60. Catches of *Agonum dorsale* in a hedge and the neighbouring fields for one month before the beginning of the grain harvest at the beginning of August *(black columns)* and for one month thereafter *(white columns)*. From Thiele, 1964b

4. Can the Ground Beetles be Used in Biological Pest Control?

The investigations cited above show that ground beetles are well capable of destroying injurious insects. Scherney's varied and meticulous studies involving extensive material have provided us with an accurate quantitative picture of the predatory action of carabids on the potato-beetle larva. Here again, only the simultaneous fulfillment of two rarely encountered conditions, i.e. a low degree of infestation and maximum occurrence of carabids, would render additional measures unnecessary.

The hope that the wealth of carabids in the surrounding hedges, woodlands and shelter belts would prove to infiltrate into the adjacent fields and act as a kind of biological control mechanism on the crop pests has not been fulfilled. Present knowledge suggests that such an effect cannot be expected since the carabids of the hedgerows are forest animals and thus bound to the forest-like microclimate of these habitats. They are in fact incapable of existence in the near-lying fields.

The idea of breeding carabids in the laboratory as biological weapons for pest control does not seem to be feasible. Since carabid larvae are cannibalistic they have to be reared in isolation, which entails a large number of staff as well as extensive space and funds. In fact, the breeding of most of the species that might be useful in pest control has not yet been mastered.

The most likely successful exploitation of carabids would seem to be in combatting pests that have accidentally been introduced into foreign faunal regions. Importation of their natural enemies from their land of origin provides a natural and efficient means of bringing them under control. A classically successful case of this nature involved the importation of carabids (Burgess, 1911; Holste, 1915) to combat the gold-tail moth (*Euproctis chrysorrhoea* L.) and the gipsy moth (*Lymantria dispar* L.), both of which had accidentally been introduced into North America. Burgess systematically imported the European larval predator *Calosoma sycophanta* into New England and established a large field laboratory for breeding the considerable populations required. This is still the largest experiment of its kind on record, in which ground beetles have been used as biological tools in pest control. In the years between 1905 and 1910, 4046 living individuals were imported into America from Europe. Of their progeny, 18,000 larvae and adults, offspring of the imported individuals, were set loose in 1906–1910, mainly in forests afflicted with the gipsy moth. As a rule colonies of not less than 200 larvae were released, and 80% of all colonies set out up to 1910 showed an increase. In the autumn of 1908, 105 adult males and 110 adult females were set out as two separate colonies about 2.7 km apart. By August 1910 the animals had spread out over 11.37 square miles. Today, *Calosoma sycophanta* is well established in large sections of the eastern U.S.A., and together with many other predators and parasites of the two harmful moths, just as in Europe, it succeeds in keeping their numbers at a tolerable level. Dr. Terry L. Erwin kindly informed me in 1974 that several individuals caught in 1962 in New Jersey are in the possession of the National Museum of Natural History in the Smithsonian Institute, Washington, D. C. Dr. Erwin has also seen two individuals collected in 1974, one in Virginia and one in Maine. He believes that "the species is still established in the USA and is rather widespread in the northeast". In 1974 Weseloh published investigations on enemies of the gipsy moth according to which *Calosoma sycophanta* occurs in large numbers in Connecticut.

Even though little hope exists of combatting pests biologically with ground beetles this does not detract from their value in keeping down numbers and even destroying large quantities in the event of a heavy attack. Investigations already reported on the role of carabids in the ecosystem (see Chap. 3.H.) and studies aimed at elucidating their effect on injurious insects confirm their considerable ability to reduce the population density of pests. On their own, however, carabids and other entomophages are unable to prevent, or check rapidly enough, the damage caused by pests to agricultural monocultures or forests. They should rather be considered as valuable natural "auxiliaries" in pest control. Unfortunately, however, carabids appear to be much more susceptible to the effects of pesticides than other insects. An intensification of research on the bionomics of ground beetles on agricultural land might render it possible

(1) to curb the unavoidable use of pesticides to a level at which the populations of useful species, in this case carabids, suffer as little damage as possible, or; (2) even to limit the use of pesticides to cases where the natural enemies of the pests prove inadequate. Carabids would then be able to play a useful part in an "integrated pest control" programme.

II. Damage Done by Ground Beetles

1. Ground Beetles as Crop Pests

Many carabids have proved to be omnivores or at least partial phytophages, thus inevitably causing occasional damage to cultivated plants. All that was known on the subject up to 1954 was reviewed by Blunck and Mühlmann (1954). Although rarely encountered, damage by carabids has been reported due to their devouring blossoms, ripening seeds and seedlings, as well as leaves and succulent shoots. Berries and Gramineae, including grain crops, are particularly vulnerable. "As a rule the damage usually remains within tolerable limits". An exception, however, is provided by the grain ground beetles, *Zabrus tenebrioides*, the older larvae of which are a particular danger to young cereal plants in spring. The species is frequently found in eastern Europe and in the Near East. Severe damage was reported from Moravia and the Ukraine where 70 km^2 of grain crops were damaged or destroyed in 1932. The species is relatively rare in central Europe and only causes damage in drought years in eastern central Europe. In the dry year 1959 damage was centred in the district of Halle in the German Democratic Republic, in the dry region of central Germany. Further to the west the species is so rare that it is no longer a pest. In Bavaria Scherney (1955) found only seven *Zabrus tenebrioides* among a total of 31,937 carabids caught in fields, and Kirchner (1960) found a single individual in his total catch of 11,617 field carabids near Cologne on the Rhine. An exceptional case of damage in the Netherlands (Hendrix, 1951) is on record for 1948/49.

Damage to strawberries is relatively frequent; Blunck and Mühlmann (1954) list 14 guilty species and Briggs (1965) mentions a few more. The latter author has studied the biology of some of the more troublesome strawberry pests, including *Harpalus rufipes* which is known as the "strawberry seed beetle". Whereas this species and *H. aeneus* eat the seeds, *Pterostichus vulgaris* and *P. madidus* eat into the fleshy part of the fruit. According to Briggs an average of 15% of the strawberry crop on the East Malling Research Station in England was damaged by carabids.

Some damage, however, can be attributed solely to water shortage, which forces the carabids to attack the fresh parts of plants as in a case reported by Brandt (1949), in which unripe ears of grain were devoured by *Carabus auratus*.

Dick and Johnson (1958) found reports in the older literature of exceptional cases where *Harpalus* destroyed up to 80% of the conifer seeds in nurseries. Their own experiments with Douglas Fir seedlings, carried out in Canada, revealed less than 2% damage to batches of 20 seeds left in the open for 10 days in June (due to *Harpalus cautus*).

With a few rare exceptions, damage done by carabids is of little economic significance. Apart from *Zabrus*, mainly members of the genera *Harpalus* and *Amara*, which consume large quantities of plant matter, are responsible for the damage (see Chap. 3.F.VI.).

2. Ground Beetles as Conveyors (Vectors) of Disease

Kullmann and Nawabi (1971) showed that *Pterostichus madidus* is able to ingest trichinous carrion. The muscle trichina (*Trichinella spiralis* Owen) could usually be isolated from the connective tissue capsule of the insect intestine. Larvae and capsules of the worms were excreted undigested in the faeces. It was possible to transfer the infection to mice and hedgehogs experimentally by feeding them with *P. madidus* that had previously been fed on trichinous meat, if not more than 22 h were allowed to elapse between uptake of meat by the beetles and their being eaten by the small mammals (the length of time required for the trichinae to traverse the intestine). This suggests the possibility that carabids in nature may indeed act as carriers of trichinous infections to insectivorous animals. The ingestion of trichinae by *Carabus coriaceus* and *Pterostichus (Platysma)* was also observed by Merkushev (1955, cited by Kullmann and Nawabi).

Mobedi and Arfas (1971) infested carabids artificially with a nematode, *Capillaria hepatica*, the eggs of which could be found in the faeces of the ground beetle for 15 days. In the opinion of these authors, therefore, certain carabids can be held responsible for the distribution of such parasites, although investigations on beetles caught in the vicinity of infested rodents invariably yielded negative results.

A *Salmonella typhi* culture was isolated from dead individuals of *Pterostichus vulgaris* by Blokhov and Mukhin (1961, cited by Dubrovskaya, 1970).

The conclusion to be drawn from the paucity of information contained in the literature is that ground beetles are of little significance as carriers of diseases that are of interest to man.

B. The Influence of Man on Ground Beetles

I. The Influence of the Methods of Husbandry Employed on Cultivated Land

1. Methods of Cultivation on Agricultural Land

Opinions differ as to the effect of the processes involved in soil cultivation on the carabid fauna. It is known, for example, that the various methods employed for loosening the soil have different effects on earthworms, a rotary hoe causing a much more drastic reduction than a plough (Tischler, 1955).

To what extent does this apply to carabids? "Since many carabid species are indigenous to the littoral region they ought therefore to be adapted to periodic changes in their substrate. It is therefore not surprising that most of them suffer little loss due to turnover of the soil" (Tischler, 1965; see Chap. 2.C.II.). Heydemann (1953) pointed out, however, that differences in numbers of spring and autumn breeders on fields with root crops and winter grain fields might be connected with the fact that the cultivation processes on the two kinds of field coincide with different stages in the two types of annual rhythmicity of carabids. In fields with root crops the soil structure is radically disrupted by hoeing in spring, when the spring breeders are present as adults, whereas the autumn breeders are represented by growing larvae and pupae. Heydemann assumed that the larvae would be less affected by soil tillage than the adults, thus offering an explanation for the occurrence of fewer spring breeders in root crop fields than among cereals. He was led to this assumption by observing a reduction in numbers of *Carabus auratus* following soil cultivation in May. On the other hand, as Kirchner (1960) pointed out, a large proportion of the adult *C. auratus* die off in the second half of May anyway. He regarded the preponderance of spring breeders, that on the whole show a marked preference for warmth, among cereals, as being due to the warmer microclimate in these less shady fields. In direct contrast to Heydemann, Skuhravý (1958) stated that spring cultivation of root crops decimates the larvae and pupae of autumn breeders whereas the spring breeders, present as adults, are unaffected. His conclusions were based upon the observation that large numbers of *Pterostichus vulgaris* (overwinters as larvae) adults hatch in June on winter crop fields where the larvae and pupae have been able to develop undisturbed. This beetle is seldom encountered among root crops in June, however, although large numbers are found from July until September. Such animals are presumed to have wandered in mainly from surrounding areas. This interpretation is not necessarily the only one, in view of the great variability of the phenological situation from year to year and its large degree of dependence upon weather. On the other hand, the adults of *Pterostichus cupreus* (overwinters as adult) appeared at least as early and as frequently in root crops (April/May) as in cereals. With Louda's report (1968) of rather similar results for both species, Skuhravý's interpretation has gained in likelihood. Nevertheless, it should be borne in mind that *P. vulgaris*, in contrast to *P. cupreus*, requires plenty of water and therefore the differences in plant cover from season to season between the various types of fields may exert a considerable influence on the phenology of such widely differing species.

The effect of harvesting and cultivation on the carabid population has been studied in western Germany (Kirchner, 1960). Catches were made in an early potato field, and again following harvesting and replanting with cauliflower. Controls were provided by catches made in a neighbouring late potato field that remained undisturbed during the period in question. The total catch in the cauliflower field was only slightly lower than that from the control potato field.

Most species are unaffected by a change of crop. In the case of *Pterostichus vulgaris*, which was the species most commonly encountered, the numbers caught in an experimental field in the first week immediately following a change of crop were even higher than in the control field. Noticeably fewer *Agonum dorsale* (perhaps due to migration, see Chap. 4.A.I.3.) and *Carabus purpurascens* were recorded, although the numbers of *Bembidion quadrimaculatum* greatly increased, probably due to immigration by flight. Lindroth's (1945) explanation is that the relatively dry and sunny nature of the freshly cultivated field with its low vegetation fulfils the ecological requirements of the latter species.

Although the individual steps involved in cultivation or harvesting appear to have very little influence on field carabids, the employment of a different method of soil cultivation, if pursued for a considerable length of time, is reflected in the carabid fauna, whatever the crop involved. Investigations in connection with crop rotation on an experimental farm in Byelorussia included comparisons between fields that were worked over with a moldboard to a depth of 20–25 cm (controls) and others that were loosened twice in the course of crop rotation to a depth of 40 cm, and disc-harrowed to 10 cm in the other years (Dubrovskaya, 1970).

The latter method is the less disruptive for the carabid population, as shown by the following figures:

No. of field	Year	Crop	Carabid population in % of the control
4.	1958	Potato	120.5
	1959	Wheat undersown with clover	157.2
	1960	Second-year clover	138.5
	1961	Flax	143.2
3.	1959	Beet	115.2
	1960	Wheat undersown with clover	133.8
	1961	Second-year clover	93.3
2.	1960	Sugar beet	220.5
	1961	Wheat undersown with clover	144.9
1.	1961	Sugar beet	116.4

The influence of *fertilizers* on many different soil animals has been reported, but little appears to be known of any direct effect upon carabids, with the exception of Lindroth's (1949) studies. In experiments involving choice of substrate he showed that none of the artificial fertilizers in common use elicited a positive reaction from *Amara ingenua*, a species that does well on cultivated land in Scandinavia. Note at correction: In field experiments with nitrogen fertilizers applied at 80 kg, 160 kg, 320 kg and 480 kg/ha carabids, especially *Pterostichus vulgaris*, *Bembidion lampros* and *Amara plebeja*, avoided the plots with the highest concentration. They accumulated mainly in the non-fertilized control plots (Honczarenko 1975).

The effect on the carabid population of turning extensively utilized, unfertilized hay meadows into well-manured high-quality pasture land in the Erzgebirge in the GDR was observed by Hempel *et al.* (1971). The number of species sank, the composition of the fauna became more monotonous and the eurytopic carabid species common to cultivated land came to the fore, whilst typical meadow species receded. On a nonfertilized hay meadow (I) 35 species were identified, 36 on a nonfertilized pasture (II) and only 25 on the high-quality pasture (III). A few examples of the different numbers of characteristic species caught are given below:

	I	II	III
Bembidion gilvipes	6	85	162
Pterostichus vulgaris	27	265	415
Loricera pilicornis	4	7	69
Pterostichus coerulescens	22	19	3
Trechus secalis	116	29	42

2. *Methods of Cultivation in Forests*

One fact that stands out clearly is that the practice of clear-felling radically changes the composition of the carabid fauna. In Westphalia in western Germany (Lauterbach, 1964) cleared areas of forest and young spruce plantations were shown to be mainly inhabited by field species (*Carabus cancellatus*, *C. auratus*, *Pterostichus cupreus*, *Agonum sexpunctatum*, *A. muelleri* and *Amara* species), as well as the two species particularly characteristic of forest clearings, *Pterostichus angustatus* (see Chap. 8.F.) and *Carabus arcensis*. The dominant species in forest habitats in the same region, however, were *Pterostichus madidus*, *P. metallicus*, *P. cristatus*, *Abax ovalis*, *A. parallelus*, *A. ater*, *Molops piceus*, *Nebria brevicollis* and other forest species. The fauna of these two adjoining areas have almost nothing in common and their requirements as to microclimate are entirely different.

Similar observations were made by von Broen (1965) in the flat Mecklenburg region of northern Germany. Common species on the forested areas (a completely different vegetational type from those mentioned above) were *Carabus hortensis*, *C. arcensis*, *Abax ater*, *Leistus rufomarginatus*, *Calathus micropterus*, *Nebria brevicollis* and others. Typical field species such as *Carabus granulatus*, *C. cancellatus*, *Bembidion lampros*, *Pterostichus lepidus*, *Harpalus rufipes* and species of *Amara* were found to be the most common species of clear-felled areas. *Harpalus rufitarsis*, which also occurs in the sun-drenched heath forests, was especially typical of the latter areas. No convincing differences in abundance were found between, for example, *Pterostichus niger*, *P. vulgaris* and *P. oblongopunctatus*. Clear-felling thus appears to disrupt the soil fauna and affects the forest carabids in particular. A completely new fauna takes over the denuded areas and migration presents the only means by which the original forest fauna can reestablish itself in the newly regenerating forest. Lauterbach (1964) found 49% of the species in forest clearings to be capable of flight as compared with only 28% inside the forest itself.

In western Germany reforestation with exotic conifers brought with it an increase in the numbers of species and individuals of soil-inhabiting beetles (a greater variety of vegetation and diversity of microclimate), but this kind of reforestation appeared to have no effect on the carabid population (Kolbe, 1972).

II. The Effect of Insecticides on Carabids

1. Laboratory Experiments

The effect of five contact insecticides was tested by van Dinther (1963) in Petri dishes, using *Harpalus rufipes*. The scale of toxicity proved to decrease from Parathion via Dieldrin, Sevin, DDT to Toxaphene. The LD 50 and LD 90 values were determined and compared with the doses normally applied in combatting harmful insects in the field. Using an exposure time of one hour the LD 90 of Parathion amounted (expressed in kg/ha) to 0.003 (normal dose of 0.1–0.5 kg/ha). After 3 h exposure the LD 90 of Dieldrin amounted to 0.118 as compared with the normally applied dose of 0.1–2.0 kg/ha. Both insecticides would probably completely exterminate a population. Sevin, DDT and Toxaphene proved to be less toxic. Humphrey and Dahm (1976) analyzed the residues of chlorinated hydrocarbons in three species of carabids. In *Pterostichus chalcites* they found an especially high content of DDE (913.5 ug/g body weight = ppm). The LD 50 for dieldrin was 3357 μg/g of beetle (ppm) in laboratory experiments, a figure which far exceeded the residue levels found in the animals. The authors conclude from their results that *P. chalcites* has an unusually high resistance to dieldrin.

Since *Harpalus rufipes* causes damage to strawberries Briggs and Tew (1969) tried out more than 30 insecticides other than chlorinated hydrocarbons, in a search for an alternative to Aldrin as a means of destroying this beetle. Most of them had an initial toxicity of less than 40%, but a large number proved to be 80–100% toxic. Parathion was highly toxic. The important criteria for an effective agent are that it should combine persistence in the soil with the least possible traces in the crop itself. Using Diazinon or Fenitrothion damage could be reduced by over 90%. It is interesting to note that Diazinon is more effective after 24 h in damp clay soil than in dry soil, but that this effect is completely reversed after treatment lasting several days to a week.

Using a variety of substances Scherney (1958) carried out thorough experiments involving four different doses on six species of *Carabus*, *Harpalus* and *Pterostichus*. Parathion (E 605) was particularly toxic in all doses as compared with Toxaphene (= chlorinated Camphene), Multanin (= Lindan + DDT), Hortex (= Lindan) or even arsenate of lime. His data on the relative toxicity of these substances confirm the findings of other authors. A few species, however, react differently: of the *Carabus* species, *C. granulatus* is less sensitive than *C. auratus* and *C. cancellatus*, and among the smaller species *Pterostichus cupreus* is more sensitive than *P. vulgaris* and *Harpalus rufipes*. Under experimental conditions, on the other hand, E 605 (Parathion) is the least effective agent against potato-beetles, and was responsible for a mortality of only 10%.

Under laboratory conditions Thionazin was lethal to carabids in the doses commonly employed in combatting nematodes and harmful insects (2.24–8.96 kg/ha) whereas Menazon was not. The effect of Thionazin was enhanced in damp soil. Small carabids of the size of *Bembidion lampros* succumbed 12–13 times as rapidly as the females of *Pterostichus vulgaris* (Critchley, 1972).

According to Scherney carabids exhibited typical symptoms following ingestion of poisoned potato-beetle larvae (trembling antennae, rolling gait). If such larvae constituted the only source of food, mortality of *Carabus cancellatus* amounted to 40–80% (control animals 10%) and that of *Pterostichus vulgaris* 45–72% (controls 7.5%). The strongest toxic effect on carabids was elicited if the potato-beetles had been treated with E 605.

Indirect poisoning had already been observed by Klein-Krautheim (1953) in *Carabus nemoralis*, *C. auronitens*, and *C. arcensis*, in which typical atactic locomotory movements set in following ingestion of a single poisoned cockchafer.

Laboratory investigations on the effect of insecticides on carabids are apparently few and far between. Somewhat more is known about the

2. *Influence of Insecticides on Carabids in the Field*

On *cultivated land* 10% DDT in aerosol form has a destructive effect on the carabid fauna (Novák and Skuhravý, 1957). A potato field of 5–6 ha in Czechoslovakia was sprayed on July 3, 1954, a field of the same size being available as a control. The carabid population was investigated by means of pitfall traps at intervals of one to two weeks before and after treatment. In June, both *Brachinus crepitans* and *Pterostichus cupreus* were found in approximately equal numbers on test and control fields (several hundred of each per trap unit). Subsequent to spraying the following catches were made:

	Days immediately following spraying		Mid-July	
	Control	Experiment	Control	Experiment
Brachinus crepitans	69	2	20	0
Pterostichus cupreus	205	1	27	1

Although the experiment revealed that spraying drastically reduced the carabids it could not be followed up because the population begins to die off anyway between June and July, in the natural course of events.

Treatment of a sugar beet field with the systemic insecticide Systox (0.05%, 400 l/ha) was found by Novák *et al.* (1962) to cause no significant differences between sprayed and control fields, as seen in the abundance curves of the most common species, i.e. *Pterostichus vulgaris*, *Harpalus rufipes* and *Calathus fuscipes*. In this case the poison had little effect on the carabids.

In Germany, Scherney (1958) found the following differences in catches of larger carabids (*Carabus granulatus*, *C. cancellatus*, *Pterostichus vulgaris* and *Harpalus rufipes*) following treatment of a 5.6 ha potato field with Potasan (= Lindan + ester of phosphoric acid):

13. 7.	20. 7.	26. 7.		3. 8.	10. 8.	17. 8.	23. 8.
550	488	719	Treatment	44	117	298	690

Recovery of the population within four weeks was apparently achieved by immigration, because large numbers of carabids were at first found on the edges of the field.

The following insecticides drastically reduced the numbers of most carabids in field experiments in western Germany: Mercaptodimethur, Fensulfothion, Trichlorfan, Demeton, Dimethoate and Bromophos. Only a few species were unaffected. In laboratory experiments with Trichlorfan and Demeton, however, *Pterostichus vulgaris* individuals were undamaged, even though their numbers had been reduced by these agents in the field. Fensulfothion, on the other hand, killed 40 individuals of this species in the laboratory within five days. Most carabids are damaged by this substance in the field. The fact that it is employed as a nematocide is a good illustration of the harm that can be caused to a beneficial group of animals in the course of chemical measures aimed at a completely different group (Gese, 1974).

The influence of large-scale applications of insecticides in agriculture has recently been investigated (Basedow *et al.*, 1976b) in Sweden and West Germany. The treated areas measured 5–10 ha; changes in the capture rates in pitfall traps before and after treatment were determined. Complete sparing of all epigaeic predatory arthropods could be reached by none of the administered insecticides (fenithrotion, parathion-ethyl, parathion-methyl dust), but methoxychlor seemed to damage only a few species of carabids. After the application of the other insecticides, a quick recolonization of the treated fields was possible, too (within a span of 14 days). This was due to the low persistence of the insecticides. However, if they are applied to large areas, one must reckon with a long-lasting injury to the mostly univoltine predators, since the application is made at a time when most of the species are at the peak of their breeding period. These are the first results from the working group "Integrated pest control in grain culture of the West Palearctic Regional Section" of the "Organisation de la Lutte Biologique contre les animaux et les plantes nuisibles" (Basedow *et al.*, 1976b).

In *forested areas*, the use of DDT in combatting the pinebud moth (*Panolis griseovariegata* Goetze) was reported by Schindler (1958) to kill off about half of the carabids (especially *Calosoma sycophanta*), although no exact data were provided. In Czechoslovakia, Novák *et al.* (1953) followed up the results of the treatment against cockchafers on the margins of a mixed deciduous forest with HCH dust (hexachlorcyclohexan) from the air. A strip measuring 400 m × 30 m was treated with 5–7 kg/ha of 12% HCH. Soil arthropods were caught in baited traps one day before and on successive days following spraying.

Among the animals most affected were *Coleoptera*. Figures for the carabids (a fauna typical of the forest edge) were as follows:

	5. 5.		7. 5.	8. 5.	9. 5.	11. 5.
Experiment	84	Treatment	39	18	13	28
Control	55		49	26	18	32

The treatment appeared to have had some effect, even if not very dramatic, on the carabids.

Careful investigations were carried out in 1969 by Freitag *et al.* (1969) in connection with measures taken to combat the spruce budworm (*Choristoneura fumiferana* Clemens) on 293,000 acres of forest in Ontario, Canada. The area was sprayed with Sumathion and Phosphamidon, which led to a radical destruction of the carabids *Agonum retractum, Calathus ingratus, Pterostichus pennsylvanicus, Scaphinotus bilobus* and *Sphaeroderus nitidicollis*, as compared with a control area (Fig. 61), in spite of the fact that before treatment the latter area contained even fewer carabids. The ground spider *Trochosa terricola* was less affected by the poison.

The effect was maintained throughout the whole of 1969 (Fig. 62), during which period the carabid population on the test area remained below that of the control area (Freitag and Poulter, 1970).

The above survey of the situation does not claim to be exhaustive. A study of the pertinent literature reveals how astonishingly little is known about the effects of insecticides on an animal group of such ecological importance as the carabids. Far too little is known for insecticides to be used in what might be termed an integrated pest-control programme that would preserve the demon-

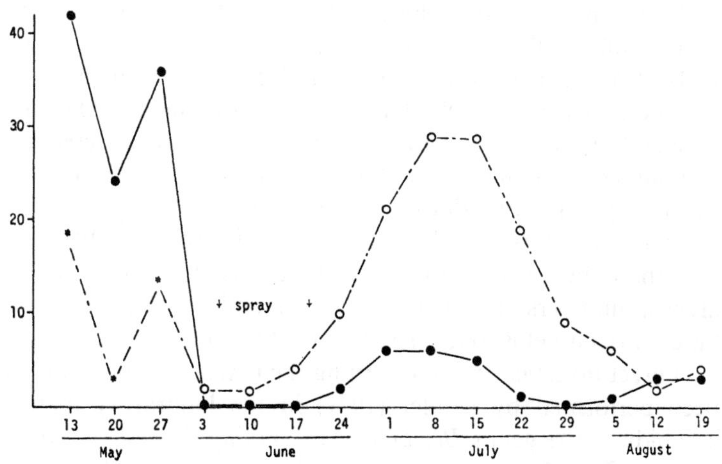

Fig. 61. Number of specimens of the carabid beetles *Agonum retractum, Calathus ingratus, Pterostichus pennsylvanicus, Scaphinotus bilobus, Sphaeroderus nitidicollis* trapped in sprayed and control stations for 15 weeks in the summer of 1968. ∗: numbers counted at two stations corrected to show expected numbers for three stations; ●: total number of specimens of three stations in the spray area; ○: total number of specimens of three control stations. From Freitag *et al.*, 1969

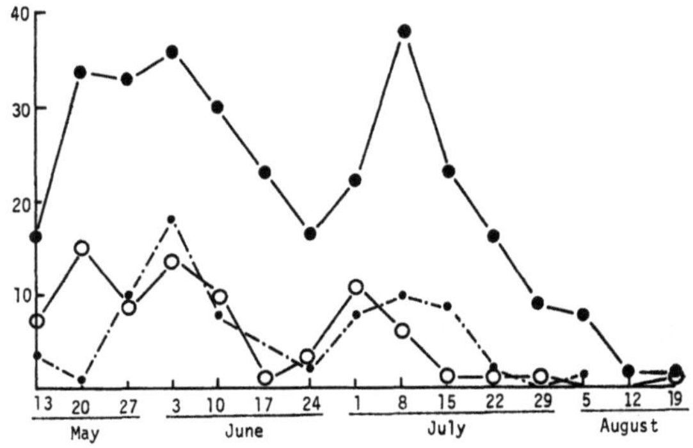

Fig. 62. Number of specimens of the same species of carabid beetles as shown in Figure 61 trapped in spray and control stations for 15 weeks in the summer of 1969. ●: total number of specimens of stations Nos. 1 and 2 in the control area; •: total number of specimens of stations Nos. 3 and 4 in the spray area; o: total number of specimens of stations Nos. 5 and 6 in the spray area. From Freitag and Poulter, 1970

strably beneficial effects of the carabids. The few publications dealt with here mostly reveal drastic damage to the carabid fauna or at least grave long-term effects.

III. The Effect of Herbicides on Carabids

As already mentioned, hedges and groups of trees in the fields constitute reservoirs for a species-rich carabid population in agricultural areas. Model experiments have been carried out on the possible effect on the carabid population of the drift of herbicides into hedges following treatment of agricultural land (Pollard, 1968a, b). In England, three treated stretches (A) of a field hedge were compared with three untreated control stretches (B). The soil flora was eliminated in the experimental stretches by repeated spraying with Preeglone (a paraquatdiquat mixture). Almost all carabid species showed a reduction in numbers on the treated areas: 93 carabids were caught in A from June 10th–21st 1965, as compared with 468 in B. From February 1965 until January 1966 304 were caught in A and 902 in B. The mean catch of *Agonum dorsale* in A was 6.4/trap and in B 26.5/trap ($p = 0.001$). Thus a reduction in numbers could also be proved for a field species that definitely overwinters in hedges. The particularly abundant species *Pterostichus vulgaris* and *P. madidus* also suffered a marked reduction in numbers. On the whole, the removal of the soil flora had "a detrimental effect on potential predators of crop pests".

It is quite possible that changes in floristic structure play an even larger role than the toxic effect of the herbicide itself. The increase on the treated areas in numbers of *Bembidion obtusum* and *Trechus quadristriatus*, both inhabitants of open fields, offers support for this view.

Direct investigations concerning the toxicity of nine different herbicides have been performed by Müller (1971) in the laboratory mainly using the small carabids of the genus *Bembidion*. In varying degrees, almost all herbicides had a toxic effect: 2.4 D and Chlorpropham (a carbamate) acted very rapidly and brought about a high rate of mortality (Fig. 63), although the dosage was above that normally used in the field. 2.4 D had a greater effect on *B. tetracolum* and Chlorpropham a greater effect on *B. femoratum*, especially if contaminated nutriment was also ingested. *B. femoratum* reacted by attempting to flee (repellent

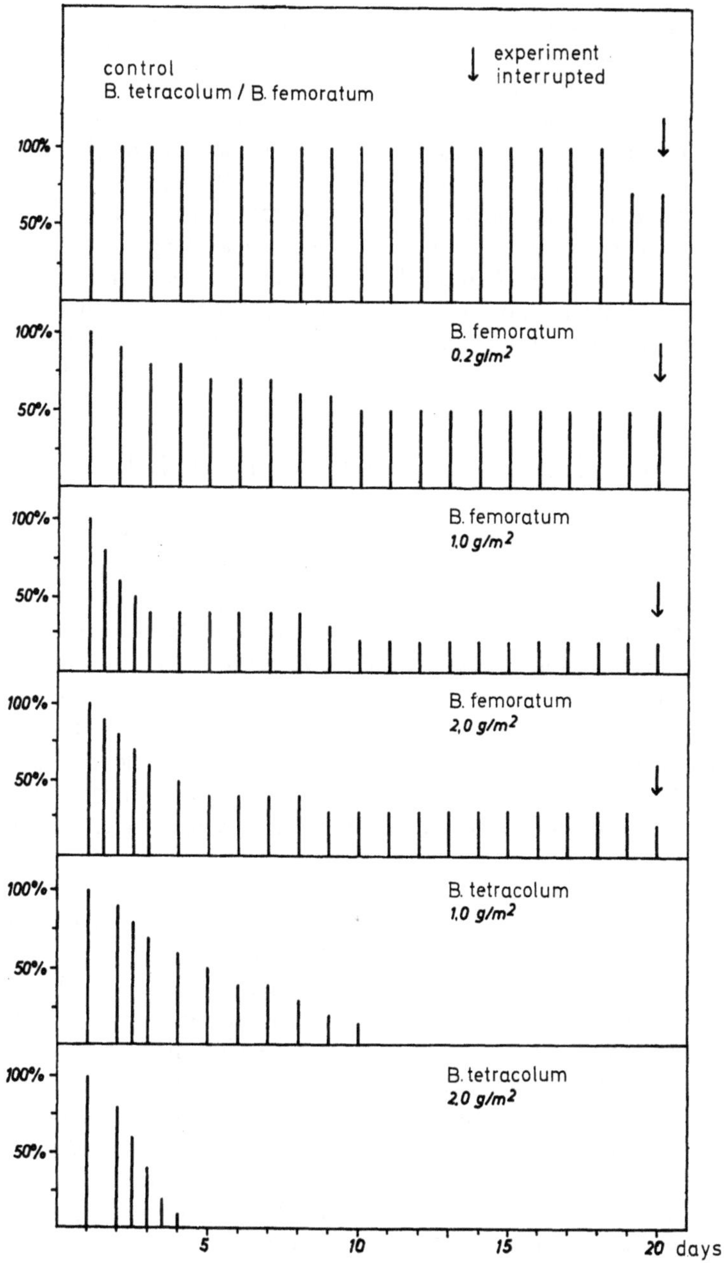

Fig. 63. Survival rate of *Bembidion femoratum* and *B. tetracolum* following treatment with a 2,4-D-substance. Normal quantities applied: 0.2 g/m^2. From Müller, 1971

action). In the open, this can lead to migration of species capable of flight, thus bringing about alterations in fauna. Field investigations with the herbicides Uvon-Kombi (II) and Elbanil (III) by Müller (1972) have not yet produced unequivocal results because attempts to prevent immigration and emigration, as well as the effects of autumn tilling, largely masked the herbicidal effect. In any case, treatment with herbicides does not seem to be very effective: "The qualitative constitution of the carabid fauna was not significantly influenced since any indirect effect of weed destruction, which might lead to changes in the hydrothermic household of the soil surface, was largely prevented in all four plots by the use of hand cultivation". With respect to the question of whether herbicides can exert any influence on carabids, the findings of Speight & Lawton (1976) may be of some interest. In a weedy field the predation pressure of carabids on an artifical prey (*Drosophila* pupae) was greater than in a "clean" one.

Under usual conditions of employment herbicides appear to have less effect on the carabid fauna than insecticides, although it should be emphasized that research into the side effects elicited by herbicides is only in its initial stages.

IV. The Influence of Industrial and Traffic Exhaust Gases on Carabids

Up to the present day scarcely any investigations have been published on possible effects exerted on the soil fauna by industrial exhalations, and it is not surprising, therefore, that only one team of authors has considered the implications for carabids (Freitag *et al.*, 1973). In Ontario, Canada, the ground beetle populations were investigated in five localities 900, 1200, 1650, 2100 and 2700 yd to the east of a kraft paper mill. A site 10 American miles to the west served as a control. All areas investigated were forest habitats. Twenty carabid species plus one species of silphid were studied. A drastic progressive reduction in number of carabids was recorded with increasing proximity to the mill (Fig. 64). Most species behaved in the same way (example, *Carabus nemoralis*, Fig. 65).

Although the agent responsible for killing-off carabids in the vicinity of the kraft mill is unknown, measurements did reveal that the reduction in their numbers ran parallel with the increasing precipitation of Na_2SO_4. In order to test whether the emissions from the mill had any effect on the growth of the beetles, measurements of the body size of the common species *Agonum decentis* were made in populations at varying distances from it. No differences were detectable. Nevertheless, the carabid fauna does provide a quantitative indicator of environmental changes connected with industrialisation.

A similar effect on the ground beetle population is produced by traffic exhaust gases (Maurer, 1974). A comparison was made between populations of meadows near a much frequented road (5000–8000 motor vehicles daily in summer) and those near a quieter road (200 vehicles daily). In each case traps were set up at distances of 2 m and 30 m from the road. Only half as many carabids were caught near the edge of the busier road as inside the field, but there was no difference in numbers trapped at 2 m and 30 m from the edge of the quieter road. Diversity was almost invariable lower throughout the year on the edge of the busier road than at a distance of 30 m from it (diversity index of Shannon and Weaver). The lead content of individuals of *Pterostichus cupreus* caught near the road was higher by a factor of 7–8 than that of individuals caught inside the field: the corresponding factor for *Carabus auratus* was 4–7.

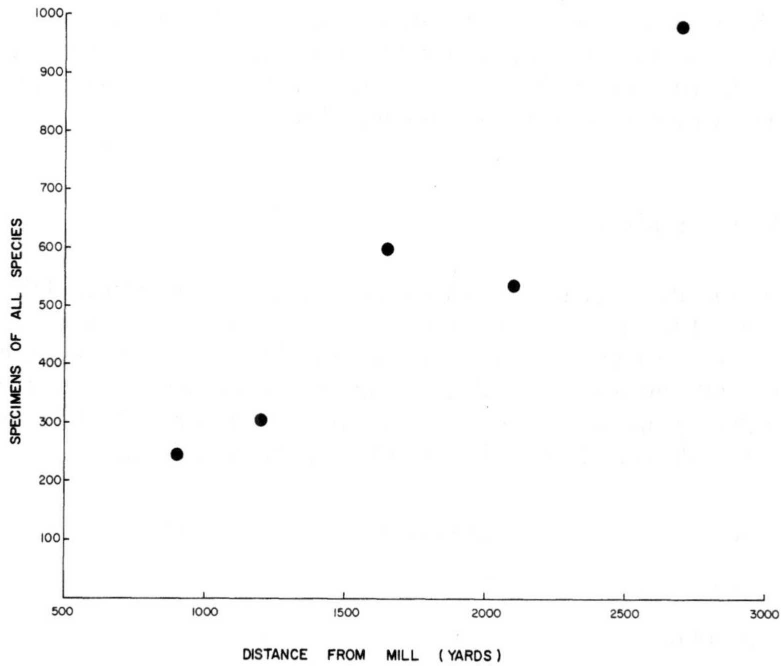

Fig. 64. Total number of specimens collected for all species against distance from a kraft mill. From Freitag *et al.*, 1973

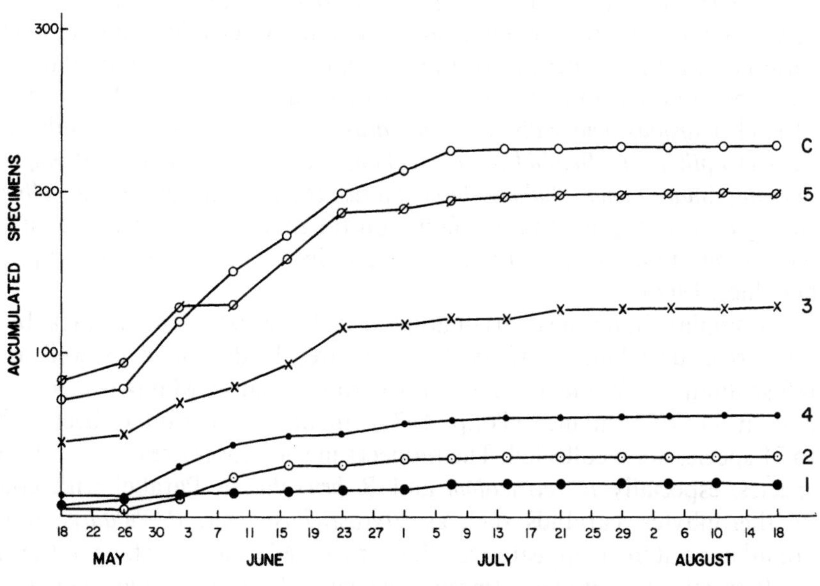

Fig. 65. Accumulated number of specimens of *Carabus nemoralis* collected in six stations near a kraft mill in Thunder Bay, Ont. over 15 weeks in summer 1971. The numbers of the stations *1–5* indicate increasing distance from the kraft mill. *C:* control station at a distance of 10 miles. From Freitag *et al.*, 1973

This does not necessarily imply, of course, that accumulation of lead, which was large in any case, accounted for the reduction in numbers of carabids at the edge of the busier road, where, in fact, the frequency of only one of the two species, *C. auratus*, was diminished.

V. Carabids in the City

Investigations carried out in Vienna offer quantitative evidence of the impoverishment of the insect fauna, especially of the carabids, in a large city (Schweiger, 1962). The figures given below indicate the number of species belonging to the different genera found in various zones starting at the city margins (still including meadows, woods and steppe), and progressing through the garden belt to the completely built-up area at the heart of the city:

Genus	Marginal zone	Garden belt	Built-up areas
Carabus	17	7	2
Trechus	6	3	1
Pterostichus	22	10	2

On rubbish heaps in the middle of the built-up area of the inner city, for example, Schweiger found *Notiophilus aestuans, Dyschirius globosus, Bembidion lampros, B. properans, B. inoptatum, Asaphidion flavipes, Trechus quadristriatus, Acupalpus meridianus, Agonum dorsale, Amara* and *Harpalus* species, that is to say, predominantly very eurytopic or distinctly xerophilic species from agricultural land, most of them also litoraea animals. Carabids of similar origin were also encountered underneath the stones of a terrace of a house on the edge of Kiel: *Carabus nemoralis, C. granulatus, C. cancellatus, Notiophilus biguttatus, Loricera pilicornis, Bembidion tetracolum, Pterostichus vulgaris, P. niger, Agonum dorsale, Amara aenea, Badister bipustulatus, Harpalus aeneus, Bradycellus verbasci.* In any case, it appears that the fauna on the margins of a city, even if connected with habitations, is richer than that found in the completely built up city centre (Tischler, 1966).

A similar fauna, also originating to a large extent in litoraea biotopes or on agricultural land, is to be found in freshly deposited as well as in older refuse dumps in Wilhelmshaven (northern Germany), Mainz (western Germany) and in Turku (Finland) (Topp, 1971). In all, 2580 ground beetles belonging to 71 species were collected. The pioneers in such cases are exclusively *Bembidion* species, especially *B. tetracolum* and *B. bruxellense.* Parallel with colonization by therophytes, carabids such as *Pterostichus niger, P. vulgaris* and *Carabus granulatus* put in an appearance. They are joined at a later stage in the succession by *Pterostichus strenuus, Amara convexiuscula* and *Agonum dorsale.* On still older deposits phytophages come increasingly to the fore *(Harpalus rufipes, Anisodactylus binotatus).* Adventive species, too, are common on refuse dumps, an example being the carabid *Dyschirius luedersi* in Turku.

VI. Carabids as Bioindicators of Anthropogenic Influences: Future Possibilities

The type of investigation already mentioned, concerning the effect of industrial emissions and the fauna of large cities, suggests a possible role for carabids as bioindicators of injurious anthropogenic environmental conditions. Carabids are easily caught with automatic, standardized traps and are not difficult to identify. Furthermore, their ecological requirements have been so thoroughly investigated that not only quantitative changes in the overall population, but the absence or presence of certain species can give us exact information as to the trend in the environmental alterations. In western Germany a team headed by P. Müller in Saarbrücken is engaged upon the first large-scale experiments exploiting such indicator effects at the periphery of a large city, in an industrial region and in areas of natural forest (Fig. 66).

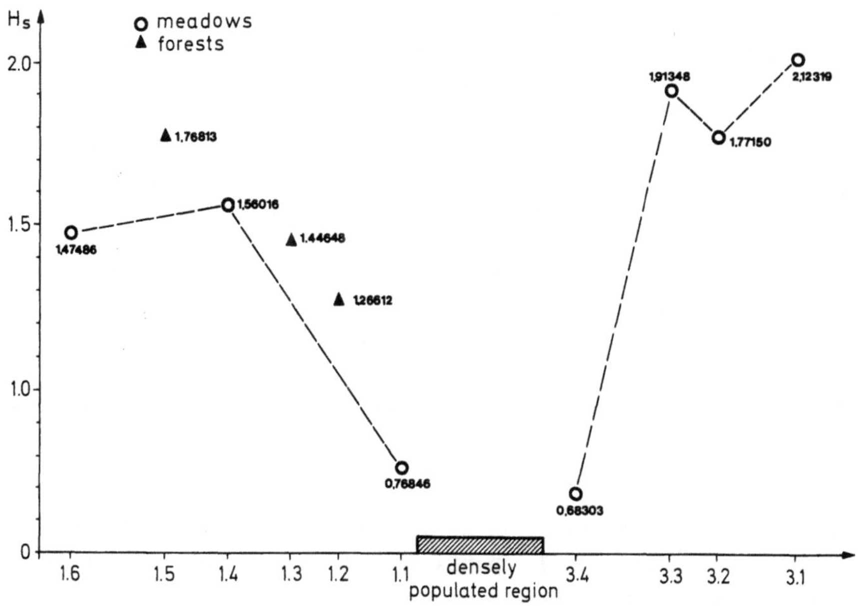

Fig. 66. Diversity values H_s of ten carabid populations in the vicinity of the densely populated region of Saarbrücken (West Germany). Increasing effects on the populations lead to a decrease in species diversity. The number of species does not necessarily decrease: it can even increase. Abscissa: Numbers of collecting sites. From Müller *et al.*, 1975

Plant protection bureaus in western Germany have begun investigations to decide whether the effect of protective agents on predators of the soil surface can be tested by standard methods. As we have seen, the sensitive and well-differentiated response of carabids to biocidal treatment renders them useful indicators of undesirable side effects.

Chapter 5

The Differences in Distribution of Carabids in the Environment: Reactions to Abiotic Factors and Their Significance in Habitat Affinity

An idea of the demands made by carabids upon the microclimate can be obtained by measuring the microclimatic factors prevailing in the habitat. As a rule, however, this does not provide conclusive information as to which specific factors determine the affinity of a species for its particular habitat. In nature, several or many factors vary in mutual dependence on one another, so that it is not immediately apparent whether the distribution of a species is governed by the combined action of all factors, or by only some of them, or even by one factor alone. "In order to recognize the individual factors governing the distribution of a species the complex of environmental factors has to be broken down experimentally into its constituents" (Thiele, 1959).

The pioneer studies in this field were carried out in the main by Scandinavian workers, of which the comprehensive studies of Bro Larsen (1936), Rolf Krogerus (1932, 1960) and Harry Krogerus (1948) are in many respects still valid today, and above all, the frequently cited standard work by Lindroth (1945–1949) on the fennoscandian Carabidae. The general section of his book (Vol. III, 1949) is indispensible to anyone seriously concerned with the influence of abiotic factors on carabids, which is to be dealt with in this chapter. The obstacle presented to some readers by the fact that Lindroth's book, consisting of over 900 pages, is in German, is to some degree circumvented by the addition of a 30-page English summary and English legends to all figures.

Lindroth (1949) clearly defines the aims and difficulties in attempting to determine the factors governing habitat affinity by means of laboratory experiments. "The experiments represent an attempt to isolate the external factors acting upon the animal in nature and to estimate their influence. However, the very isolation of such factors is an unnatural process. We cannot, for example, claim to know the absolute temperature preference of a particular species since this simply does not exist; it depends, among other things, upon air humidity... This and other considerations led me to seek only comparative results from each experiment... Preferably, species that are systematically closely related but which differ with respect to either distribution or ecology should be chosen. If, then, the experiments reveal that the species react differently it is justifiable to consider that the reason for their different behaviour in nature is to be sought within the complex of factors measured. The same applies to non-related species that more or less agree in the respects mentioned, and also exhibit similar reactions in experiment" (Lindroth, 1949).

"An experiment can show that a species may in fact be indifferent to a factor with whose intensity gradient its frequency in the field appears to be correlated. In some cases the preference shown may be the reverse of that suggested by the distribution of the species in nature. The conclusion is then justified that the distribution of the species is not influenced by this factor, but rather by another or several others which vary in parallel with the first factor. If a species shows a preference, and this agrees with the intensity of the factor at which the species lives under natural conditions, the conclusion may be drawn that the factor plays a part in the make up of the pattern of distribution, although it does not imply that it alone is effective" (Thiele, 1964a).

Despite his success in applying preference experiments to the explanation of habitat choice in carabids (see also Chap. 5.B.III.), Lindroth wrote in 1945: "...we are still far from recognizing the exact, decisive factors—indeed as far as I can see this presents an insuperable obstacle". In the meantime Lindroth's experimental approach has been copied by many workers and the method of preference experiments has undoubtedly contributed much to a causal explanation of habitat affinity in carabids. An attempt to summarize these results will be made at the end of the first section of this chapter (see Chap. 5.A.IV.). In view of Lindroth's qualifying remarks concerning the relative nature of observed preferences, it is particularly important to compare only results obtained by identical or similar methods. The following is therefore largely based on the data obtained by my own group of workers, for which such criteria hold.

A. Climatic Factors

I. Temperature and Orientation in the Environment

1. Experimental Method

By means of a piece of apparatus developed by Herter (1924) and described as a "Temperaturorgel" (= "temperature organ" or temperature gradient apparatus) it is possible to obtain information on the preferred temperature (PT) of an animal within a very short time. The apparatus consists of a straight or circular narrow cage, the floor of which is a metal band heated at one end and cooled at the other. By means of thermometers built in at intervals the temperature of the floor can readily be measured. A group of animals placed in this apparatus is free to choose its own temperature and in so doing exhibits thermotaxis. Data from such experiments give characteristic diagrams for any one species. Uniform air humidity is achieved over the entire apparatus by moistening the floor, thus eliminating the risk of orientation in humidity gradients. Investigations of this nature have been carried out on carabids by Lindroth (1949), van Heerdt (1950), Schmidt (1956a, 1957), Kirchner (1960), Kless (1961), Thiele (1964a), Lauterbach (1964), Thiele and Lehmann (1967), Paarmann (1966), Tietze (1973d) and Becker (1975).

The value of experiments in a temperature gradient and other artificial factor gradients lies in the possibility of elucidating the reasons underlying choice of habitat. In the case of carabids Lindroth (1949) formulated two basic methodological questions regarding the usefulness of the data and offered experimental solutions.

1. Many species are shown to be eurythermic in experiments. The question arises as to whether the population under investigation consists of a series of distinct physiological types, all of which are stenoplastic but in different temperature regions, or whether each individual reacts euryplastically. By marking each specimen and observing it during the experiment Lindroth was able to show that the second possibility held for *Pterostichus nigrita*. In 10 consecutive experiments the individuals chose different places over the broad range of preferred temperature and were thus all euryplastic.

2. Conclusions can only be drawn from experimental results if the reactions of at least the majority of species show a substantial degree of constancy. Lindroth distinguished here between species with stable and those with labile preferences. *Oodes gracilis* is an example of the former type, its PT being constant at about +20° C. *Oodes helopioides*, on the other hand, exhibits a labile preference that "clearly varied according to the initial temperature on the day of the experiment (or possible rose during the course of the summer, independently of the smaller fluctuations in daily temperature)" (Lindroth, 1943). In 1949 Lindroth listed many more examples of stable and labile temperature preferences. He was the first to show that there are species in which the PT of wild individuals and

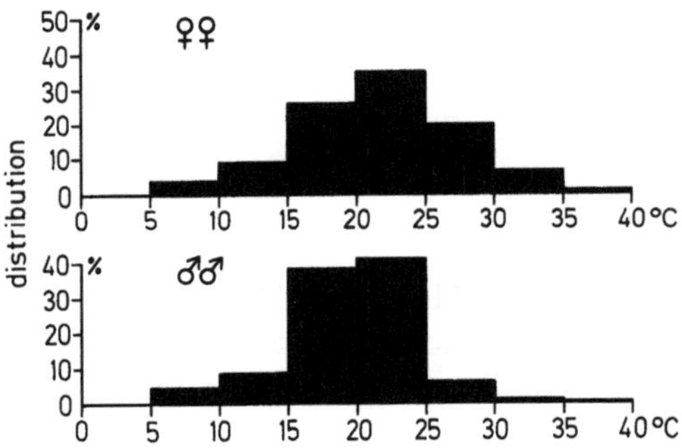

Fig. 67. Preferred temperature of *Abax ater*. 10 recordings from each of 20 males and 20 females. 400 recordings altogether. From Thiele, 1964a

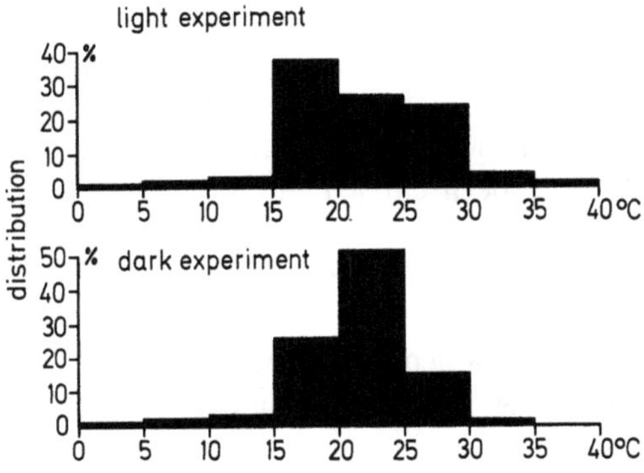

Fig. 68. Preferred temperature of *Abax ater* in light and dark experiments. 10 recordings in light and 10 in dark from each of 20 animals (10 males and 10 females). 400 recordings altogether. From Thiele, 1964a

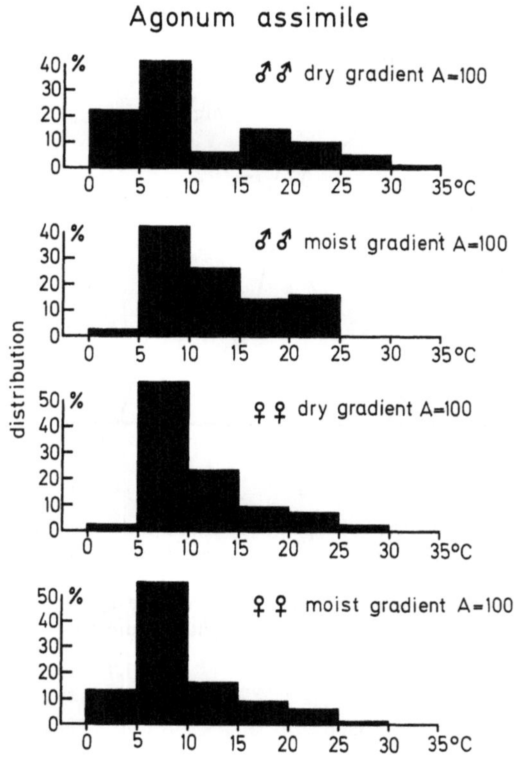

Fig. 69. Preferred temperature of laboratory bred *Agonum assimile* in summer at different air humidities. *A*: number of recordings (10 per animal). From Thiele, 1967

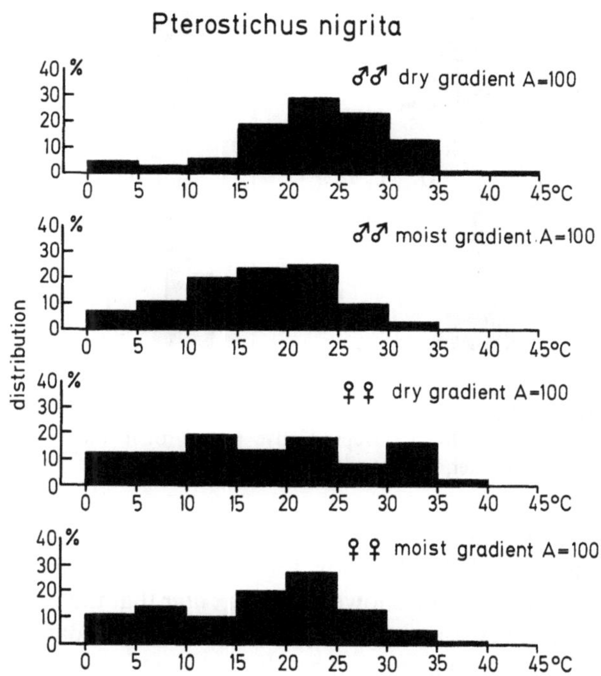

Fig. 70. Preferred temperature of laboratory bred *Pterostichus nigrita* in summer at different air humidities. *A*: number of recordings. From Thiele, 1967

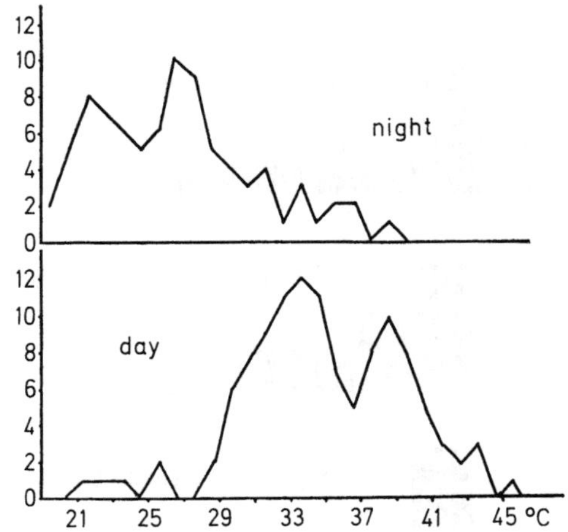

Fig. 71

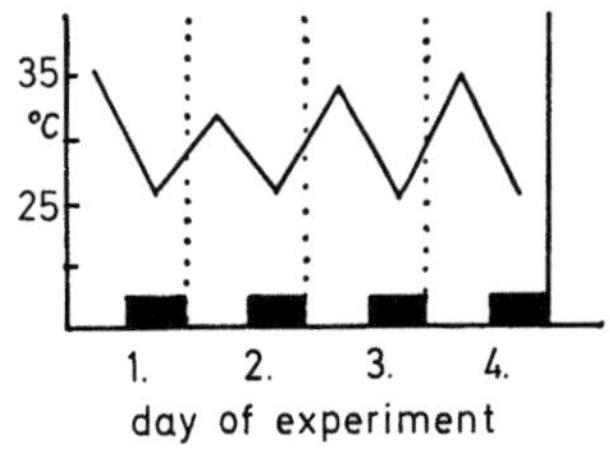

Fig. 72

Fig. 71. Preferred temperatures of two male specimens of *Cicindela campestris* at night and during the day. *Ordinate:* number of observations. From Remmert, 1960

Fig. 72. Preferred temperature of a male specimen of *Cicindela campestris* on four consecutive days. Mean values of day and night measurements. From Remmert, 1960

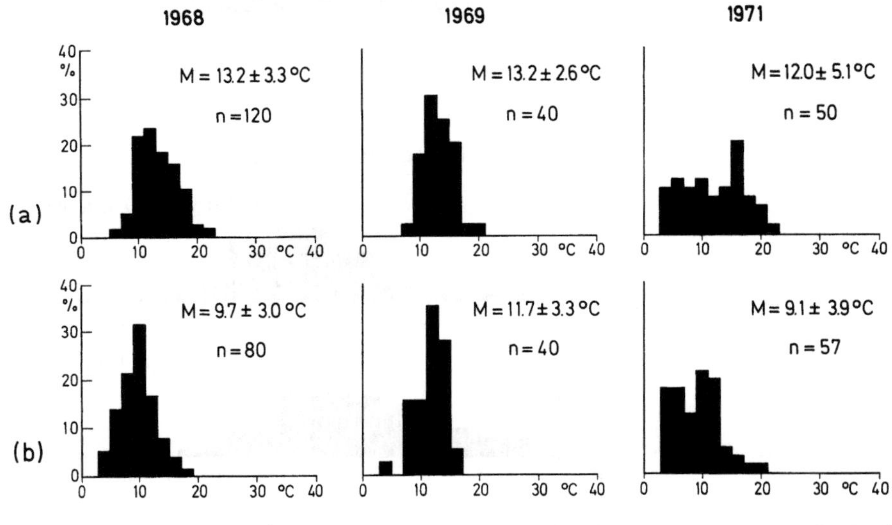

Fig. 73. Preferred temperature of *Agonum assimile* caught wild between January and March (a) and April and May = reproductive period (b) in three years. M = mean ± standard deviation. From Neudecker, 1974

those of the F_1 generation bred from them in the laboratory do not differ: The mean PT values for *Pterostichus anthracinus* were 20.17° C and 20.53° C.

Thiele (1964a) showed for *Abax ater* that the preferred temperature is identical in both sexes and is independent of light or darkness (Figs. 67, 68)[10]. Detailed comparative analyses of reactions to temperature, humidity, and light were per-

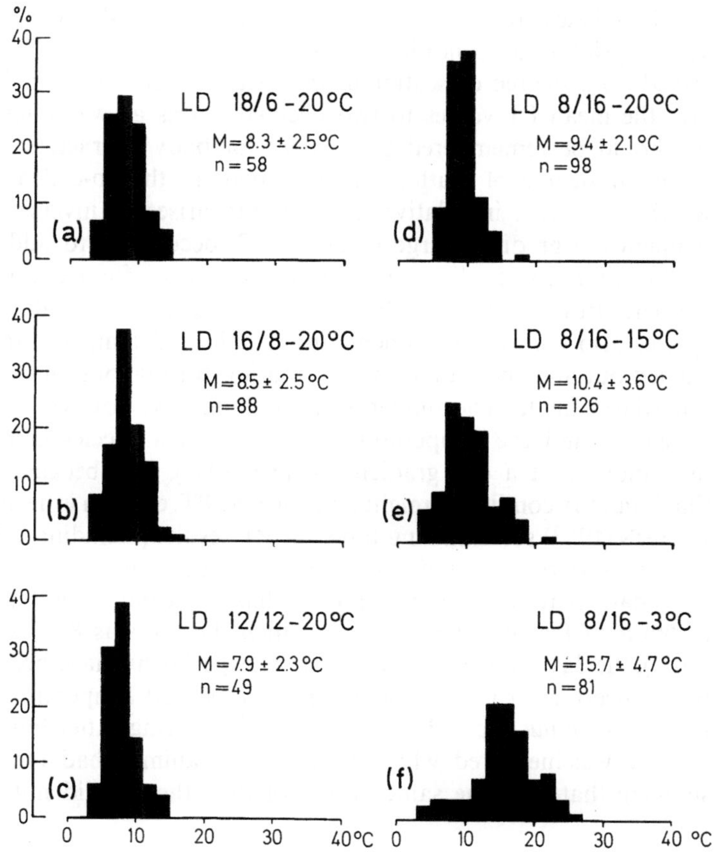

Fig. 74. Preferred temperature of laboratory-bred specimens of *Agonum assimile* following treatment with different photoperiods and temperatures. M = mean ± standard deviation. From Neudecker, 1974

formed on *Agonum assimile* and *Pterostichus nigrita* (Thiele, 1967). *Agonum assimile* is stenopotent with respect to these factors, whereas *P. nigrita* is eurypotent. If the same experiment is repeated under changed conditions the behaviour of the animals in the factor gradients is fairly constant. Males and females of a species react similarly. The differences *between the two species* were proved to be statistically significant in 73% of 22 comparisons between similar experiments. In 46 *intraspecific* comparisons of experiments (e.g. temperature preference at different seasons, different humidity, following different adaptation temperatures, between the two sexes) differences were statistically proved in only 17% of the cases. Judging from the consistency of the reactions it appears that the PT (besides the preferred humidity and preferred light intensity, see below) is an important physiological character of a species, and one which can be determined experimentally with good reproducibility (Figs. 69, 70).

[10] Lindroth (1949) was already unable to detect any consistent differences in the reaction of male and female carabids.

Nonetheless, the consistency of PT measurements should not be overestimated. The physiological condition of the individual as well as external influences introduce a degree of scatter in the measurements that makes it pointless to give the mean PT values to two decimal places as was often done previously. (It should be remembered that values for body characters may be subject to a similar degree of scatter.) In particular, in the so-called "dry gradient" in which a gradient in relative air humidity arises in inverse proportion to the temperature gradient, large errors in PT occur due to additional orientation in the humidity gradient. *Dyschirius thoracicus*, for example, chooses higher temperatures in a moist substrate than in a dry one (Palmén, 1954). For this species also control experiments in an unheated temperature organ have been carried out and show uniform distribution throughout. Such controls have not often been made. The influence of air humidity, however, varies from species to species and the temperature preference of a species can even be identical in a moist and a dry gradient (Thiele, 1967). But because of the possibility that humidity conditions might influence the PT, caution is necessary in evaluating Herter's (1953) summarizing tables of PT results (including values for carabids), since they were measured entirely in a dry gradient.

A daily pattern of differences in PT has so far only been reported by Remmert (1960) for *Cicindela campestris*, in which the PT was 8° C lower at night than during the day (Figs. 71, 72). Up to now, experiments have brought little information concerning seasonal changes in the preferred temperature in carabids. *Harpalus punctatulus* showed a PT of 27.08° C in spring. After four months' captivity 21.06° C was measured, whilst freshly caught animals had a mean PT of 28.83° C, showing that over the same period of time the PT did not change in natural

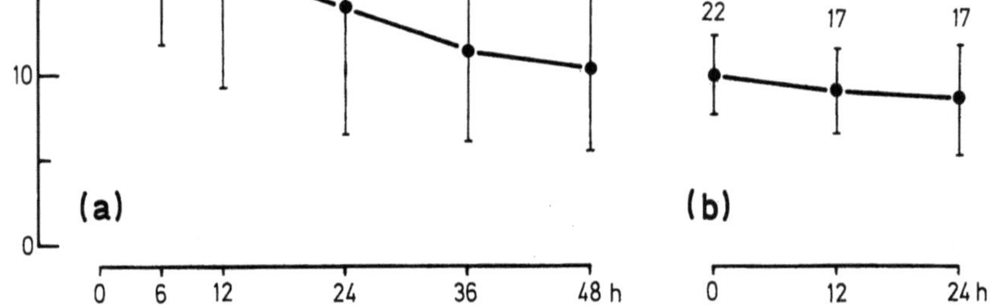

Fig. 75 a and b. Changes in preferred temperature (mean values + standard deviations) in laboratory bred individuals of *Agonum assimile* after transfer from +3° C to +20° C. *Abscissa:* Time of obervation following transfer to short day and 20° C; *numbers* indicate number of animals involved. (a) experimental animals from short day and +3°C; (b) control animals from short day and +20°C. From Neudecker, 1974

surroundings. The lowered PT in the captive animals was attributed to an enhanced need for moisture, which might have been due to the abnormal conditions under which they were kept (Lindroth, 1949).

With great thoroughness Neudecker (1974) endeavoured to show seasonal changes in PT in *Agonum assimile*. Animals caught in the wild state in three years, however, showed very similar PT values, which were at the most slightly lower in the reproductive phase (Fig. 73). Even animals pretreated at different temperatures and photoperiods showed very similar PT values (Fig. 74). Only animals pretreated with +3° C exhibited a mean rise in PT of about 5° C. This and the somewhat higher PT of animals taken from their winter quarters in the wild, may be connected with the fact that individuals kept under experimen-

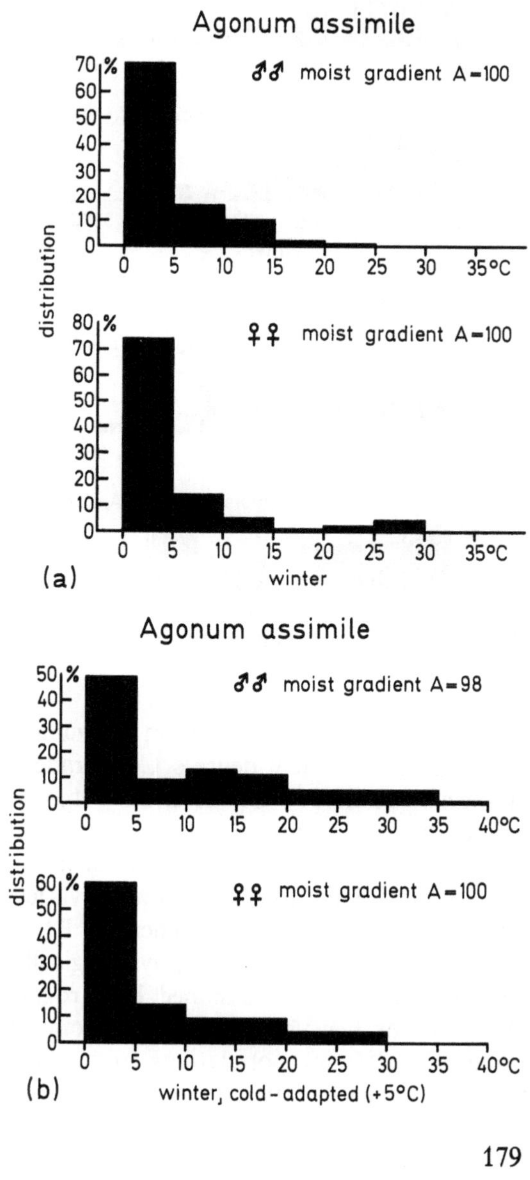

Fig. 76a and b. Preferred temperature of laboratory bred *Agonum assimile* in winter. (a) Animals had been kept at +20° C; (b) following cold adaption (1 week at +5° C). *A:* number of recordings (10 per animal). From Thiele, 1967

tal conditions at +3° C for 8 months showed a large increase in PT as compared with control animals that had been kept at 20° C. However, this effect lasted for only 36–48 h (Fig. 75).

A short-term cold adaptation (one week at +5° C) of laboratory animals kept under natural light revealed no change in PT in either *Agonum assimile* or *Pterostichus nigrita*, as compared with animals that had been kept under similar light conditions in the same winter season at 20° C (Thiele, 1967, Figs. 76, 77).

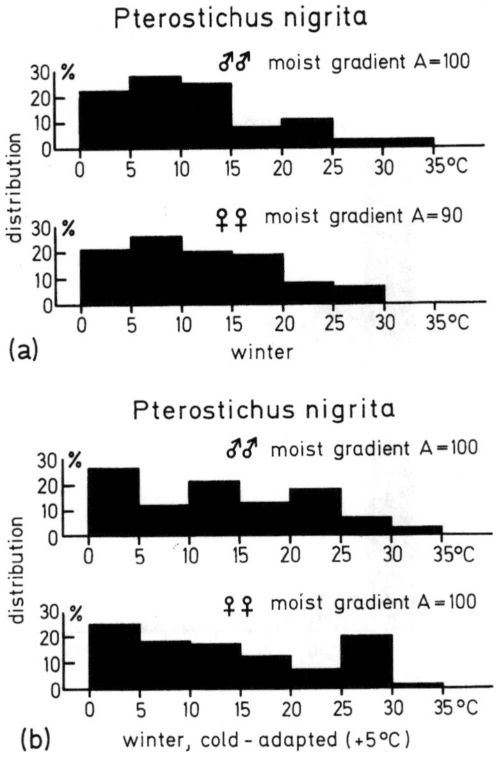

Fig. 77 a and b. Preferred temperature of laboratory bred *Pterostichus nigrita* in winter. Conditions as in Figure 76. From Thiele, 1967

The PT values determined in my undergraduate ecological courses varied at the most by a few degrees Centigrade from year to year in the majority of species (Table 39). Of these, only *Agonum dorsale* proved to have a labile temperature preference.

In view of the different measurements obtained in experiments, where the physiological condition of the animal varied, it would be risky, merely on the basis of relatively small differences in PT between populations from different sources, to conclude that ecophysiologically different subspecies are involved. Nevertheless, this was assumed by Krogerus (1960), who found a mean PT of 15.7° C for *Bembidion bruxellense* from about 66° N in Finland, and a mean value of 18.2° C for populations from about 60° N. Such differences can also be found in repeat experiments with one and the same group of animals.

Table 39. Mean values for preferred temperatures (in °C) and humidity (% R. H.) of carabids, determined by students in ecology courses

Preferred temperature	1975	1974	1973 1. course	1973 2. course	1972	1971	1970	Largest difference
Agonum assimile	8.3°	8.5°	8.1°	7.8°	8.7°	8.1°	12.5°	4.7°
Agonum dorsale	19.8°	24.6°	17.4°	16.6°	—	—	—	8.0°
Pterostichus angustatus	—	21.9°	19.5°	—	—	22.0°	20.0°	2.5°
Abax parallelus	22.8°	19.9°	—	—	20.6°	—	—	2.9°
Nebria brevicollis	—	19.4°	—	—	—	22.2°	—	2.8°
Molops piceus	16.6°	—	—	—	17.8°	—	—	1.2°
Preferred humidity								
Agonum assimile	50 %	49 %	45 %	45 %	54 %	50 %	47 %	9 %
Agonum dorsale	58 %	66 %	59 %	65 %	—	—	—	8 %
Pterostichus angustatus	—	57 %	58 %	—	—	60 %	56 %	4 %
Abax parallelus	81 %	79 %	—	—	85 %	—	—	6 %
Nebria brevicollis	—	69 %	—	—	—	69 %	—	0 %
Molops piceus	—	79 %	—	—	74 %	—	—	5 %

In determinations of PT of Westphalian *Carabus problematicus*, Schmidt (1956b) found two groups of individuals with PT maxima at 15° and 25° C, respectively. He, too, assumed that they were two physiologically different forms. On the other hand, the PT values measured by Thiele (1964a) on animals of the same species from the Rhineland revealed a uniform eurythermicity in all individuals. Regarding the above-mentioned reservations concerning the degree of consistency to be expected from PT measurements it can be said that no conclusive evidence of the existence of "temperature races" in carabids on the basis of PT measurements has so far been obtained. This would also require evidence of the heritability of such PT differences in the various populations. The most extensive study on this point is still that of Krumbiegel (1932), who found a successive rise in PT from populations of *Carabus nemoralis* from East Prussia to Spain, but using methods that are nowadays considered to be out-of-date, and with very different numbers of individuals in the various populations.

We should bear in mind the words of Krogerus (1960) concerning conclusions from temperature preference experiments as to the significance of temperature in ecological behaviour: "...one has constantly to be aware of the fact that comparative and not absolute values are the aim of the experiments. These must be performed simultaneously on two or more species that have been collected at the same time. If the experiments then reveal different reactions the conclusion is justified that the underlying cause of the different behaviour of these species in nature depends upon the factor tested". Lampe (1975), however, observed that the temperature prevailing at the time of day (night) and season (spring and autumn) at which the adults are active is, for *Abax ovalis*, even in absolute terms almost identical with the temperature optimum for this species as determined in the temperature gradient.

2. *Preferred Temperature and Habitat Affinity*

Results obtained using the temperature gradient apparatus have certain advantages over field observations. Although it is possible in the field to study the numbers and distribution of a species in different habitats and at the same time measure the temperature on the spot, numerous other factors vary with temperature. In the experimental temperature gradient, on the other hand, climatic factors such as humidity and light intensity can be kept nearly constant, so that only the effect of temperature on the animals is observed. Carabids that spread over a wide range of temperature are termed eurythermic, whereas those that concentrate in a very narrow temperature range are called stenothermic. Warm and cold stenothermic species can be distinguished. An experiment may reveal that the prevailing temperature is immaterial to a species even though, in nature, it is only found in habitats of a particular temperature. In such a case it is justifiable to conclude that temperature influences the distribution of the species little if at all, this depending rather upon other factors. If a pronounced temperature preference is detectable and if it coincides with the temperature sought in nature, it is permissible to assume that temperature plays a part in determining the distribution spectrum of the species.

Comparative investigations on the PT of a large number of carabid species have been carried out by Lindroth (1949: 15 "limestone" species), Krogerus (1960: 12 species), Kless (1961: 15 species), Lauterbach (1964: 21 species), Thiele (1964a: 23 species) and Becker (1975: 10 species).

The PT range is fairly narrow (Thiele, 1964a). For most species more than 70% of the experimental choices fall within a range of 15°, and only occasionally 20°. For the large majority of woodland species and field animals investigated this range only rarely commences at 5° C (usually at 10°) and extends to 30° C. The percentage of choices above 30° to 40° is very small, and only a few of the species so far investigated exceed this range. In two cases of alpine-subalpine species and species from the far North Krogerus (1960) found mean PT values below 10° C: *Nebria nivalis* 5.0°, *N. gyllenhali* 8.0°. Other species with this type of distribution exhibited mean PT values below 15°, which are also observable in central European montane or sub-montane species. A peak between 40° and 50° C was exceptional and was only seen in the dry grassland species *Callistus lunatus* (Becker, 1975). Thiele (1964a) found only one cold-preferent species among ten field animals, the rest being either eurythermic or preferring warmth. Of 13 forest species, too, only four preferred cold (see Chap. 7). The PT obviously plays quite a considerable role in the geographical and ecological distribution of carabids, but it is not the most important response to microclimate.

A most thorough analysis of the significance of PT and other reactions to the microclimate for the distribution of two closely related species (*Pterostichus oblongopunctatus* and *P. angustatus*) was undertaken by Paarmann (1966), including experimental observations on the reactions of larval stages. *P. oblongopunctatus* is a eurytopic forest species of palaearctic distribution, whilst *P. angustatus* occurs in the continental and sub-continental regions of Europe, inhabiting mainly warm, dry forest clearings, bare-felled areas and patches left desolate by forest fires. The temperature preference of the adults of the two species is almost identical in a temperature gradient. In a combined light-temperature gradient, however, ranging from light-cold to dark-warm, *P. angustatus* chooses a much warmer temperature region than *P. oblongopunctatus* on account of its dark-preference, although the latter, too, is dark-preferent (Fig. 78). At least

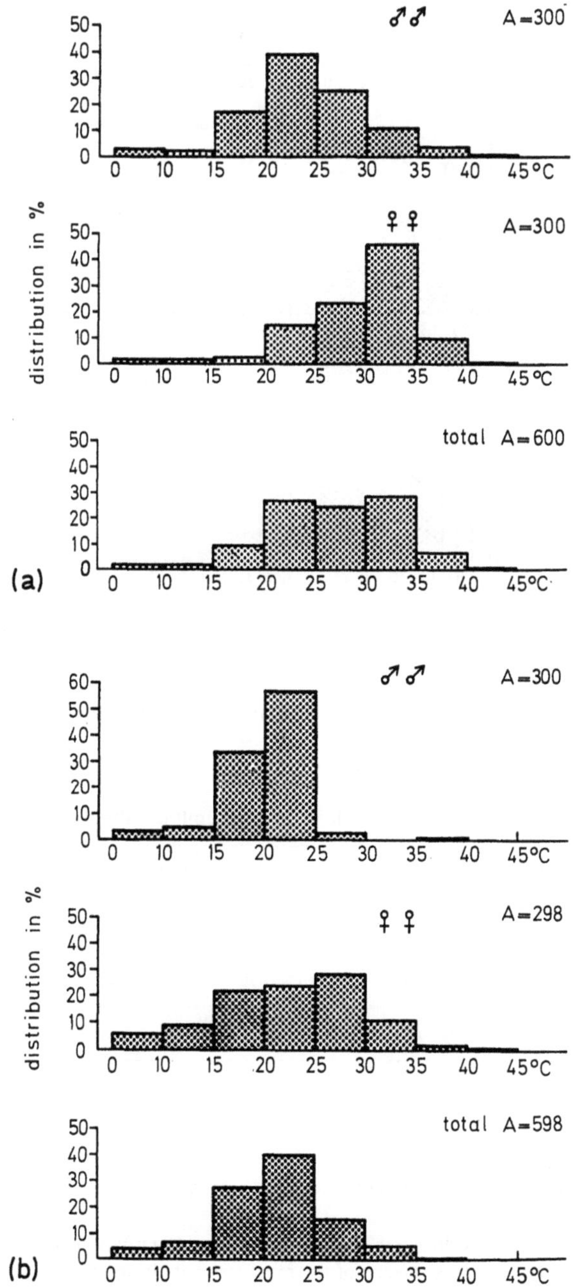

Fig. 78a and b. Distribution of two *Pterostichus* species in a combined temperature-light intensity gradient. The gradient extended from 5°C at 1750 Lux to 45°C at 2 Lux. (a) Distribution of *P. angustatus;* (b) distribution of *P. oblongopunctatus.* A: number of registrations (usually 10 per animal). From Paarmann, 1966

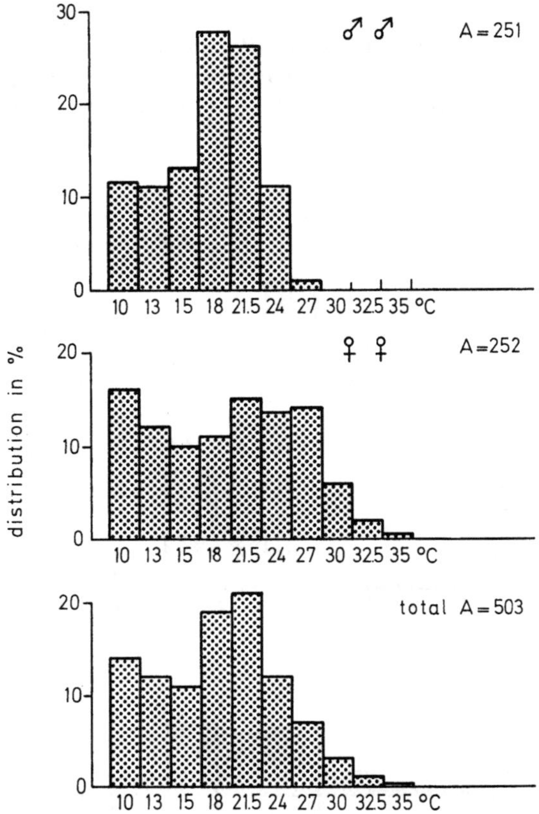

Fig. 79. Distribution of 4-week-old adults of *Pterostichus angustatus* over the temperature gradient in the substrate-temperature gradient apparatus. *A*: number of animals investigated. From Paarmann, 1966

Table 40. Preferred and repellent temperatures of the larvae of two *Pterostichus* species. (According to Paarmann, 1966)

	Preferred temperatures of the larvae			
	P. oblongopunctatus		*P. angustatus*	
	Mean values	(Limiting values)	Mean values	(Limiting values)
Larvae I	18.74° C	(16.95–20.50° C)	22.52° C	(20.95–23.55° C)
Larvae II	19.07° C	(17.90–21.00° C)	22.69° C	(20.80–24.50° C)
Larvae III	18.61° C	(16.70–20.35° C)	23.23° C	(21.65–24.75° C)
	Repellent temperatures of the larvae			
	P. oblongopunctatus		*P. angustatus*	
	Mean values	(Limiting values)	Mean values	(Limiting values)
Larvae I	30.72° C	(29.55–32.15° C)	33.50° C	(32.70–34.45° C)
Larvae II	32.26° C	(31.55–33.85° C)	33.55° C	(33.30–33.85° C)
Larvae III	32.88° C	(32.20–34.15° C)	34.91° C	(33.80–35.85° C)

The mean values are calculated from ten animals in each case (ten measurements per animal). Limiting values: means of the measurements from animals showing extreme behaviour

Fig. 80. Distribution of sexually mature adults of *Pterostichus oblongopunctatus* over the temperature gradient in the substrate-temperature gradient apparatus. *A*: number of animals investigated. From Paarmann, 1966

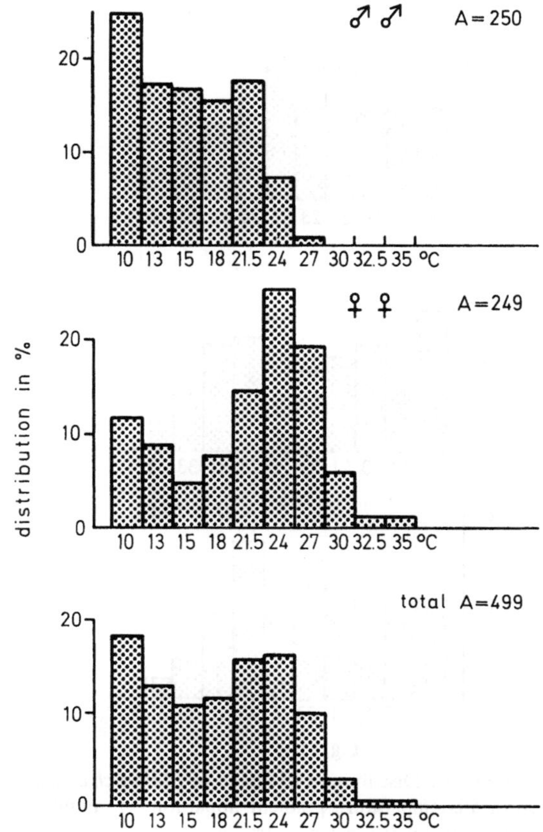

P. angustatus tolerates higher temperatures (Figs. 79, 80). The PT of larvae of *P. angustatus* is definitely higher than that of the larvae of *P. oblongopunctatus*. This is also true for the repellent temperatures (Table 40). In a substrate temperature gradient apparatus the females of both species seek higher temperatures than the males (in contrast to their behaviour in a temperature gradient with a smooth track). In so doing, they choose a temperature that corresponds to the PT of the larvae (Figs. 81, 82). In *P. oblongopunctatus* this higher PT in the females only becomes obvious at sexual maturity. Paarmann regards this physiological idiosyncrasy of the females as being of adaptive advantage in guaranteeing that they lay their eggs in a site thermally suited to the needs of the hatching larvae. The moisture requirements of this closely related and very similar species pair are at least as decisive in determining habitat affinity as their proven differences in temperature requirements.

3. *The Physiological Basis of Thermotaxis*

Thermoreceptors have neither been directly demonstrated nor localized by means of exclusion experiments in carabids. An interesting approach to an explanation of the temperature preferences of carabids was made by Schmidt in a series of investigations.

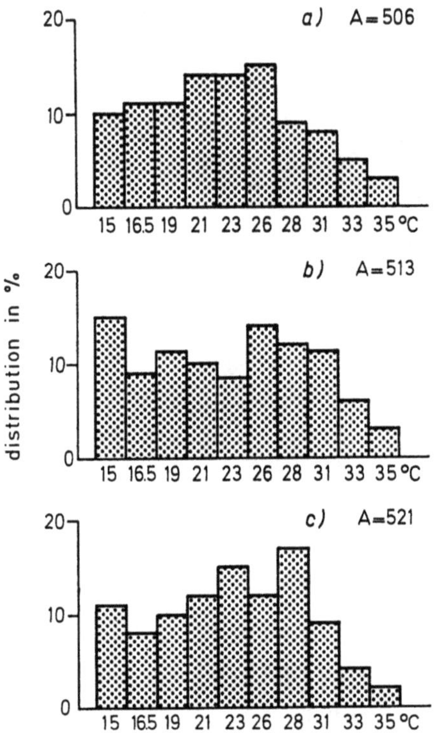

Fig. 81

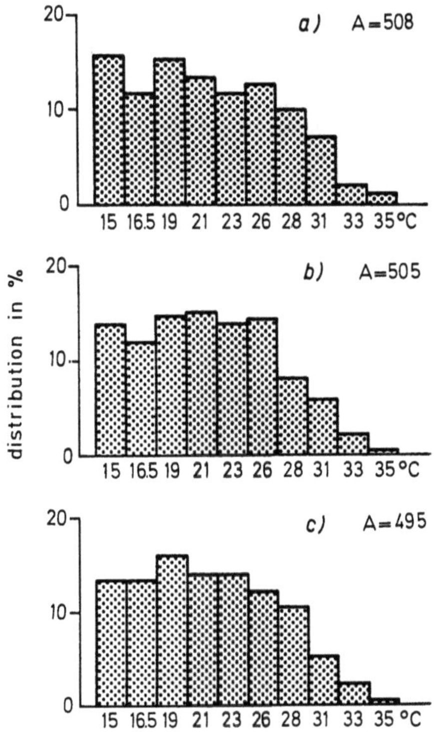

Fig. 82

Fig. 81 a–c. Distribution of larvae of *Pterostichus angustatus* over the temperature gradient in the substrate-temperature gradient apparatus. *A*: number of animals investigated; (a) 1st instar; (b) 2nd instar; (c) 3rd instar. From Paarmann, 1966

Fig. 82 a–c. Distribution of the larvae of *Pterostichus oblongopunctatus* over the temperature gradient in the substrate-temperature gradient apparatus. *A*: number of animals investigated; (a) 1st instar; (b) 2nd instar; (c) 3rd instar. From Paarmann, 1966

Schmidt (1956a) found that the transpiration rate in species of the genus *Carabus* did not rise steadily with increasing temperature, but remained constant within a certain narrow temperature range, which differed from species to species. Schmidt concludes from this that *Carabus spp.* are able to regulate their transpiration rate actively over this range, which coincides more or less with the PT range determined in a dry temperature gradient. Schmidt therefore assumes that it is not so much the perception of temperature as its own transpiration rate and the search for an optimal transpiration region that constitute the physiological basis of temperature preference. The need for finding this optimal transpiration region is all the more vital to members of the genus *Carabus* since their transpiration rate increases with decreasing water content.

In the course of measuring the dependence of oxygen consumption on temperature Schmidt (1955/56) discovered that no mere physical dependence exists, and that over certain ranges the oxygen consumption, in fact, remains constant and is independent of temperature. This suggests the existence of a regulatory mechanism for metabolism and a temperature optimum connected with it. Kühnelt (1955) also demonstrated similar regulatory regions in investigations including some of the same species of Carabus.

However, perception of its own transpiration rate cannot be the only factor underlying PT. *Agonum assimile* chooses very low temperatures combined with dryness. It seeks out the driest region in a moisture gradient at constant temperature (at 40–50% R. H.), that is to say, a region where its transpiration rate must be high. In both wet and dry temperature gradients *Agonum assimile* collects in the cooler region. This means that in a dry temperature gradient the warm and dry region with its attendant high rates of transpiration is avoided, in favour of the cool, wet region where transpiration is lower. Results of this nature point to a direct perception of temperature stimuli and speak against maintenance of a constant transpiration rate (Thiele, 1964a).

4. *Cold Resistance*

Miller (1969) found large seasonal fluctuations of freezing tolerance in *Pterostichus brevicornis* in Alaska. Beetles collected in summer died at −6.6° C, whereas those collected in winter survived temperatures below −35° C and 67% of a group of 17 beetles even tolerated a 5-h exposure to −87° C.

A cold resistance of this magnitude is far beyond that measured for carabids from the northern temperate zone. In nature, the species is exposed to temperatures of −60° C for brief periods and has to tolerate frost temperatures of −40° C lasting for several weeks (Baust and Miller, 1970). Seasonal changes in cold resistance are only in part associated with alterations in glycerol content. The glycerol content of the haemolymph is zero in summer and rises to as much as 23% in winter, in parallel with an alteration in the freezing point of the haemolymph, which in winter may sink as low as −5° C. "The plot of mean freezing points was a near mirror image of the glycerol curve." However, deviations from this rule were seen in mid-winter: glycerol could decrease without a corresponding alteration in cold resistance. Further, the supercooling temperature measured for the entire body of the living animal did not strictly follow the glycerol content, so that other protective mechanisms must be assumed to be effective. The lowest supercooling temperature at which spontaneous freezing was observed was −11.5° C in winter (Fig. 83)[11].

Accordingly, *P. brevicornis* in its winter quarters at −20° C was not supercooled but frozen (Miller, 1969). The findings indicate intracellular freezing. In Japan, *Pterostichus orientalis* was also found in a frozen condition in its winter quarters. In contrast to the extreme temperature tolerance of *P. brevicornis*, the Japanese species only tolerated −10° C for longer periods of time (40 days) without suffering damage (Ohyama and Yasahina, 1972). *Agonum assimile* from central Europe showed a somewhat lower cold tolerance (Neudecker, oral communication). In Norway, *Pelophila borealis* only survived the winter under a protective

[11] Recordings made from motoneurones in the trochanter of the hind legs showed an absolute neural extinction at −11.7° C if the animals were cold-adapted. Coordinated movements could still be observed up to about −12° C (Baust, 1972).

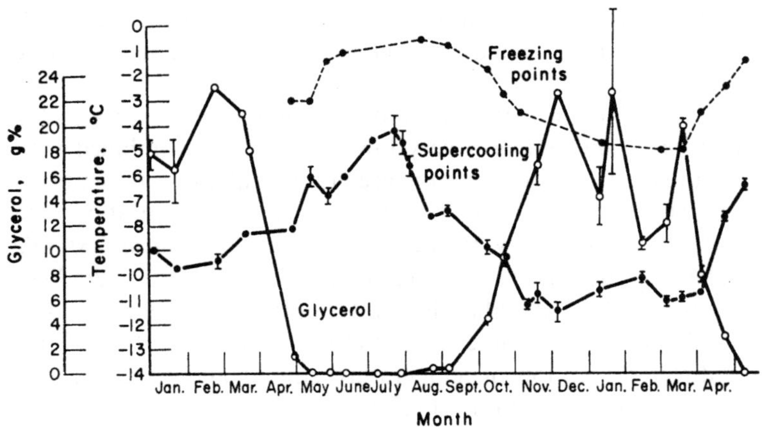

Fig. 83. Seasonal variations in haemolymph glycerol content, super-cooling points and haemolymph freezing points of *Pterostichus brevicornis*. Super-cooling points are whole body determinations. Values are mean ± S. E. From Baust and Miller, 1970

covering of snow: temperatures below about −10° C were lethal to this species. Its supercooling point sank in the course of the year from about −5° C in June/July to below −20° C in winter (although the animals died after the latter experiments). *P. borealis* survived undamaged 127 days at a temperature of 0° C in an anoxybiotic condition (nitrogen atmosphere of 99.9%). Subsequently the animals showed a large rise in oxygen consumption for about one week and an increase in lactate accumulation in the haemolymph of about 10 times the initial value. Since both phenomena were also recorded in animals collected from their winter quarters, anoxybiosis is apparently also possible under such conditions. Anoxybiosis permits the animals to overwinter in the soil beneath a covering of ice (Conradi-Larsen and Sømme, 1973a, b). However, carabids usually only awake to an active life when the temperature rises above freezing point. Sixteen species of *Harpalus* were exposed to gradually increasing temperatures and the point at which supine animals righted themselves was noted. The air temperature in the vicinity of the animals was recorded. The mean values obtained for individual species varied between the fairly narrow limits of +3.4° C and +8.2° C (Lindroth, 1949). The animals began to move, nevertheless, at much lower temperatures, e.g. the eurytopic species *H. aeneus* at −6° C (Lindroth, 1949). Krogerus (1932) found values of a similar order of magnitude for the commencement of activity in *Cicindela*, *Dyschirius* and *Amara* species.

5. *Heat Resistance*

Just as the lower temperature point at which carabids awaken to activity shows very little variation even between species of widely differing ecological types, so is the upper limit of temperature tolerance extremely uniform (more so even than to cold). In the 16 species of *Harpalus* investigated by Lindroth (1949), already mentioned in the preceding section, the temperatures at which partial paralysis set in (usually of the hind legs) were recorded as well as those at which total heat paralysis occurred. The mean temperature for the latter lay between 47.4° C and 51.7° C for the 16 species investigated, i.e. a range of only 4.3° C. Again, the values obtained by Krogerus (1932) for *Cicindela, Dyschirius* and *Amara* species were in much the same region. In view of the great uniformity shown by such reactions it is not feasible to make an ecological differentiation between various species on the basis of their heat tolerance.

6. *The Influence of Temperature on the Developmental Stages*

For overwintering larvae cold is often obligatory in order to overcome their larval dormancy (see p. 257).

In spring breeders with summer larvae development is accelerated in a medium temperature range by rising temperatures. The curve for the duration of development of central European *Pterostichus nigrita* is of the type described by Janisch as a "Kettenlinie" (chain line). From 10° to 25° C this curve approximates well with Blunck's hyperbola (cf. Tischler, 1949). The corresponding values for speed of development thus lie on a straight line and it is possible to calculate that the zero point for development is 6.8° C, which is in excellent agreement with the observed data. Mortality is at a minimum between 15° and 25° C, and pupal weight is at a maximum at 20° C (Ferenz, 1973; Figs. 84, 85).

It could be shown for *P. angustatus* and *P. oblongopunctatus* that the temperature requirements of the individual stages differ considerably and represent adaptations to the temperature conditions in the corresponding environments (Paarmann, 1966; Fig. 86; for distribution of the species see p. 182). In both species

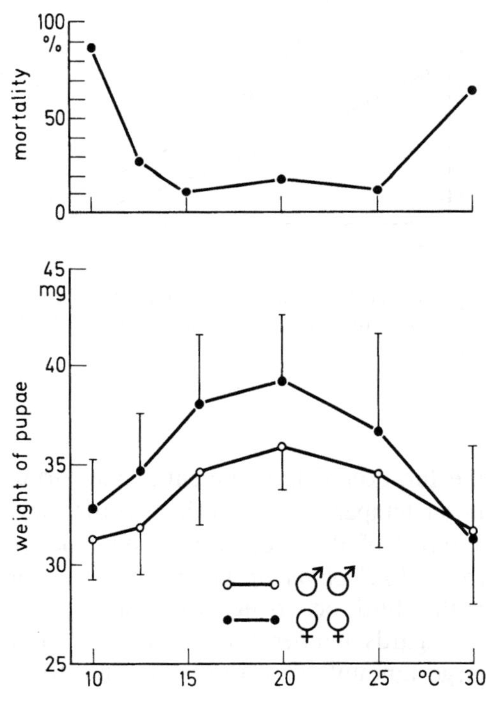

Fig. 84. Mortality of all stages of development and pupal weights of *Pterostichus nigrita* following exposure to various temperatures. From Ferenz, 1973

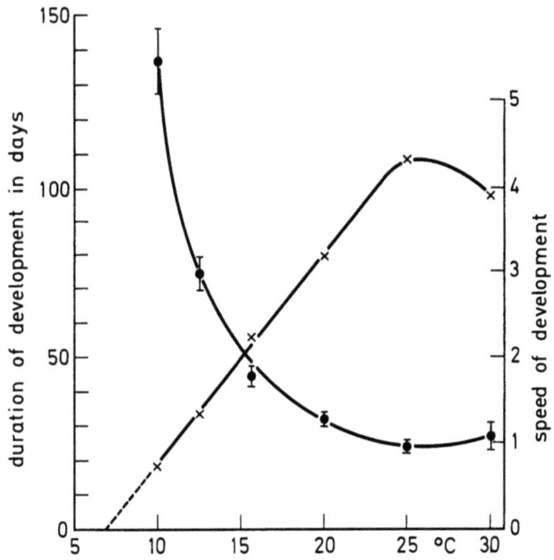

Fig. 85. Duration of development (–•–) and speed of development (–×–) of *Pterostichus nigrita* at various temperatures. From Ferenz, 1973

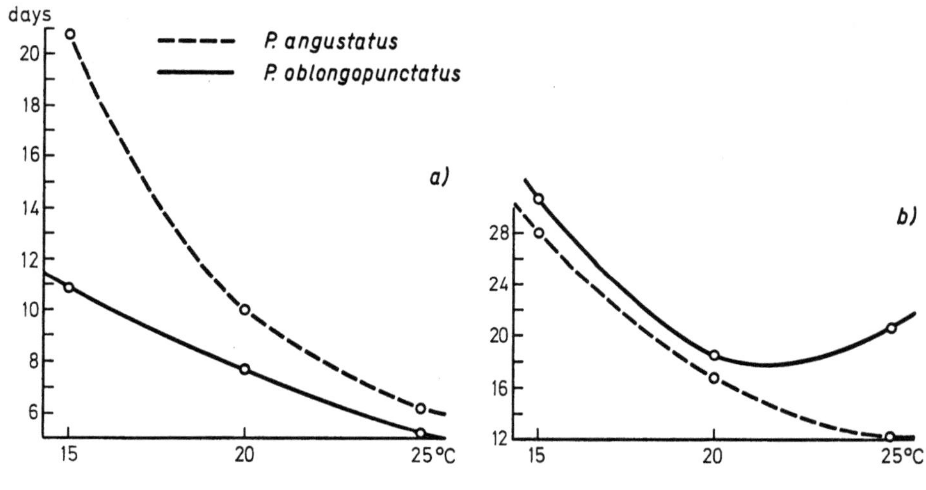

Fig. 86a and b. Dependence of duration of development of the 1st larval instar (a) and the 3rd larval instar (b) of *Pterostichus angustatus* and *P. oblongopunctatus* on temperature. From Paarmann, 1966

the duration of development of one particular larval stage is especially dependent upon temperature. In other words, development at a speed essential for the survival of the species can only take place over a very narrow temperature range. In *P. oblongopunctatus* the stage most highly dependent upon temperature is the third, occurring in late spring which is a particularly favourable season as regards temperature. In *P. angustatus*, however, the first stage requires a large amount of warmth, but due to its appearing very early in spring the species is limited to exceptionally warm habitats such as clear-felled areas and localities left bare by forest fire.

Whereas in the laboratory insects are usually kept under constant temperature conditions, in nature they are exposed to daily fluctuations in temperature. Kaufmann (1932) made the theoretical deduction that alternating temperatures in the lower temperature regions raise the speed of development but decrease it at higher temperatures.

"If the zero point for the development of a species is +10° C it will not develop at all at constant temperatures of this magnitude. If, on the other hand, the temperature fluctuates over the course of the day between +5° and +15° C, about a mean value of 10° C, the species is able to develop in the part of the day in which temperatures are above +10° C" (Thiele, 1973a). In the region of the upper limits for development an inverse effect can be observed. Alternating temperatures have the required influence, even if a small one, on *P. nigrita*: development is accelerated at 15° ± 5° C and 15° ± 10° C by four days, whereas at 25° ± 5° C it is retarded by two days and by five days at 20° ± 10° C as compared with development at constant temperatures. At 20° ± 5° C the animals develop at the same speed as under constant temperatures (Ferenz, 1973). In *Pterostichus angustatus* as well no difference in duration of development was observed under constant 20° C and 20° ± 5° C (Paarmann, 1966).

II. Humidity and Orientation in the Environment

1. *Experimental Method*

As with the investigations of older authors, more recent comparative studies on different species have brought only relative conclusions as to moisture requirements. The animals are offered a choice of substrates soaked with successively larger quantities of water. Investigations of this kind have been made on carabids by Lindroth (1949), Kless (1961) and Krogerus (1960). The results of various authors can be better compared if the animals are permitted to move about freely in an apparatus in which a stepwise humidity gradient has been established over sulphuric acid or a series of hygroscopic salts. A ring-shaped apparatus is preferable to a straight one since it avoids the thigmotactical effect which might otherwise introduce errors into the results (this holds for preference studies involving any factor).

The results of humidity preference experiments can, as a rule, be well reproduced. *Abax ater* and *A. ovalis* caught in the wild have already been cited as examples of the fact that the preferred humidity is independent of season and temperature (if this is not too far above the optimum, Fig. 23). Laboratory-bred individuals of *Agonum assimile* invariably chose dryness within the temperature range of 9°–29° C, thus differing from *Pterostichus nigrita*, which was euryhygric over this range. No differences could be detected between the reactions of males and females of the two species. Seasonal variations were slight and in no case obliterated the differences between the two species. Cold adaptation (one week,

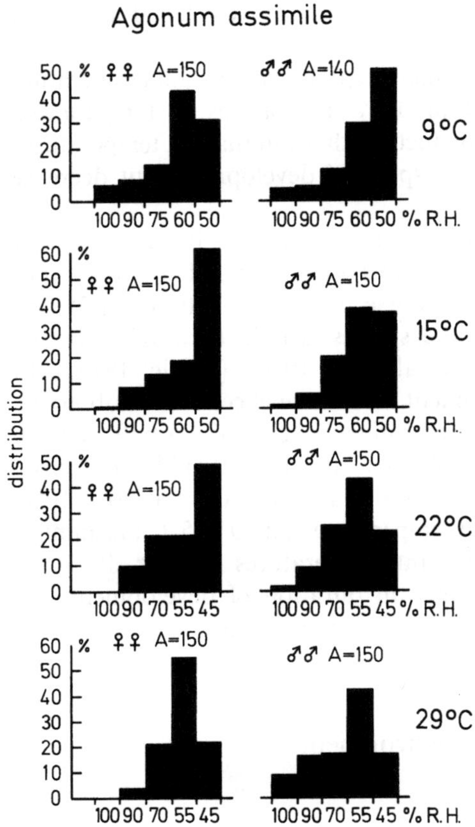

Fig. 87. Preferred humidity of laboratory bred *Agonum assimile* in summer at different temperatures. *A*: number of registrations (10 per animal). From Thiele, 1967

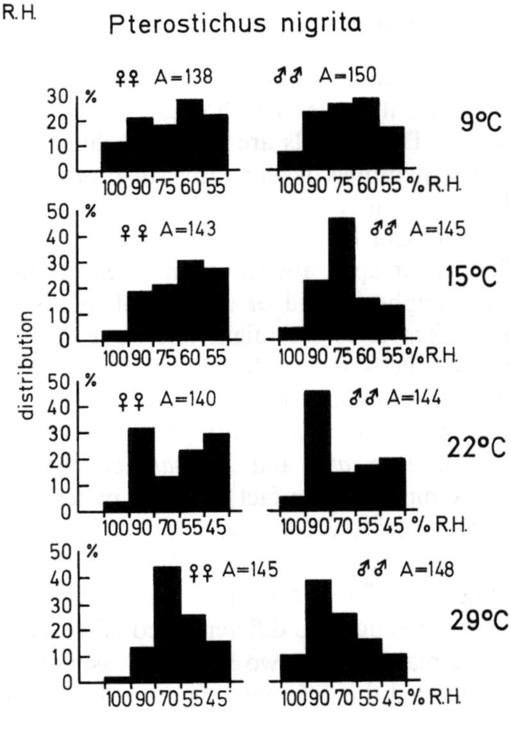

Fig. 88. Preferred humidity of laboratory bred *Pterostichus nigrita* in summer at different temperatures. *A*: number of registrations (usually 10 per animal). From Thiele, 1967

+ 5° C) had no effect in humidity preference (Thiele, 1967; Figs. 87–89). Unfortunately we know little about the influence of pretreatment with various degrees of soil or air humidity on the humidity preferences of carabids.

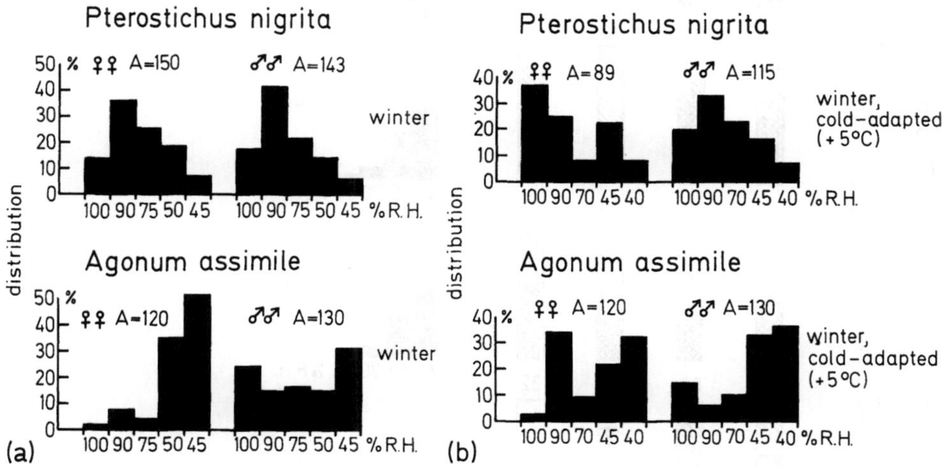

Fig. 89a and b. Preferred humidity of laboratory bred *Pterostichus nigrita* and *Agonum assimile* in winter, observed at +22° C. (a) Animals previously kept at +20° C; (b) following cold adaptation (1 week at +5° C). *A*: number of registrations (usually 10 per animal). From Thiele, 1967

Neudecker (1974) investigated 551 individuals of *A. assimile*, pretreated with different temperatures and photoperiod, factors known to influence the state of maturity. He came to the conclusion "that the humidity preference of *A. assimile* is largely independent of age, sex, photoperiod, temperature and state of maturity of the animals. The preferred humidity of *A. assimile* thus has an almost constant character" (Fig. 90).

Results obtained by students in my ecology courses (Table 39) on a series of species also confirm that the preferred humidity is a fairly constant character. Here, too, only *Agonum dorsale* was shown to have a labile humidity preference. In his experiments using a substrate humidity gradient, Lindroth (1949) found, as with temperature preference, that some species had a stable and some a labile humidity preference. The preference of *Harpalus serripes* for dryness was fairly stable, whereas, in contrast, individuals of the species *H. punctatulus* reacted variously according to the season at which they were caught or the nature of their pretreatment. Although this species, too, usually prefers dryness, it can swing completely to moisture preference, e.g. following two days in water-saturated air. Exactly the same behaviour is exhibited by *Agonum dorsale*, a species that usually prefers dryness in the gradient apparatus, but after being exposed for 18 h to about 50% R. H. chooses moisture-saturated air (Kreckwitz, personal communication).

Although the results of such experiments are usually well reproducible and the factor involved is of considerable importance for habitat affinity, air humidity

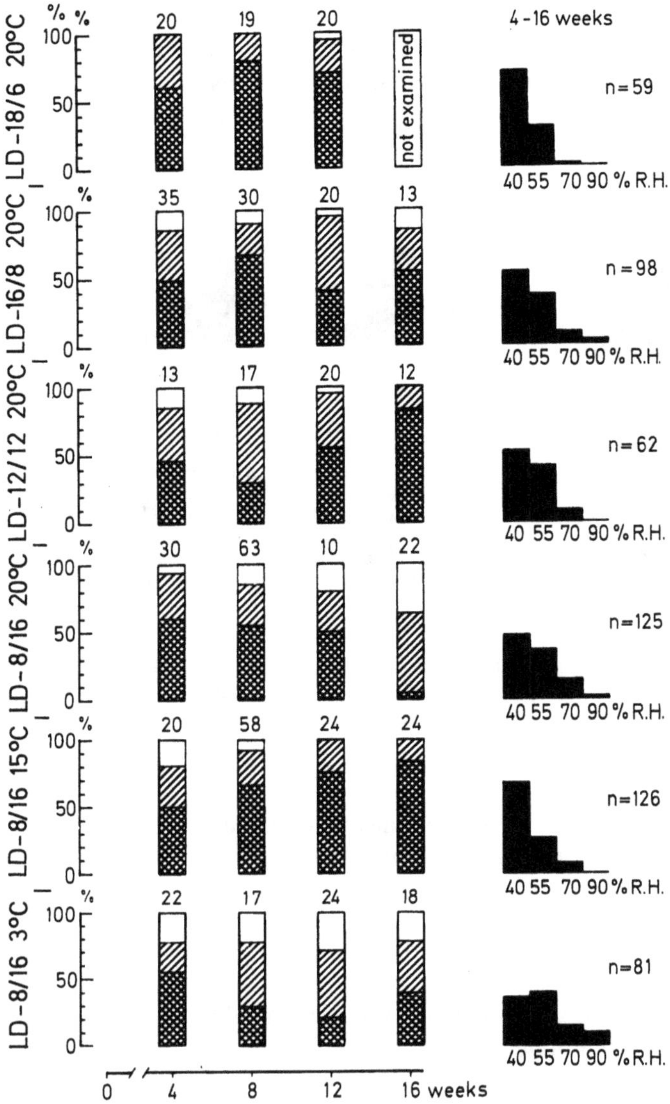

Fig. 90. Preferred humidity of laboratory bred *Agonum assimile* following treatment with different photoperiods and temperatures. *Abscissa:* duration of treatment and age of adults. *Figures above the columns:* number of animals investigated. Percent beetles preferring 40% R.H. *(cross-hatched)*, 55% R.H. *(hatched)*, 70 and 90% R.H. *(white)*. Summarizing diagrams on the *right*. From Neudecker, 1974

preference experiments with carabids have been performed by relatively few authors (von Heerdt, 1950; van Heerdt *et al.*, 1956, 1957; Perttunen, 1951; Thiele, 1964a, 1967; Lauterbach, 1964; Weber, 1965b; Paarmann, 1966; Tietze, 1973d; Becker, 1975).

The most reliable conclusions as to habitat affinity can be drawn from experiments that last no longer than a few hours. Weber (1965b) found that all central European *Carabus* species investigated were strictly hygrotactic after one or two days in a humidity gradient. Only the Mediterranean species *Carabus morbillosus* required up to five days to achieve this state. "These results are not in agreement with those of Leclerq (1947), who found that *Carabus auratus* chose a region with 41% relativé air humidity for several days... The great need for moisture exhibited by *Carabus* spp. was explained by Schmidt's (1955) experiments which revealed that they are not adapted to low air humidities, their transpiration rate rising with increasing loss of water" (Weber, 1965).

Carabids possess a very well developed ability to distinguish between small differences in air humidity, e.g. *Carabus granulatus* and *C. morbillosus* can readily distinguish between 90% and 95% R. H.

2. *Preferred Humidity and Habitat Affinity*

A larger number of species has been subjected to comparative studies by Lindroth (1949: 15 "limestone species"), Perttunen (1951: eight species), Kless (1961: 15 species), Lauterbach (1964: 21 species), Lehmann (1962: eight species), Thiele (1964a: 23 species), Weber (1965b: 10 species) and Becker (1975: 10 species).

All authors found that the large majority of species from wet habitats preferred moisture and the majority of those from dry environments chose dryness or were euryhygric. This was established by Perttunen in comparative investigations on species from dry and wet habitats in open country, by Thiele and Weber in comparisons between forest and field species, by Lauterbach in studies on species preferring forests on the one hand, and forest clearings on the other, by Lehmann in species from river banks and meadow forests and by Becker from an analysis of the behaviour of species from forests and grasslands. In some cases inhabitants of moist environments exhibit a well-developed preference for dryness. *Agonum assimile* (Thiele, 1967) overcompensates for this both in nature and in experiment with a cold preference. Their preference for cold induces the animals to seek out cold-wet habitats where they are prepared to accept the prevailing humidity (see p. 278).

Adult overwinterers that spend the winter in tree stumps and felled timber rather than in the soil are more or less euryhygric to dry-preferent in experiment (*Agonum assimile, Carabus granulatus, Pterostichus oblongopunctatus, Loricera pilicornis*). Perhaps the soil is too wet for them in winter (Thiele, 1967). In any case the humidity factor is undoubtedly one of the most important factors governing choice of habitat.

3. *The Physiological Basis of Moisture Preference*

Hygroreceptors have not so far been demonstrated in carabids. Nevertheless, by means of experiments in which distal segments of the antennae were successively removed it has been possible to localize such receptors. In the first such experiments Perttunen (1951) showed in two species that amputation of the antennae leads to a loss of the animals' ability to orientate themselves in a humidity gradient.

Weber (1965b) improved upon these experiments and demonstrated that *Carabus granulatus* was unable to orientate itself in a humidity gradient with only 0–4 antennal segments, whereas the behaviour of individuals with at least five segments was not significantly different from that of animals with 11 segments. This narrows down the position of the receptors on the antennae.

4. Resistance to Desiccation

Not only the experimentally demonstrated ability to orientate rapidly in gradients of relative humidity, but also the property of tolerating dry air for longer or shorter periods of time is significant for an affinity to certain habitats. Experiments involving exposure of the experimental animals to dry air have been carried out by Lindroth (1949), Kless (1961), Thiele (1964a, 1967) and Paarmann (1966). The ability of carabid species to survive in dry air fluctuates over very wide limits. At 20%–30% R. H. and 20°C the survival time varied from 18 to 97 h (in 13 species, Thiele, 1964a, 1967). Using very similar methods an extremely short survival time of 15–16 h was found for a genuine cave beetle *(Laemostenus navaricus)* by Boyer-Lefèvre (1971).

Lindroth (1949) tested resistance to drying in 16 species of *Harpalus*. The animals were kept in small glass dishes without water in a dry room (R. H. was not measured). The mean survival time for each species varied enormously, ranging from 492.5 h in *H. serripes* to only 50 h in *H. rupicola*. Since both species live in xerothermic habitats it can be concluded that resistance to desiccation is not obligatory (*H. rupicola* is highly thermophilic, see Chap. 5.C.). "In many cases, but by no means all, resistance to drying agrees well with moisture preference. Strongly xerophilic species are also very resistant (particularly *Agonum assimile*, further *Harpalus rufipes* and *Pterostichus oblongopunctatus*)" (Thiele, 1964a). Very small species such as *Pterostichus nigrita* and *Patrobus atrorufus* can be euryhygric, at the same time having a low resistance to drying. This suggests that other modes of behaviour than humidity preference enable the animals to avoid lethal desiccation, e.g. the dark preference of the two above-mentioned species, which ensures that they are only active at night. However, body size and resistance to drying are not correlated. Kless observed his animals at 52%–54% R. H. and found the longest survival time of 57 h in the dry grassland species, *Harpalus dimidiatus*. It is difficult to compare results obtained by various methods. A striking example of differences in resistance was studied by Paarmann. The two roughly equal-sized species *Pterostichus oblongopunctatus* and *P. angustatus*, although very similar and closely related, differ vastly in their resistance to dryness. *P. angustatus* is confined to hot, sunny forest clearings and is much more resistant to dryness than the eurytopic forest inhabitant *P. oblongopunctatus*. The survival times of the two species if kept over calcium chloride at 21°C were 67.0:35.6 h, and 28.3:18.0 h at 28°C. *P. angustatus* has a slightly higher preference for dryness than *P. oblongopunctatus*.

The individual range of variation in resistance is narrow in less resistant species and considerably wider in more resistant types e.g. from 43 to 102 h in *Agonum assimile* (mean 74) and from 65 to 126 h in *Harpalus rufipes* (mean 97).

The water content of the body appears to be uniform, even in carabids from very different environments. In 13 species the mean varied between 56% and 64%, with no recognizable connection to the environment (Thiele, 1964a and 1967). *Laemostenus oblongus*, a species that lives in bat guano at the entrance to caves, had the very low water content of 48.5%. A close relative, *L. navaricus*, a genuine cave-dweller, also has a low water content with 52.7%. *Laemostenus terricola*, which although it inhabits cellars is not troglophilic, was found to have a water content of 58.4%, a figure well within the region of the above-mentioned values (Boyer-Lefèvre, 1971).

The degree of water loss at which death ensues lies between 40% and 58% and scarcely deviates from 50% in most species (Thiele, 1964a). These figures are referred to the initial body water content; referred to fresh weight they correspond to a loss of weight amounting to 25%–34% (Thiele, 1964a, 1967). Cave carabids, too, survive a water loss of about 50% of their total body water (Boyer-Lefèvre, 1971).

III. Light and Orientation in the Environment

1. Experimental Method

Observation of behaviour in a gradient of diffuse light produced by a variety of grey filters (light intensity gradient) is more reliable in judging the significance of the light factor than those made with the use of a single point light source.

The fact that light not only varies in intensity but also in quality (=spectral colour), makes the investigation of its effect on behaviour more complicated than studies involving temperature or humidity. For an experiment to provide valid information regarding the role of the light factor in habitat choice in nature, the spectral constitution of the light used ought to resemble that in the animal's natural habitat. Artificial "white light" from different sources may vary considerably in its spectral make-up, according to the source used. For this reason it is essential that make and spectral type of the light source employed be given in publications concerning experiments on light. For the sake of simplicity the light intensity is usually measured as intensity of illumination in Lux, and not as the absolute energy released by the light source. For this purpose a luxmeter or photometer is employed, the sensitivity of which varies greatly from one spectral region to another. The course taken by the sensitivity curve is approximated to that of the human eye. It should be borne in mind that, as a rule, the insect eye has a different sensitivity curve. Hasselmann (1962) carried out electrophysiological investigations on the spectral sensitivity of the carabid eye using *Carabus auratus* and *C. nemoralis*. The sensitivity maximum for both species was found to be at about 500 nm, whereas the sensitivity maximum of the selenium cells used in luxmeters is about 550 nm. If "white light" is used in the experiment then this degree of coincidence is adequate and the beetles perceive the illumination offered as being bright light. Above 570 nm, however, the sensitivity of the *Carabus* eye drops sharply (Fig. 91). Experiments on the light intensity preferred by carabids have been performed

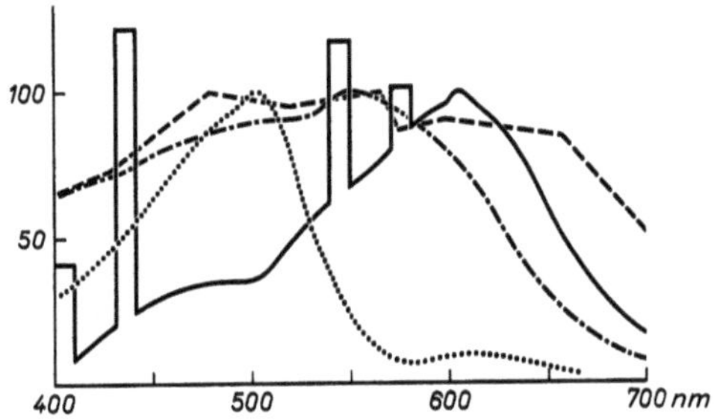

Fig. 91. *Ordinate:* relative spectral intensity (——) in the light gradient apparatus. Fluorescent light employed: Osram HNW = white light); ----- daylight on the edge of a forest; relative spectral sensitivity ····· of the eye of *Carabus nemoralis* (from Hasselmann, 1962); -·- of the selenium cell of a luxmeter. All values given in relative units; maxima of the curves are taken as 100. *Abscissa:* wave length. From Thiele, 1964a

by Thiele (1964a, 1967), Lauterbach (1964), Paarmann (1966) and Neudecker (1974). The gradients offered by these authors (except Neudecker) were chosen more or less at random or were approximately linear.

According to the Weber-Fechner law, the strength of the reaction is logarithmically and not linearly dependent on the intensity of the stimulus. It would therefore seem to be more logical to construct the light intensity gradients on an exponential rather than on a linear basis, since the beetle would then perceive an equal increase in light intensity from step to step. Neudecker used gradients of this kind in the experiments published in 1974.

Since the Weber-Fechner law holds for reactions to all kinds of stimuli the same criteria ought, in principle, to be applied to experiments involving gradients in any type of factor. So far, however, it has proved not to be practicable in experiments involving temperature and humidity, in contrast to investigations on reactions to light.

As far as is known from physiological investigations, there seems to be no absolute memory for light-intensity. Degrees of light can only be perceived comparatively, i.e. as being lighter or darker than others. This ought to mean that the animals would not be able to show a preference in the middle region of a light-intensity gradient, whereas photophilic animals would always congregate in the lightest, and photophobic animals in the darkest portion of the gradient. This is so in the majority of carabids, apart from the fact that some species are indifferent over a wide range of light intensity. In a light gradient ranging from 20–1650 Lux Thiele (1964a) found, nevertheless, for *Agonum dorsale* and *Loricera pilicornis*, well-developed peaks of preference in an intermediate region between 550 and 700 Lux. This appears at first to be paradoxical in view of what has already been said, and should perhaps be interpreted as implying that a photophilic species, although avoiding darkness, find light intensities above a certain value unpleasant (this also applies to humans). Thus, in a light gradient it is "caught" in an intermediate region due to its phobic reactions. Such an interpretation also fits in with the observed ecological behaviour of the species (see below).

As far as is known, the preferred intensity seems to be a fairly constant characteristic. Excellent agreement between the reaction of the two sexes was found by Thiele (1967) in *Pterostichus nigrita* and *Agonum assimile*. Neudecker (1974) found a more or less strongly developed preference for darkness in *A. assimile*, but says: "In the months of April to May *A. assimile* shows a greater preference for light than in the winter period of rest, which is to say that less individuals are found in the two darkest steps of the gradient." This applies to animals caught in the field. In laboratory-bred animals, however, the darkest steps of the light-intensity gradient (<1 Lux) were chosen with varying frequencies by three experimental groups as follows: by 60% of the animals treated at 3°C with short day, by 40% of those kept at 20°C and short day, and by 30% of the animals kept at 20°C and long day. This appears to indicate that the temperature at which the animals were pretreated influences the choice of light intensity. If subsequently kept in long day and at 20°C the differences between the three groups of animals were eliminated. The experiments demonstrated that the light intensity preferences of animals ready to reproduce were identical with those of animals in a state of dormancy. "It can be concluded from studies with laboratory-bred animals that the seasonal changes in choice of light intensity shown by animals caught in the wild are probably merely dependent upon temperature."

2. *Preferred Light Intensity and Habitat Affinity*

The light intensity chosen by a species in an experimental gradient is not necessarily comparable with that prevailing by day in its natural habitat (in contrast, e.g. to experimental temperature preference and habitat temperature). Besides influencing their distribution, light also plays an important part in the daily pattern of activity of carabids. In the chapter on daily rhythmicity in carabids light is shown to be the most important zeitgeber for daily rhythmicity. Species that are found to be dark-preferent in a light-intensity gradient are almost without exception night-active; the light-preferring species are day-active and species indifferent to light intensities (euryphotic) are active both night and day. This was found in 20 of 22 species studied by Thiele (1964a), the two exceptions being those mentioned above, *Agonum dorsale* and *Loricera pilicornis.* Both of these are photophilic and according to investigations so far available (only field experiments, Kirchner, 1964) *A. dorsale* is night-active and *L. pilicornis* can be either day- or night-active. As has already been said, neither species shows an extreme preference for light. Both inhabit fields and their margins, and overwinter in the less dense type of forest, which conforms well with their weaker light preference. Their seasonal change of habitat suggests a corresponding change in photophily, a familiar phenomenon in other beetles.

Statistical comparison reveals a particularly close correlation between preferred light intensity and habitat affinity. According to Thiele (1964a), 12 out of 13 forest species preferred darkness, as compared with only four of 10 field species. Lauterbach (1964), in a comparison between inhabitants of forests and their clearings, obtained similar results. Observations on a large number of species revealed that 87% of the forest animals but only 53% of the field species preferred darkness (see Table 43). Nevertheless, there seems to be no direct connection between behaviour and light intensity during the day in the habitat. Instead, it becomes increasingly obvious that forest carabids are almost exclusively night-active species, whereas the field inhabitants are in the main day-active or day- and night-active. Most species living in forests, in contrast to those of the fields, show a much stronger preference for moisture and require much more of it than the field animals. It is probably of selective advantage to them to be active only in the moister night hours. These questions will be considered in more detail in Chapter 7.

A comprehensive analysis of the reaction of two species of *Elaphrus* to irradiation was undertaken by Bauer (1974). The dependence of sharpness of vision on light intensity was tested in an optomotor drum with tracks of equal width. The visual acuity generally rose with the logarithm of the intensity of illumination. *E. cupreus*, a shade-seeking animal, attained its full sharpness of vision at a lower intensity then *E. riparius* (Fig. 92). In keeping with its habitat preference,

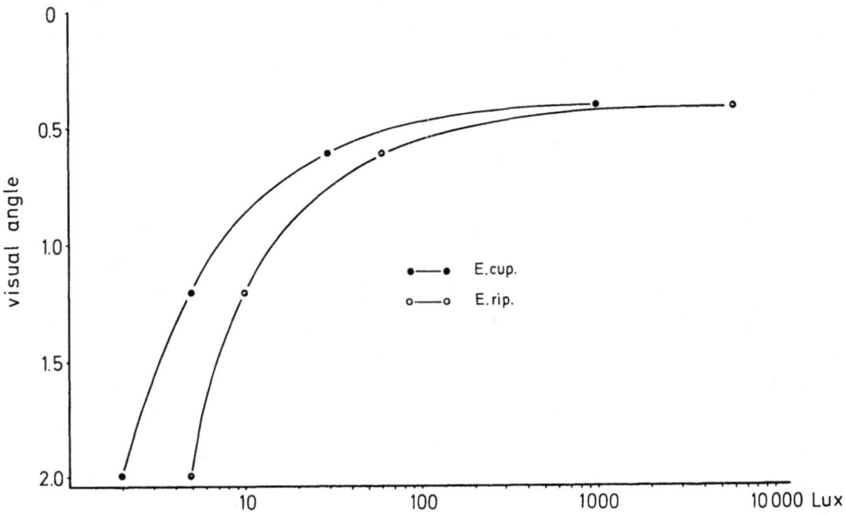

Fig. 92. The dependence of visual acuity on the strength of illumination in *Elaphrus cupreus* and *E. riparius*. The visual acuity was tested with striped patterns of various widths in the optomotor drum. From Bauer, 1974

E. riparius sought out the lightest sector in a gradient (10,000 Lux), and *E. cupreus* chose intermediate regions between 600 and 2000 Lux (Fig. 93). This need not necessarily mean that there is a "preferred light intensity". In a gradient of infrared light involving intensities that are definitely not optically detectable to the animals, *E. riparius* chose the highest intensity and *E. cupreus* once again sought out intermediate regions (Fig. 94). Bauer considers it not impossible that the animals were, in fact, warmed up to differing extents in the various sectors of the light intensity gradient, despite the apparatus being cooled with a water bath. The results of infrared preference experiments suggest that *E. riparius*, being a sun-loving animal, might have a somewhat higher temperature preference and for this reason expose itself to stronger irradiation. This suspicion could not be confirmed in a temperature gradient. Since the body temperature of *E. cupreus* rose more rapidly in the experiment with a cooled light than that of *E. riparius*, it can be assumed for the former that its integument offers less protection from radiation. This indicates that a "radiation preference" is an important factor in determining habitat affinity.

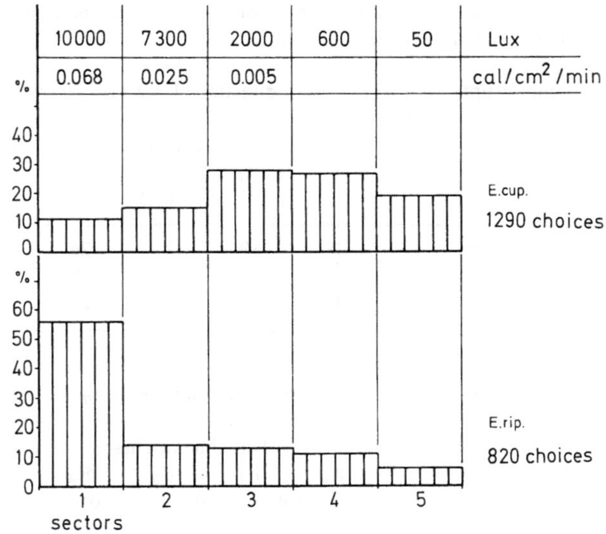

Fig. 93. Light intensity preferred by *Elaphrus cupreus* and *E. riparius* in a five-level gradient. From Bauer, 1974

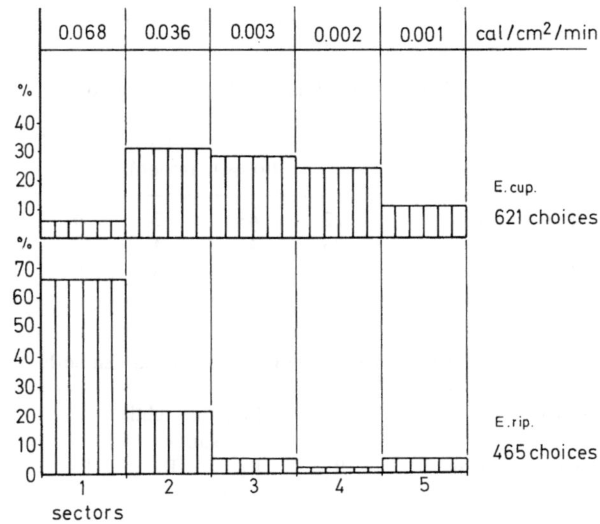

Fig. 94. Infrared intensities preferred by *Elaphrus cupreus* and *E. riparius* in a five-level gradient. The energy of radiation in the first sector was adjusted so as to correspond to that in the light intensity gradient. The infrared lamp employed had a maximum radiation flow at 950 nm. From Bauer, 1974

3. *Orientation Using Silhouettes on the Horizon*

This type of orientation, as for example towards the silhouette of a forest, has long been recognized in flying beetles (cockchafers). Lauterbach (1964) was the first to demonstrate it in the forest carabids that only move on the ground. The latter are capable of running towards dark silhouettes on the horizon even if the light falling vertically from above is not of graduated intensity. Lauterbach performed numerous experiments with the forest species *Abax ater*, both in the open and in the laboratory. Whereas *Abax ater* is attracted by forest silhouettes and artificial outlines on the horizon, the field-dwelling species

Table 41. Choice of direction of carabids in choice quadrats. (According to Lauterbach, 1964.) Each experiment involved 100 animals

a) Uninterrupted horizon in every direction. Time 11–12 h; sunshine. Area 1 m^2, smooth.

	Abax ater				*Carabus auratus*			
	R_1	R_2	R_3 (sun)	R_4	R_1	R_2	R_3 (sun)	R_4
1. Experiment	48	19	7	26	29	18	25	28
2. Experiment	53	20	9	18	32	30	15	23
3. Experiment	41	27	3	29	22	17	29	32

b) Forest edge. Recapture in three days and three nights. Area 4 m^2; covered with vegetation.

	Day-time catches				Night-time catches			
	R_1	R_2 (plantation)	R_3	R_4 (forest)	R_1	R_2 (plantation)	R_3	R_4 (forest)
Abax ater	3	1	1	7	17	4	11	42
Carabus auratus	14	29	12	7	1	11	6	3

c) Forest edge. Time 1–2 h, full moon. Area 1 m^2, smooth.

Abax ater	R_1	R_2 (forest)	R_3 (moon)	R_4
1. Experiment	22	43	7	28
2. Experiment	26	45	4	25
3. Experiment	24	49	6	21

d) Experiments with artificial silhouettes. Area 1 m^2, smooth.

Abax ater	14–16 h, slightly sunny				0–1 h, moonlight			
Angle of horizon[a]	R_1	R_2 (Silh.)	R_3	R_4	R_1	R_2 (Silh.)	R_3	R_4
10°	30	47	21	2	33	38	22	7
20°	11	71	14	4	16	51	29	4
30°	4	94	—	2	8	77	14	1

e) Laboratory experiments involving horizons differing in degree of brightness. Area 1 m^2, smooth.

Abax ater	R_1	R_2	R_3	R_4	R_1	R_2	R_3	R_4
Lux	85	6	2	1	9	4	2	1
Choice of direction	—	9	26	65	—	12	23	65

[a] Angle of elevation of silhouette above the horizon

Carabus auratus turns away from them under identical conditions. Not only does *A. ater* move in the direction of forest outlines during the daytime but also on nights with a full moon, although if the sky is overcast the animals are disorientated (Table 41). The attractiveness of a silhouette to forest carabids rises in parallel with the angle, from the animal's position, at which it is elevated above the horizon. In field experiments involving recapture of marked animals over large areas U. Neumann (1971) was able to show that a forest outline with an elevation of 10° (corresponding to a distance of 64 m) above the horizon was still a considerable aid to *Carabus problematicus* in its orientation, even if not to the extent provided by an angle of 30° (corresponding to a distance of 17 m). *Abax ater* was not so well orientated as *C. problematicus*, but was still well able to find its way to the forest at a distance of 17 m. This kind of orientation, which depends upon form perception, is certainly a significant factor in the environmental affinities of forest carabids (Fig. 95).

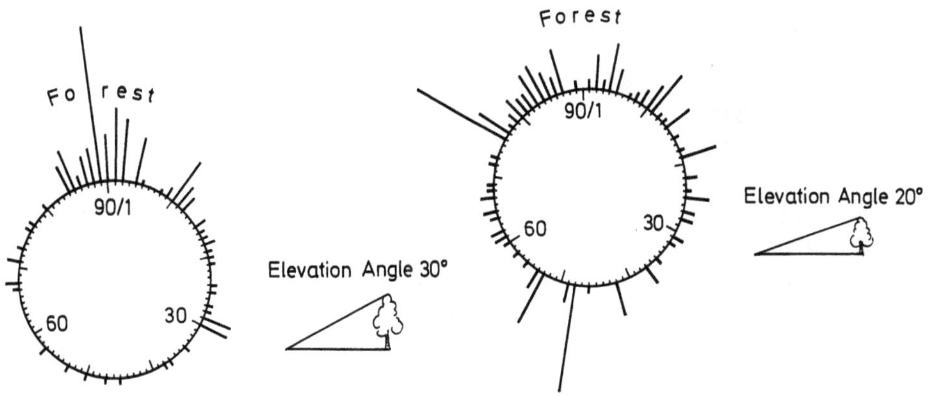

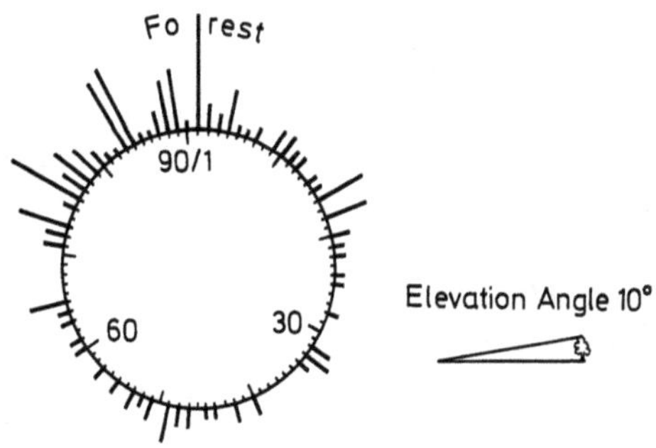

Fig. 95. Choice of direction *Carabus problematicus* set loose within a circle of traps (radius 7 m, 90 traps about 0.5 m apart around the circumference), at various distances from the edge of a forest. The animals were marked and recaught for checking. Distance from edge of forest 17 m (=elevation 30°), 34 m (=elevation 20°) and 64 m (=elevation 10°). *Shortest column on a circle:* 1 animal. *Numbers within the circle:* trap numbers. From Neumann, 1971

4. The Role of Form Perception in the Search for Living Quarters

Telotactical orientation to dark shapes of a certain form is exhibited by numerous carabid species in experiment (Bathon, 1973, 1974). Nearly all species investigated chose the shape with the largest horizontal dimension, e.g. *Carabus problematicus, Abax ater* and *Agonum assimile*. It has also been shown for these species that

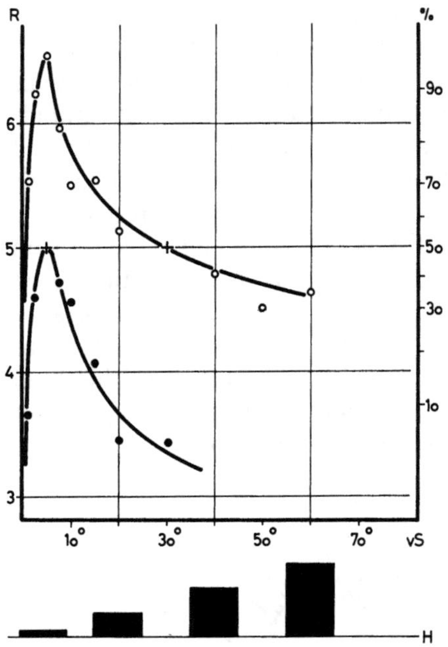

Fig. 96. Frequency of choice *(ordinate)* by *Notiophilus biguttatus* of black rectangles 30° wide and of different heights (these are given along the abscissa in degrees as they appear to the beetles). From Bathon, 1974

they are attracted in the open towards dark outlines on the horizon. *C. problematicus* and *A. assimile* chose shapes with the largest vertical dimension (at constant width); in *Abax ater*, on the other hand, there was an upper limit of 30° to 40° for the receptive field for black shapes (Bathon, 1973). Due to their ability to distinguish forms these forest-dwelling species are attracted, firstly, from open country into forests, and secondly, to dark living quarters such as felled trees, broken branches, tree stumps and so on. "The most complicated releasing mechanism among carabids of the forest floor is seen in the day-active *Notiophilus biguttatus*, a species living and hunting its prey in the litter layer. It, too, is attracted most by very broad shapes. If, however, the various shapes offered present an angle higher than 6° it avoids them with increasing size of this angle" (Fig. 96; Bathon, 1974). Horizontal patterns consisting of a number of stripes each presenting an angle of 6° are more attractive than a single stripe (Fig. 97). "The triggering of the releasing mechanism by narrow horizontal stripes enables the beetle to find a suitable hiding-place in the spaces in the litter layer as quickly as possible after its sortie into light" (Bathon, 1974).

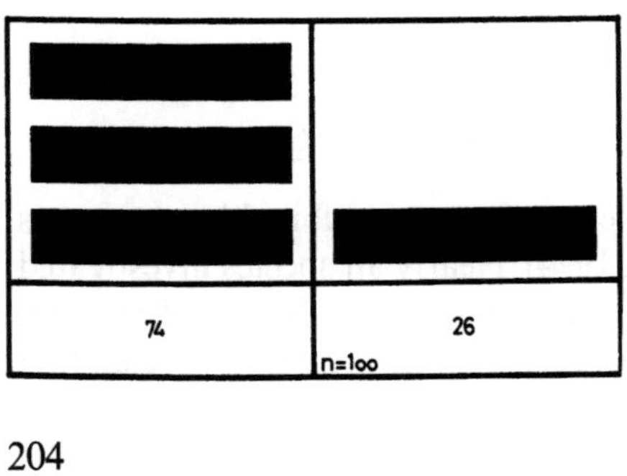

Fig. 97. Relative frequency of choice *(below)* by *Notiophilus biguttatus* of two different patterns. *Left:* three 6°-high black strips separated by two white gaps each 3° high. *Right:* one black strip 6° high. From Bathon, 1974

Dromius quadrinotatus, a tree and shrub dweller, was found to choose vertical in preference to horizontal stripes, which conforms with its mode of life. In contrast to that of the night-active inhabitants of the forest floor, the behaviour of *Dromius* differed in that the animal was, firstly, able to distinguish at all between horizontal and vertical stripes and, secondly, that a vertically striped pattern was just as attractive as a plain black shape of equal area (forest and night animals always preferred the plain shapes).

5. *Astronomical Orientation*

Many animals are known to exhibit a so-called sun-compass orientation. Their inner clock enables them to calculate the sun's position and thus run in the right direction. For beach and shore dwellers this has the special advantage of enabling them to orientate themselves towards the safety of dry land in times of flood. The instinctive direction taken must, of course, be peculiar to the special topographical conditions of the environment of any one population. According to Papi (1955) the shore carabids *Omophron limbatum*, *Scarites terricola* and *Dyschirius numidicus* possess an orientation mechanism that helps them maintain a particular azimuthal angle by reacting to stimuli arising in movements of the firmament. If placed in water beneath a clear sky *O. limbatum* and *S. terricola*, even if previously kept for a considerable time in the dark, invariably move in a direction of an azimuth approximately coinciding with "inland", as seen from their original dwelling place. Experiments with mirrors (according

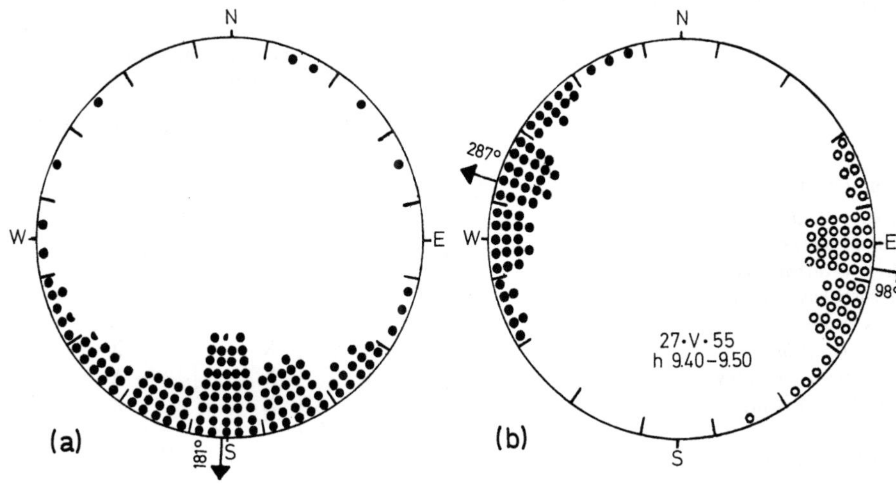

Fig. 98a and b. Direction of escape taken by *Omophron limbatum* in experiments under different conditions. (a) direction of escape taken by a population from the south bank of the river Morto Nuovo, Italy. Each point represents four observations. The animals were able to see the sun. Southerly direction of escape (inland). (b) Direction of escape taken by a specimen of *O. limbatum* from a habitat where the theoretical inland escape direction is 280–290°. *Black points:* results from an experiment in the sun (9.40–9.45 a. m.). *White points:* direction of escape of the animal in the mirror experiment according to Santschi (9.45–9.50 a. m.). *Each point* represents one observation. From Papi, 1955

to Santschi) and under overcast skies have shown that the orientation of the animals must depend upon an internal clock. *Dyschirius numidicus* can also orientate itself in the shade, and it can be assumed that the direction of the oscillations of polarized light plays an important role in orientation (Fig. 98).

IV. A Survey of the Microclimatic Requirements of Carabids from Different Habitats

Experimental investigations concerning reactions to microclimate (temperature, humidity and light intensity) have been made in my laboratory on a total of 47 carabid species in apparatus offering climatic gradients. The advantage of results of this nature is that, having been obtained by a single group of workers employing uniform methods, they can safely be compared, and they therefore enable us to draw general conclusions regarding the behaviour of carabids from different habitats (Thiele, 1964a; Kirchner, 1960; Lehmann, 1962; Lauterbach, 1964; Paarmann, 1966; Becker, 1975; Tables 42 and 43).

Let us first take a look at the behaviour of all 47 species, irrespective of habitat (Table 42). Cold-preferring species are apparently in the minority (only 8), whereas the majority are eurythermic or prefer warmth. Two-thirds of the species are dark-preferent, but the numbers of hygrophilic, xerophilic and euryhygric species are fairly equal.

Table 42. Behaviour of 47 carabid species in experiments with climate gradients. For explanation of the numbers 1, 2 and 3 in the headings, see Table 43

	Behaviour towards								
	Temperature			Humidity			Light		
	1	2	3	1	2	3	1	2	3
Stenotopic forest species									
Molops elatus	X			X			X		
Molops piceus	X			X			X		
Cychrus attenuatus	X			X			X		
Pterostichus metallicus	X			X			X		
Abax ovalis	X			X			X		
Abax parallelus		X		X			X		
Pterostichus cristatus		X		X			X		
Patrobus atrorufus			X	X			X		
Calathus piceus			X		X		X		
Agonum assimile	X					X	X		
Eurytopic forest species									
Abax ater		X		X			X		
Carabus problematicus			X	X			X		
Carabus coriaceus			X			X	X		
Carabus monilis		X			X			X	
Pterostichus oblongopunctatus		X				X		X	
Total out of the 15 forest species	6	5	4	10	2	3	13	2	0

Table 42. (continued)

	Behaviour towards								
	Temperature			Humidity			Light		
	1	2	3	1	2	3	1	2	3
Species of the river banks									
Bembidion tetracolum		X		X			X		
Bembidion femoratum		X		X				X	
Bembidion punctulatum			X	X			X		
Species of forest clearings									
Carabus arcensis			X			X		X	
Pterostichus angustatus		X				X	X		
Eurytopic species									
Carabus nemoralis		X			X		X		
Carabus purpurascens		X			X		X		
Carabus granulatus			X			X	X		
Pterostichus vulgaris		X		X			X		
Pterostichus niger			X	X			X		
Pterostichus madidus		X		X				X	
Pterostichus nigrita		X			X		X		
Loricera pilicornis	X				X				X
Nebria brevicollis			X		X		X		
Asaphidion flavipes		X			X			X	
Total out of the 10 eurytopic species	1	6	3	3	6	1	7	2	1
Species of cultivated fields									
Nebria salina	X				X		X		
Stomis pumicatus		X		X			X		
Calathus fuscipes		X				X	X		
Harpalus rufipes			X			X	X		
Broscus cephalotes			X			X	X		
Carabus auratus			X		X			X	
Pterostichus cupreus			X			X		X	
Agonum dorsale			X			X			X
Species of dry grassland									
Harpalus dimidiatus		X			X			X	
Harpalus rubripes			X			X	X		
Pterostichus interstinctus		X				X		X	
Brachinus crepitans		X				X	X		
Harpalus distinguendus			X		X			X	
Harpalus froelichi			X			X		X	
Pterostichus dimidiatus			X		X		X		
Callistus lunatus			X			X	X		
Cicindela hybrida			X		X			X	
Total out of the 17 field species	1	5	11	1	6	10	9	7	1
Total	8	19	20	17	14	16	32	13	2

Table 43. Behaviour of carabids from different habitats to important microclimatic factors as seen in preference experiments

		Temperature			Humidity			Light		
	n	Cold-preferent (1)	Eury-thermic (2)	Warm-preferent (3)	Hygro-philic (1)	Eury-hygric (2)	Xero-philic (3)	Dark-preferent (1)	Eury-photic (2)	Light-preferent (3)
Forest-dwelling carabid species	15	6 40 %	5 33 %	4 27 %	10 67 %	2 13 %	3 20 %	13 87 %	2 13 %	0 0 %
Eurytopic carabid species	10	1 10 %	6 60 %	3 30 %	3 30 %	6 60 %	1 10 %.	7 70 %	2 20 %	1 10 %
Field-dwelling carabid species	17	1 6 %	5 29 %	11 65 %	1 6 %	6 35 %	10 59 %	9 53 %	7 41 %	1 6 %
Total	42	8 19 %	16 38 %	18 43 %	14 33 %	14 33 %	14 33 %	29 69 %	11 26 %	2 5 %

Table 44. Behaviour of carabids from various types of meadows in preference experiments involving temperature, humidity and light. (According to Tietze, 1974)

		Temperature			Humidity			Light		
	n	Cold-preferent and meso-thermic	Eury-thermic	Warm-preferent	Xero-philic	Eury-hygric	Hygro-philic and meso-philic	Dark-preferent	Eury-photic	Light-preferent
Species of dry grassland	7	2	—	5	6	1	—	5	1	1
Species of mesophilic green land	6	2	—	5	4	3	—	4	3	—
Species of moist to wet green land	6	2	2	2	1	3	2	3	2	1

If we now consider the species according to their ecological type, many connections between experimental behaviour and habitat affinity are revealed (Table 43). The percentage of cold-preferent species decreases from the forest-dwellers via the eurytopic species to the field species, with the figures of 40%, 10% and 6%. The corresponding values for warm-preferring species, in contrast, rise from 27% via 30% to 65% of the field species. Far more forest inhabitants are hygrophilic than cold-preferent (67% compared with 40%). Of the forest species only 20% are xerophilic as compared with 53% of the field species. In keeping with the rather loose kind of habitat affinity shown by eurytopic species, they contain a large proportion of eurythermic and euryhygric forms.

A singular preference for darkness is encountered in forest species (87% of the species). Although it sinks to 70% in the eurytopic species it is still relatively high (53%) in field species. Species with a distinct preference for light are very rarely encountered, but the percentage of euryphotic forms rises from 13% of the forest species to 20% of the eurytopic and to as much as 41% in the field species. Certain species from extreme habitats were not included in the synopsis in Table 43. The three *Bembidion* species that are fairly frequently met on river banks (Table 42) are all moisture-loving and at the same time thermophilic or eurythermic, in accordance with their choice of habitat. Equally in keeping with their habitats, which are extremely sunny forest clearings or burned areas, is the preference of *Carabus arcensis* (also found on bogs, see Chap. 9) and *Pterostichus angustatus* for dryness (and warmth).

A comparison of the preferenda of carabid species from various types of one and the same habitat even reveals differences in behaviour, as Tietze (1973d) showed for meadows (Table 44). The same distribution between light and dark preferring species of carabids is seen in dry, moist and wet meadows. Wet meadows, however, harbour far more hygrophilic species than the drier types, and the latter harbour more eurythermic to warm-preferent species.

Thus, preferences with regard to microclimate prove to be an important factor in determining habitat affinity, although this fact only becomes recognizable if the response of a number of species from various environments to a variety of factors is tested.

Widely varying combinations of microclimatic requirements can be encountered, however, in species of one and the same distributional type. "The many nuances in physiological demands... account for the manifold nuances in distributional pattern. The climatic adaptations of different species in a biocenosis present at least as complex a picture as the morphological. Each species is, as regards its climatic requirements, the result of a quite specific evolutionary process. In a complex environment the species united in a single biocenosis are not the result of any *one* selective factor of paramount importance" (Thiele, 1964a).

Extreme climatic adaptation in a single direction is more the exception than the rule in central Europe. Only extremely eurytopic inhabitants of cool, wet forests exhibit a simultaneous preference for coolness and moisture, and are night active on account of their preference for darkness. On the other hand, only a few stenotopic species of warm, dry fields are day active and prefer light, warmth and dryness (Thiele, 1964a). Similar conclusions were arrived at by Becker (1975) in an investigation involving carabids of extremely warm grasslands. "Oddly, no species could be found that was at the same time thermophilic, xerophilic and photophilic. One of these factors fulfils a regulatory function in the individual species, preventing spread to regions with an unsuitable microclimate; even the carabids of the dry grasslands which, as compared with forest species, are adapted to higher temperatures and lower air humidity, would quickly dry out under too extreme climatic conditions. Xerothermophilic species like *Callistus lunatus*, *Harpalus azureus*, *H. rubripes* and *Pterostichus cupreus* probably counter this danger with their preference for darkness".

The factors that from the statistical aspect seem to be of special importance for the individual ecological groups will be more closely inspected in Chapter 7, after other traits of behaviour, such as annual and daily rhythms, brood care and reaction to soil factors have been considered.

B. Chemical Factors

The ground beetles, as their popular English name implies, are confined to the ground. Logically, the influence of soil properties on their distribution early became the subject of interest and experimental study. Some of the especially important chemical properties of the soil which might exert an influence on the distribution of carabids include pH value, sodium chloride and calcium content. These will be considered more closely in the following.

The effect of physical characteristics of the soil, such as particle size, will be dealt with separately in a following section concerning environmental structure.

I. pH Value of the Soil

We possess little information concerning the effect of pH value and electrolyte content of the substrate on carabids. Correlations have often been established between hydrogen ion concentration in the soil and the distribution of soil animals, but most authors are of the opinion that a causal relationship does not exist (see also Schwerdtfeger, 1963). The only preference experiments to have brought positive results are those of Krogerus (1960). The carabids (usually 20–30 animals) were offered a choice of ten boxes, arranged linearly in a metal container 1 m in length, and covered with a wire mesh. *Sphagnum fuscum* was used as substrate, its pH having been gradually adjusted to an alkaline region by the addition of $Ca(OH)_2$. An experiment to determine an animal's preferred pH region lasted two to three days. According to the diagrams in Krogerus' article the bog species *Agonum ericeti* and *Agonum munsteri* (a form of *A. consimile*) showed a strong preference for acid regions (up to pH 3.6). *Agonum consimile* chose the weakly alkaline regions around 8.0, but the choice of *Agonum thoreyi* and *A. versutum* was not so well defined. Unfortunately, the diagrams indicate neither the exact number of experimental animals, nor the nature of their distribution in pH gradients. No information is given regarding a possible influence of other factors such as light on the distribution along gradients. Krogerus found that the haemolymph was more acid in acidophilic species than in other carabids: 6.2–6.6 (e.g. *Agonum ericeti* 6.2–6.3) as compared with 6.6–7.2 (e.g. *Agonum sexpunctatum* 7.1).

II. The Sodium Chloride Content of the Soil

Many carabids are known only from areas with soil containing sodium chloride, whether on sea coasts or inland. Such species are "topographically" halobiontic. Species that merely show a peak in distribution on salty ground are called "topographically" halophilic. The additional connotation "topographical" indicates that this is the only connection so far known, and says nothing about a causal connection with NaCl content. According to Horion (1959) salt beetles in Europe are of continental origin, and may be connected with salty habitats for climatic reasons. Heydemann (1967a) states that all halobiontic species are at the same time either hygrobiontic or hygrophilic.

The halobiontic species *Dichirotrichus pubescens* exhibited a poor ability to choose between salt concentrations in an experimental gradient (Heydemann, 1968). Nevertheless, the ill-defined maximum in preference at 20‰ agrees with the steep maximum shown in the field at this value. Orientation towards weak salt concentrations in the halophobic species *Agonum dorsale* was more pronounced, but, again, not so striking as that shown in the open.

A. dorsale, in contrast to *D. pubescens*, consumed energy for osmoregulation if the soil was covered by 2 mm of 30‰ salt water (Heydemann, 1968), its oxygen consumption rising. If flooded with salt water, however, its resting metabolism sank to very low values (about one-sixth of the normal O_2 consumption), whereas *D. pubescens* lowered its O_2 consumption by only about 50%. From this, Heydemann concludes with respect to *A. dorsale:* "Insufficient O_2 is available for the mobilization of energy for osmoregulation in the face of flooding with 30‰ salt water". If flooded with sea water containing 94 γ Na^+/10 mg the halobiontic species *D. pubescens* was capable of maintaining a value of 14γ/10 mg body weight for 30 h. In *A. dorsale*, however, the Na content rose over the same period of time from a similar starting value of 12.5γ/10 mg to 30γ/10 mg.

Thus even a halobiontic species possesses only limited powers of osmoregulation and resistance to NaCl, apart from which the mere possession of such properties would not necessarily mean a connection with a salty habitat. Heydemann (1967a) challenged the view that relatively resistant forms of this type have been crowded out of other habitats and into salty areas by competition from stronger rivals, and have developed successfully because of their relatively high resistance to NaCl. Salty ground is in fact densely populated and in no way "lacking in competitors". Heydemann tends rather to favour the idea of some direct connection between the species and NaCl, although as yet physiologically unexplained. *D. pubescens*, very common on the North Sea coast, is lacking on the Baltic coast of Finland (as are other halobiontic carabids e.g. *Pogonus chalceus*, *P. luridipennis* or *Bembidion aeneum*). The fact that the species is again frequent on the Arctic coast of Finland proves that it is not the temperature that is responsible for its absence, but rather the low salt content of the Baltic.

Lindroth (1949) performed simple experiments showing that the halophilic species *Bembidion aeneum* and *B. minimum* prefer a substrate containing salt. *B. minimum* lives, almost without exception, on sea coasts. The distribution of *B. aeneum* in Sweden is very unusual: in the southern and central parts of the country it occurs exclusively on clay soils that have been deposited from the Yoldia Sea and thus are salty. Further east, however, the animal is absent from the clay soils deposited by the fresh water Ancylus Lake, whilst in western Norway the species is only found on the sea coast. In choice experiments involving two sorts of sand, one of which had been soaked with tap water and the other with 1% NaCl, both species chose the salt-containing substrate twice as often as the salt-free. These experiments are further evidence in favour of a direct dependence of the distribution, of some species at least, on sodium chloride.

III. The Calcium Content of the Soil

The existence of plants and animals that serve as "limestone indicators" because of their regular occurrence on limy ground—if not over their entire distributional area, at least over a large part of it—has long been recognized and studied. Some carabids also exhibit this tendency, and to them Lindroth (1949) devoted a large section of his book, giving us at the same time a model for the analysis of the dependence of the distribution of an animal species upon an environmental factor.

For his investigations Lindroth chose the genus *Harpalus*. Cambrio-Silurian lime in the form of bed rock is of very limited occurrence in Scandinavia, being confined to the southerly Baltic islands of Öland and Gotland and small isolated regions of central Sweden. This, then, ought to determine the distributional pattern of the "limestone" species, a large number of which belong to the genus *Harpalus*.

1. Limited to Gotland and Öland:
   Harpalus azureus | *Harpalus rupicola*[12]
   Harpalus punctatulus
2. Northernmost Scandinavian limit on Öland and Gotland:
   Harpalus hirtipes | *Harpalus puncticeps*
   Harpalus melleti | *Harpalus serripes*
   Harpalus neglectus
3. Occur as relics in the central Scandinavian lime regions:
   Harpalus anxius | *Harpalus rufitarsis*
4. Northernmost limit of distribution in northern Uppland:
   Harpalus rubripes | *Harpalus smaragdinus*

---

[12] This species has later on also been detected near Stockholm (Lindroth, personal communication).

5. Northernmost limit in southern Norrland:
   Harpalus seladon *Harpalus tardus*
6. Almost ubiquitous in Scandinavia:
   Harpalus aeneus

The ten species of groups 1–3 are thus the "limestone" species, and the remaining five can be taken for purposes of comparison.

Since *Harpalus* species are to a greater or lesser extent phytophagous initial feeding experiments were carried out from which it transpired that the animals were in fact polyphagous. They are thus not secondarily confined to calciferous soils on account of their affinity for a particular limestone-dependent plant.

Using five limestone species a choice of substrate was offered, ranging from pure silicate gravel through eight mixtures containing increasing quantities of $CaCO_3$ (0.78%–50%). A number of variations on this experiment were made but in no case did the calcium prove to attract the species tested.

Since limestone soils show a high pH value the five species were confronted with soils of difffering pH value, i.e. 7.5, approx. 6.0 and 4.8. Again, no preference could be detected for the pH value of 7.5 at which the animals live in the field.

The overall conclusion arrived at was that "limestone" species are not bound to their substrate on account of its chemical properties.

Eight of the *Harpalus* species mentioned were offered the choice of silicate or lime gravel: five of them, "limestone species", showed a definite preference for the calcium-containing substrate, i.e. 62–83% of the individuals of *H. azureus, anxius, melleti, rufitarsis* and *rupicola*. Since obvious physical differences exist between the two types of gravel (the silicate particles are well rounded and those of the lime gravel more flattened) a choice of lime or slate gravel (which also has flat particles) was offered. None of the above-named species any longer chose the lime gravel.

This seems to dispose of the idea that the distribution of "limestone" species is in any way connected with the calcium itself. Other factors appear to be linked with the calcium content of the soil and it must be these that are responsible for the occurrence of the animals. By means of meticulous measurements in habitats where limestone and silicate occurred contiguously, it could be shown that limestone soils have a characteristic microclimate on account of their dryness and high temperature minima.

Laboratory experiments on the temperature and moisture requirements of the 15 species provided a basis upon which they could be arranged according to their degree of thermophily on the one hand and xerophily on the other (Table 45).

Each of the ten "limestone species" is shown in bold-faced type in the table in which it occupies the highest position: "Limestone species" were found in the three uppermost places in each case and in neither table does one of these species occupy a position lower than six.

In this way it has been possible to unmask the limestone species. They are either particularly thermophilic or xerophilic animals. Thus, it is their climatic

Table 45. 15 species of the genus *Harpalus*, arranged according to their thermophily and xerophily rank. (According to Lindroth, 1949)

Thermophily rank	Xerophily rank
1 ***azureus L***	1 ***hirtipes L***
2 ***puncticeps L***	2 ***serripes L***
3 ***punctatulus L***	3 ***neglectus L***
4 ***rufitarsis L***	4 *tardus*
5 ***melleti L***	5 *smaragdinus*
6 ***rupicola L***	6 ***anxius L***
7 *rubripes*	7 *punctatulus* L
8 *tardus*	8 *azureus* L
9 *hirtipes* L	9 *rubripes*
10 *serripes* L	10 *rufitarsis* L
11 *aeneus*	11 *melleti* L
12 *seladon*	12 *puncticeps* L
13 *neglectus* L	13 *aeneus*
14 *anxius* L	14 *rupicola* L
15 *smaragdinus*	15 *seladon*

Each of the 10 limestone species (indicated by L) is printed in bold-faced type in the column in which it holds the higher position

requirements which are responsible for their being bound to the particular microclimate that, at least in Sweden, is solely encountered on limestone.

This is yet another illustration of the paramount importance of climatic factors in the ecological distribution of carabids.

C. Distribution of Carabids and Environmental Structure

The term "environmental structure" (Raumstruktur) is used here to cover the structural elements of the soil and vegetation in the immediate surroundings of the carabids, as factors which could exert an influence on their distribution.

I. The Substrate

It seems highly probably that the movements of carabids are determined to a large extent by the structural nature of their substrate. Experiments have been carried out to ascertain whether species differences could be detected in this respect (Table 46).

Groups of as a rule ten beetles were placed in glass cylinders of 30 cm diameter, the bottoms of which had been covered with a sand gypsum mixture to produce a rough surface that could be kept uniformly moist. Exactly half of this surface was covered to a depth of 1–2 cm with a layer of moist peat and fallen leaves. The number of animals choosing the uncovered half was

Table 46. Percentage of individuals of different species of carabids choosing a litter-free surface in experiments. (The figures indicate percent values of the largest possible number of observations.) Observations made during the hours of darkness; further remarks in text

Species	Laboratory exp 1.	Laboratory exp 2.	Field exp	Running speed in cm/s at 20° C
Largest possible number of observations	130	120	450	
Forest species				
Abax ovalis	19	31	22	9.7
Abax ater	34	18	20	12.2
Abax parallelus	0	8	10	11.0
Molops elatus	n. d.	19	21	n. d.
Molops piceus	n. d.	5	7	3.9
Pterostichus madidus	n. d.	11	19	9.1
Pterostichus oblongopunctatus	n. d.	6	7	9.4
Pterostichus nigrita, newly caught	n. d.	4	9	7.2
Pterostichus nigrita, caught wild; after overwintering in the laboratory	n. d.	4	6	7.2
Pterostichus nigrita, 1st lab. gen.	n. d.	4	6	7.2
Pterostichus cristatus	38	n. d.	n. d.	13.0
Pterostichus niger	42	n. d.	n. d.	16.1
Nebria brevicollis	40	n. d.	n. d.	12.4
Patrobus atrorufus	0	n. d.	n. d.	5.4
Agonum assimile	3	n. d.	n. d.	9.9
Field species				
Agonum dorsale	9	(14[a])	n. d.	8.1
Pterostichus cupreus	0	(1[a])	n. d.	10.6

[a] Observations made during the hours of daylight
n. d.: not determined

recorded as a percentage of the total observations possible (e.g. ten animals observed ten times gives a possible 100). Laboratory experiments were performed using artificially alternating day-night (200–500 Lux, LD 16/8) at 19° C; experiments in the field were carried out in May/June. Four observations were made daily in the laboratory, 30 min to 1 h before and after light on and 30 min to 1 h before and after light off. In the field large numbers of recordings were made both day and night for two weeks. Almost all species investigated were predominantly night-active and thus only their night data are given. Daytime activity was small, only the two field species *Agonum dorsale* and *Pterostichus cupreus* behaving conversely.

All species were found predominantly in or on the litter, i.e. were thigmotactic. The percentage of animals entering the uncovered half varied, however, from 0 to 42%, notably the larger species frequenting it most *(Abax ater, A. ovalis, Molops elatus, Pterostichus madidus, P. cristatus, P. niger)*. Nevertheless, the relatively small species *Nebria brevicollis*, too, gave a high value of 40%.

The speeds of the various species, shown for the sake of comparison, indicate that a higher rate of occurrence in the exposed half of the vessel is in no way connected with greater activity on the part of the animal. Of two species equal in size and speed, *Abax parallelus* and *A. ovalis*, the former obviously avoids the exposed portion. The two *Molops* species (of which *piceus* is, however, smaller and slower) also behave very differently. It is striking that the animals that seldom enter the uncovered area are largely forest species of the water-meadow woodlands and the moist oak-hornbeam forests, where a hard, bare substrate does not occur *(Abax parallelus, Molops piceus, Patrobus atrorufus, Pterostichus nigrita, Agonum assimile). Abax ovalis, Pterostichus cristatus* and *Molops elatus*, on the other hand, occur in mountain beech forests, e.g. on rendzina soils, and *M. elatus* also ventures into the open on rocky ground. *Pterostichus madidus, P. niger, N. brevicollis* and *A. dorsale* are more or less confined to cultivated fields where the covering of vegetation is thinner. The behaviour of *P. oblongopunctatus* and *P. cupreus* cannot be fitted into this pattern.

It thus seems that substrate structure may be an important factor in the distribution of carabids in their natural surroundings, although its significance and mode of action are as yet poorly understood. Since wild catches were generally used in these experiments it is impossible to say whether the observed differences in behaviour were inherited or had been acquired in the environment. Nevertheless, wild and laboratory bred individuals of *P. nigrita* behaved identically. The influence of soil particle size on the distribution of carabids has already been dealt with in connection with sandy soils (p. 36) and "limestone species" (Chap. 5.B.III.).

II. Environmental Resistance

The term "Raumwiderstand" or "environmental resistance" was coined by Heydemann (1957) to describe the varying degrees of resistance met with by epigaeic arthropods in their habitats due to the nature of the surface of the substrate, and especially the vegetation. *Carabus auratus* covered about 6–8 m/min at 22° C on a smooth, bare and firm substrate (such as a path). It hunts its prey along a trail of about 3 cm width and can therefore cover about 700×3 cm $= 2100$ cm^2 = about 0.2 m^2/min, or, assuming a distance of 50 m in 24 h, 10 m^2. The following recorded and calculated values were obtained from areas where environmental resistance is high:

	Speed	Area covered
Harrowed fields	2–3 m/min	3–3.5 m^2
Winter rye crop	1.8 m/min	2.5 m^2
Meadow	0.5–1 m/min	0.8 m^2

"*Carabus* set free in a meadow behaved like animals in a strange environment, moving clumsily among the blades... For this reason meadow formations which are difficult for the larger carabids to penetrate contain only a few *Carabus* species such as *C. granulatus* and *C. clathratus*... Both forms are able to clamber about among the blades of grass and even feed in this position (*C. granulatus*, in any case)".

Novák (1971b) tested the behaviour of carabids confronted with artificial obstacles. The animals were placed in experimental vessels on the bottom of which pins had been stuck at various densities. The animals were free to move about within the pin area or to remain outside it. Preliminary results showed with statistical significance that *P. vulgaris* was more often found inside the pin area than *Harpalus rufipes* and that the males of both species were more often found in the pin area than the females. These results, too, were statistically significant. The findings are no more than a starting point from which to develop the use of the method and the author justifiably suggests that "preference experiments involving microclimatic factors should also offer the animals physical obstacles in combination with the climatic gradient".

Heydemann (1957) attempted to calculate relative values for environmental resistance. First, he calculates what percentage of a vertical section is horizontally impermeable on account of plant stems or tree trunks. This figure is then divided by the distance between the plant stalks or trunks in metres, since it is also relevant for permeability. A good degree of correlation was found between these relative values and the speeds of the animals:

Speed	Habitat	Environmental resistance
	Carabus cancellatus	
4.80 m/min	weed patch	40
1.70 m/min	field of winter cereal	160
0.60 m/min	meadow	900
	Pterostichus niger	
6.50 m/min	smooth woodland path	0
2.10 m/min	woodland grass	110

The observations of Dubrovskaya (1970) support the above findings: "There were always fewer carabids in the same field when it was occupied by row crops." She also points out that, "In the opinion of most authors, carabids prefer soil in a crumbly state, and their abundance and the number of species represented are greater in cultivated fields, especially fields with row crops, than in fields of perennial grasses and other fields in which the soil is not cultivated". We cannot, however, necessarily generalize from these observations, "The direct opposite, higher abundance on clover and uncultivated land, is, however, noted in some papers...". Scherney (1959) also states that carabids are more abundant in some instances on clover than in other fields although they are as a rule more plentiful in grain fields and fields of row crops.

The forest litter layer, too, offers considerable resistance and thus favours the long slender life form as seen in the staphylinids (body length:width = 1:4–1:9). "Their individual and species density are exceedingly high as compared with those of other families in the same size class. Their activity in zones with a higher environmental resistance is hindered to a much lesser degree than that of other beetles with similar requirements... Thanks to their suitable body structure with respect to environmental resistance staphylinids have at their disposal an ecological niche that is not only of widespread occurrence but to a large extent unoccupied. According to Ludwig this may have contributed to the development of such a large number of staphylinid species" (Heydemann 1957). These observations warrant special consideration in view of the fact that the staphylinid niche, as will be discussed further on (Chap. 10), only partially overlaps that of the enormous wealth of more epigaeic species of carabids (body index according to Heydemann, 1:2–1:3), which, in turn, are in almost uncontested possession of a large niche.

III. The Sense of Gravity

Although organs connected with this sense have not been found in carabids experiments on *Dyschirius nitidus* showed that the animals must in fact possess an effective sense of gravity (Bückmann, 1955). To find its prey *Dyschirius* digs vertical ducts in the sand of the seashore (see Chap. 3.III.). The direction of these ducts was observed when the animals were offered a series of substrates sloped at different angles to the horizontal (Table 47). At an angle of 5° or more the direction of the duct is vertical with a statistical significance of $p<0.01$. Only at an angle of 3° is the significance somewhat lower, with $p=0.02$, but still suggests that even such a slight slope is detectable to the animals.

Table 47. Behaviour of *Dyschirius nitidus* with respect to gravity. (From Bückmann, 1955)

Angle formed by substrate surface with the horizontal	Number of ducts	Number directed upwards
20°	11	10= 91 %
10°	7	7=100 %
5°	19	16= 84 %
3°	25	18= 72 %
0° (control)	28	26= 93 %

D. The Behaviour of Carabids to Water and Their Resistance to Inundation

I. Carabids Hunting in Water

Some species of *Carabus* which exclusively inhabit the margins of inland waters voluntarily enter the water and clamber about among the water plants to hunt their prey beneath the surface (water snails, small crustaceans, leeches, insect larvae, amphibian larvae, small fishes), e.g. *Carabus clathratus, alyssidotus, variolosus, melanckolicus, galicianus* (Sturani, 1962). Sturani observed much longer maximum submersion times than previous authors i.e. 21 min in *C. variolosus*, 17 min 30 s in *C. clathratus* and 15 min in *C. alyssidotus*. The following series involving longer periods of submersion alternating with shorter periods on the surface was recorded for a single specimen of *C. clathratus:* 13 min/30 s/17.5 min/30 s/10 min/25 s/10 min/15 s/7 min, after which it remained on the surface. The ratio of total time submersed to total time above water was 57 min, 30 s (five dives) to 1 min, 40 s (four surfacings). The same individual returned to the water after a pause of 30 min.

These *Carabus* species can apparently renew their respiratory air within a very short time, and store it beneath the elytra at the tip of the abdomen (Fig. 99). In order to replenish its air supply *C. clathratus* crawls onto a rock or similar object, just far enough out of the water for the front part of its body, including the thorax, to be in the air (Fig. 99). A bubble of exhausted air is then expelled from the subelytral space and by means of ventilation of the stigmae in the joint cavities of the prothorax an exchange of respiratory

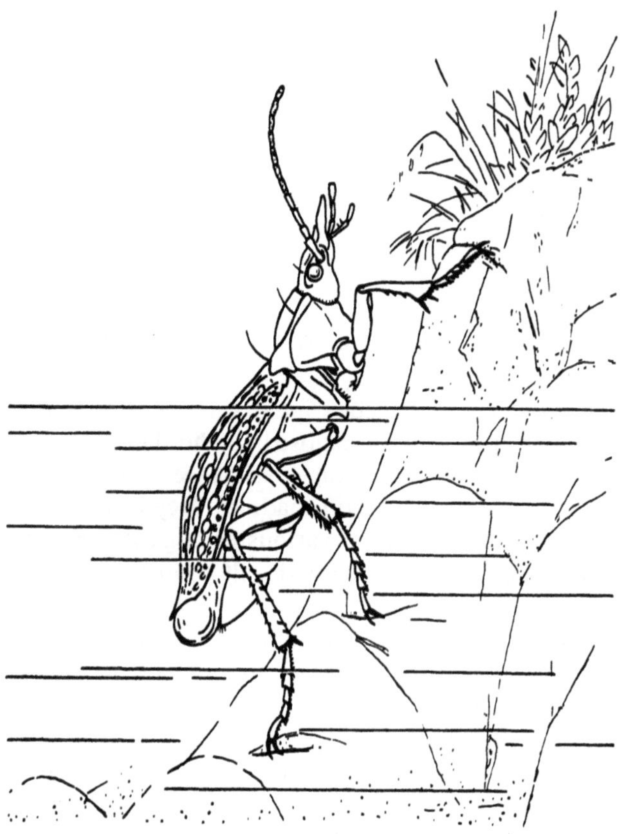

Fig. 99. *Carabus clathratus* climbs half out of the water to renew its supply of respiratory air (bubble at the end of the abdomen beneath the elytra). Explanation in text. From Sturani, 1962

air can be effected. Fresh air permeates the entire tracheal apparatus and from this the air bubble in the subelytral space can be renewed.

The larvae of *C. variolosus* are able to swim and by means of "dabbling" with the upper part of their body submerged they catch live prey beneath the water surface (Fig. 100).

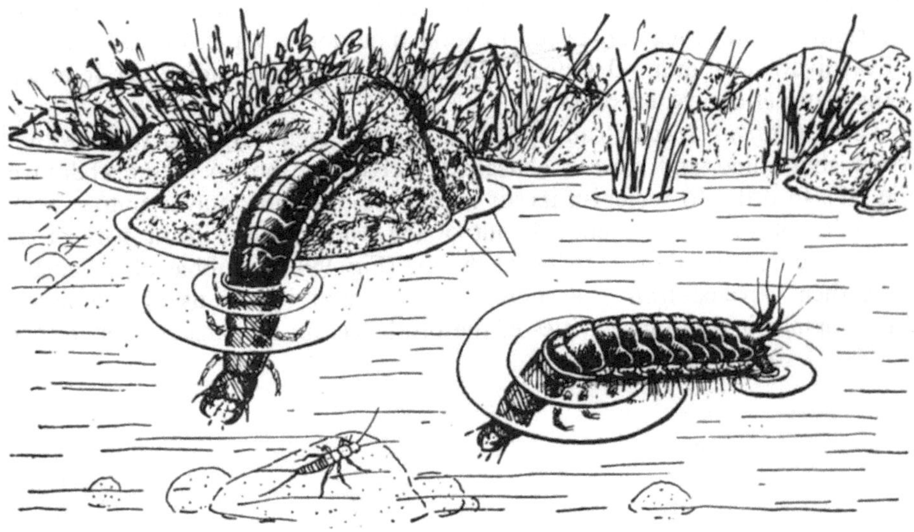

Fig. 100. Two larvae of *Carabus variolosus* searching for prey with their heads under water. From Sturani, 1962

II. The Resistance of Carabids to Inundation

Stenotopic carabid species inhabiting coastal regions or the margins of inland waters are particularly exposed to the dangers of flooding. Have such species developed an unusually high resistance to submersion? This question was the subject of an investigation made by Heydemann (1967a) to determine how the behaviour of carabids of various ecological types faced with flooding by fresh or salt water, is connected with salt content, O_2 content, temperature and water movements.

Heydemann chose a halobiontic coastal species, *Dichirotrichus pubescens*, which occurs in large masses in the salt meadows on the North Sea coast. It showed a substantially higher survival time than inland species under continuous flooding with non-aerated fresh-water. Whereas *D. pubescens* survived as long as 32 days at 10–13° C, two inland species (*Agonum fuliginosum* and *A. dorsale*), tested for the sake of comparison, survived only a maximum of 17 days at 8° C under otherwise similar conditions.

The resistance of *D. pubescens* to inundation is influenced by a number of factors. From the wealth of available data the following conclusions can be drawn: (1) Low water temperatures (5–8° C) prolonged the survival time considerably as compared with higher temperatures (21–23° C). (2) In salt water (32‰) survival time was lower than in fresh water. In moving, brackish water (15‰) it lay between the two. (3) If flooding occurred more or less in a tidal rhythm (6 h ebb, 6 h flood) the survival time in brackish water was several times that under constant flooding.

The greater survival time in fresh water is explained by the fact that salt water at the same temperature contains less oxygen. Further, Heydemann assumes that more energy is required for osmoregulation in salt water than in fresh water. "The dependence of the halobiontic species *Dichirotrichus pubescens* on the salt content of the substrate in no way corresponds with its ability to meet the osmoregulatory demands placed upon it by intense contact with water. Even salt beetles tolerate flooding with rain water better than with salt water." (See Chap. 5.B.II.)

Evans *et al.* (1971) observed that the O_2 consumption of *D. pubescens* sank to minimum values during experimental flooding with salt water. The longer the submersion, the longer the subsequent period of recovery required before normal activity was resumed. However, if the animals had been collected from the mud at the end of a period of flooding no such phase of recovery was needed. From this it was assumed that the animals were not in direct contact with the salt water during flooding but survived in channels filled with air.

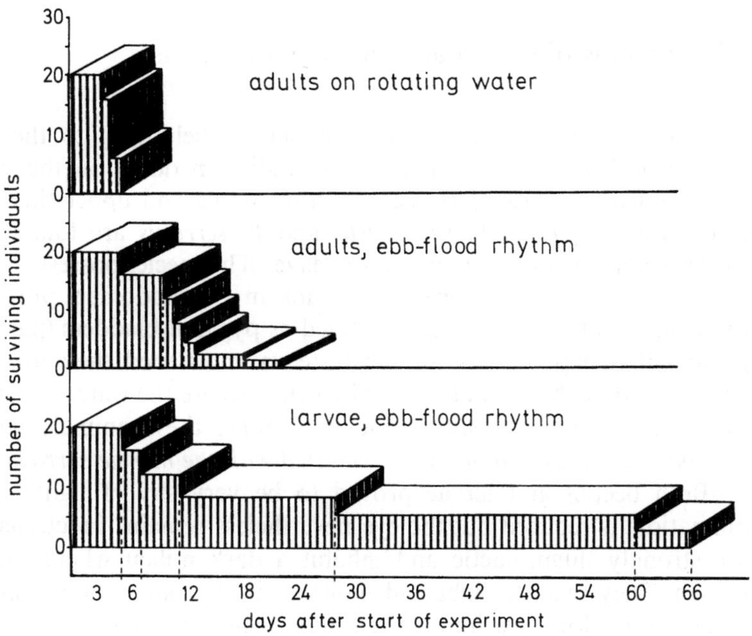

Fig. 101. Various influence of rotating water *(above)* and still water (rhythmical submersion, 10 h ebb/2 h flood) on adults and larvae of the halobiontic ground beetle *Dichirotrichus pubescens*. Temperature: 15–18° C. Salt content of the water 32‰. From Heydemann, 1967

Survival times similar to those found by Heydemann for *D. pubescens* were also found in cold salt water for *Bembidion doris*, *B. unicolor*, *Acupalpus dorsalis* and *Agonum obscurum* by Palmén (1949). The longest survival time in the face of flooding with *fresh water* was found by Palmén (1945) to be 50–70 days in *Agonum fuliginosum*. He considers it impossible for carabids to overwinter under water (important in view of remarks on shore fauna, Chap. 6.B.VI.). In contrast, he feels that hibernation of carabids frozen in ice is possible. This was confirmed by recent observations (see Chap. 5.A.I.4.).

It is interesting to note that the larvae of *D. pubescens* survived flooding much longer than the adults. If continuously submersed with aerated salt water of 32‰ adults survived a maximum of eight days at 15–18° C whereas the larvae survived 83 days. In an ebb and flood rhythm the adults survived a maximum of 22 days at 16–19° C and the larvae 65 days (Fig. 101). Similar powers of resistance in the larvae could also be demonstrated in *Amara convexiuscula* and *Pogonus chalceus*—as well as in *Pterostichus vulgaris*—which shows that the phenomenon is not solely confined to shore carabids. A clue to the reason behind the better survival powers of the larvae was provided by experiments in which larvae of the third instar of *A. convexiuscula* survived eight times as long in moving as in still water. Larvae breathe exclusively through their skin and are not enveloped by a film of air under water, so that moving water provides them with more oxygen. Adults, in contrast, are at a disadvantage in moving water since it tends to wash away the air film that, in their case, serves as a physical gill.

III. Carabids of the Tidal Zone of Rocky Coasts

Regions such as these are chosen by species belonging to the Trechini tribe, e.g. *Aepopsis robinii*. The species was studied in detail on the French Atlantic coast by Richoux (1972). It occurs in the middle and upper tidal zones, regions where *Fucus spiralis*, *F. vesiculosus* and *F. serratus* are found. It is confined to crevices in the rocks, which it never leaves. The species possesses no morphological adaptations to submersion. Organs in the abdomen previously thought to be air sacs have now been identified as pygidial bladders (part of the defence glands). If, in experiments, the substrate was moistened, the animals immediately withdrew to the bottom of artificial crevices where the water could not penetrate. Thus, although living within the tidal zone, these animals *inhabit an aerial environment which is in no way connected with the marine surroundings.*

Both beetles and larvae proved to be very shy of light. The number of ommatidia is reduced but the eye nevertheless remains functional. The animals are strongly thigmotactic and inhabit a dark milieu where the air humidity is high. They react to light and moisture in the same manner as the terrestrial and cavernicolous Trechini of the genus *Aphaenops*. *Aepopsis* shares the unusual habit of the *Aphaenops* species of laying only one relatively large egg. It is impossible to say whether the ecological similarity to the cavernicolous Trechini is due to their being related, or to convergence.

IV. Swimming Carabids

Apart from the few species that enter water voluntarily the ability to swim is of vital importance for the survival of carabids living on the water's edge. The possibility of distribution by means of drifting has also to be borne in mind in estimating their powers of dispersal (see Chapter 8.C.).

Bembidion minimum, a species of the foreland of the North Sea coast, proved to be a swimmer of much greater endurance than *Amara convexiuscula*, a species found mainly in regions not endangered by floods (Heydemann, 1967a). In still, fresh water *B. minimum* was able to survive by swimming for a maximum of 35 days at 18–20° C. If the water's surface was subject to movement then the figure sank drastically to 19 days. The corresponding values for *Amara convexiuscula* were four and three days. *Pterostichus madidus* can occur in large numbers in the river bank zone of the Rhine, although its distributional peak is in forests and cultivated fields. Of 20 specimens of this species ten were able to swim for 24 h in still water at 19–20° C, and only one specimen achieved a maximum of six days (Lehmann, 1965).

The behaviour of *Pterostichus madidus* in water was different from that of a species typical of the banks of the Rhine, *Bembidion femoratum* (Lehmann, 1965). 50 *B. femoratum* or 45 *P. madidus* were placed in an aquarium 1.20 m in length, with a sloping floor of gravel and sand. Water was then let in slowly. *B. femoratum* behaved in the same way in a number of experiments. "As the water slowly rose, the animals left their retreats and fled to the highest point of the artificial bank. If stranded on a stone they swam the shortest distance to the nearest dry land... They invariably sought to escape in the direction of the sides of the aquarium along which there was still a continuous dry area. Once there, the animals then tried to escape by flying... *Pterostichus madidus* was also driven out of its hiding-places by the water but in contrast to *B. femoratum* its efforts to escape were undirected".

Depending upon whether it was on land or on water *Bembidion litorale* reacted very differently to striped patterns. It preferred horizontal stripes on land but vertical ones on water (Fig. 102; Bathon, 1973). Thus, on land, the beetle can quickly recognize hideouts such as small self-made holes in the sand, or beneath rocks or small clumps of plants, and creeps into them if necessary.

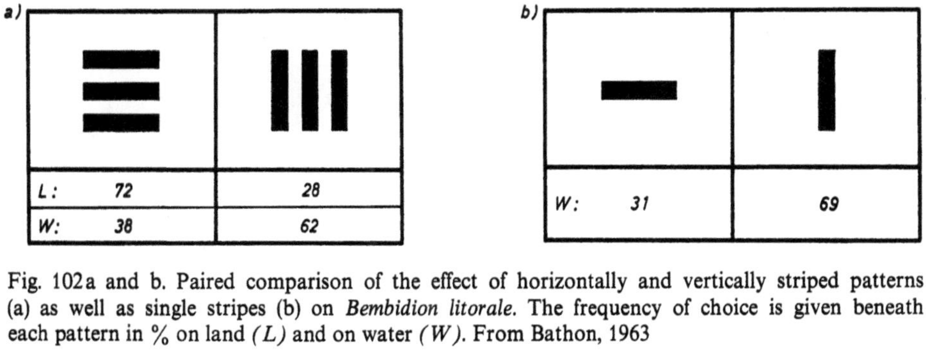

Fig. 102a and b. Paired comparison of the effect of horizontally and vertically striped patterns (a) as well as single stripes (b) on *Bembidion litorale*. The frequency of choice is given beneath each pattern in % on land *(L)* and on water *(W)*. From Bathon, 1963

The preference for horizontal stripes may be partly due to its biotopic affinity for shores. The significance of the preference for vertical stripes on water (reed belts as shore indicators?) remains unsettled. It is interesting to note that the pattern preference changes as soon as the species reaches water.

At lower temperatures *Dichirotrichus pubescens* can swim much longer (Heydemann, 1967a). On salt water at 6°–8° C half the population was still alive after 56 days if the surface was undisturbed. At 21° C the same proportion of animals had died after only 11 days. Survival times were lower in moving salt water.

At the same temperature *D. pubescens* did not live so long on *salt water* as on fresh water. If at 21° C half the population was still alive after 12 days in moving fresh water, the entire population would already be dead under the same conditions in salt water. Despite possessing certain powers of osmoregulation (Chap. 5.B.I.) the animals were not particularly well adapted to salt water. Experiments involving flooding and swimming revealed that carabids of coastal and shore regions show a particularly high resistance to water. This resistance increases with sinking water temperature so that the autumn, winter and spring floods cause smaller losses of individuals than the summer floods. Fresh water is in general better tolerated than salt water, although amazingly long periods of swimming have been recorded, suggesting that drifting over long distances might be possible. As early as 1944 Palmén showed experimentally that the ability of insects to survive in water by swimming is inversely proportional to salt content. For this reason only few carabids were able to colonize Newfoundland from the North American mainland by the sea route, as Lindroth convincingly showed in 1963. Of 152 individuals of 21 carabid species washed up in one case, all were dead or dying. He considers that a small number of carabids might survive drifting on ice-floes. On the other hand, where the salt content is low, such as in the inner Baltic Sea, swimming could play a larger role in distribution (see Chap. 8.C.). This type of transport is also possible across large inland lakes such as Lake Erie. The colonization of oceanic islands in the tropics involves yet another type of transport: "'rafting', or drifting on objects floating on the sea surface—from small pieces of wood to veritable islands—happens most often in the wet tropics" (Lindroth, 1962).

Chapter 6

Ecological Aspects of Activity Patterns in Carabids

The preceding chapters were concerned with the "discontinuity in spatial distribution" of carabid species, to use the term coined by Strenzke (1964). However, organisms are also subject to a discontinuity in their temporal distribution. They are neither incessantly active, nor is their activity of a completely random nature: in most cases a temporal architecture can be recognized in their behaviour. Carabids in particular nearly always exhibit a highly developed rhythmicity in the form of peaks in activity which recur at regular intervals and are separated from one another by periods of inactivity.

This rhythmicity is manifest as a daily rhythm on the one hand and an annual rhythm on the other. The two periodicities are of considerable adaptive importance and are closely interconnected.

A. Daily Rhythmicity in Carabids

Daily rhythmicity in the activity of carabids has been recognized in numerous field investigations involving traps, and in laboratory studies using actographs, under both artificial and natural light conditions (Greenslade, 1963a; Grüm, 1966; Heydemann, 1967c; Kirchner, 1964; Lauterbach, 1964; Novák, 1971a, 1972, 1973; Paarmann, 1966; Scherney, 1961; Skuhravý, 1956; Thiele, 1964a; Weber, 1965d; Williams, 1959).

A comprehensive review of the entire field was published by Thiele and Weber (1968), incorporating their own results and those then available from the authors mentioned above (Table 48). Of the 67 carabids that had been investigated at that time 60% were shown to be strictly night-active and 22% predominantly day-active, whilst 18% occupied an intermediate position, behaving sometimes as diurnal, sometimes as nocturnal animals e.g. according to geographical locality. In some cases individuals exhibiting very different types of behaviour can be found side by side in one and the same population, as for example nocturnal, diurnal und indifferent *Carabus auratus* individuals in one population (Thiele and Weber, 1968; Fig. 103). During the reproductive period diurnal activity is found in *C. cancellatus* in addition to the normal nocturnal activity (Weber, 1966c, d, e; Fig. 104). In experiments with *Pterostichus angustatus* and particularly *P. oblongopunctatus* the proportion of the population exhibiting diurnal activity increased considerably with decreasing strength of illumination in the light phase (Paarmann, 1966; Figs. 105, 106). This might

Table 48. Percentages of Day Activity (DA), habitat affinity and annual rhythmicity in the carabid species so far investigated

Species	Habitat affinity	Annual rhythmicity	DA (%)			
			0–15	15–30	30–45	>45
Abax ater	F	u	+			
Abax ovalis	F	SB	+			
Abax parallelus	F	SB		+		
Agonum dorsale	O	SB	+			
Agonum ruficorne	O	u	+			
Agonum assimile	F	SB	+			
Amara lunicollis	e	SB				+
Amara communis	O	SB				+
Asaphidion flavipes	O	SB				+
Badister bipustulatus	e	SB			+	
Bembidion lampros	O	SB				+
Bembidion femoratum	O	SB	+			
Bembidion quadrimaculatum	O	SB				+
Calathus fuscipes	O	AB	+			
Calathus melanocephalus	O	AB	+			
Calathus piceus	e	AB	+			
Carabus nitens	O	SB				+
Carabus auratus	O	SB				+
Carabus cancellatus	O	SB				+
Carabus auronitens	F	SB			+	
Carabus granulatus	e	SB		+		
Carabus alpestris	O	?		+		
Carabus irregularis	F	SB		+		
Carabus hispanus	F	?	+			
Carabus rutilans	?	?	+			
Carabus glabratus	F	AB	+			
Carabus nemoralis	e	SB	+			
Carabus depressus	e	?	+			
Carabus silvestris	F	SB	+			
Carabus problematicus	F	AB	+			
Carabus purpurascens	e	AB	+			
Carabus violaceus	e	AB	+			
Carabus coriaceus	e	AB	+			
Carabus clathratus	O	SB	+			
Carabus variolosus	F	?	+			
Carabus hortensis	F	AB	+			
Carabus lusitanicus	O	?	+			
Carabus splendens	F	?	+			
Carabus morbillosus	O	?	+			
Carabus arcensis	F	SB				+
Cychrus caraboides	F	AB	+			
Dyschirius globosus	O	SB				+
Harpalus rufipes	O	AB	+			
Harpalus aeneus	O	SB		+		
Loricera pilicornis	O	SB	+			
Laemostenus terricola	?	AB	+			
Molops piceus	F	SB		+		
Molops elatus	F	SB		+		
Pterostichus cupreus	O	SB				+
Pterostichus vulgaris	e	AB	+			

Table 48. (continued)

Species	Habitat affinity	Annual rhythmicity	DA (%) 0–15	15–30	30–45	>45
Pterostichus madidus	e	AB		+		
Pterostichus angustatus	F	SB	+			
Pterostichus oblongopunctatus	F	SB			+	
Pterostichus coerulescens	O	SB				+
Pterostichus niger	e	AB	+			
Pterostichus cristatus	F	AB	+			
Pterostichus metallicus	F	SB	+			
Pterostichus nigrita	e	SB	+			
Nebria salina	O	AB	+			
Nebria brevicollis	e	AB	+			
Notiophilus substriatus	e	SB				+
Notiophilus rufipes	e	SB				+
Notiophilus biguttatus	e	SB				+
Stomis pumicatus	O	SB	+			
Trechus quadristriatus	O	AB	+			
Trichotichnus laevicollis	F	AB		+		
Zabrus tenebrioides	O	AB	+			
Species of the open countryside (total: 24)			12 50%	2 8%	0	10 42%
Eurytopic species (total: 17)			10 59%	2 12%	1 6%	4 24%
Forest species (total: 20)			12 60%	5 25%	2 10%	1 5%
Autumn breeders (total: 21)			19 90%	2 10%	0	0
Spring breeders (total: 36)			12 33%	6 17%	3 8%	15 42%

O: species of the open countryside; F: forest species; e: eurytopic species; SB: spring breeders; AB: autumn breeders; u: unstable breeding habits. In the case of *C. nemoralis, clathratus, arcensis, violaceus, L. pilicornis* and *P. madidus* only DA values from central European populations were used. Where various DA values were obtained for the other species, means were calculated. In species where the habitat affinity and reproductive phase varied geographically, the conditions prevailing in the populations investigated were used. The mediterranean species *C. morbillosus, lusitanicus, splendens, rutilans* and *hispanus* are not included in the analysis of daytime activity according to habitat affinity and seasonal activity (at the bottom of the table)

explain the widely differing figures given by various authors for the extent of daytime activity, particularly in *P. oblongopunctatus.*

Periodicity in larval activity has been little investigated. Whereas the adults of *Carabus problematicus* are strictly nocturnal, van der Drift (1951) caught larvae during the daytime. His material is small, however. Heydemann (1967c) found the majority of carabid larvae to be nocturnal, with the exception of those of *Nebria brevicollis* (adults strictly nocturnal) and *Notiophilus substriatus* (adults also diurnal: according to Greenslade, 1963a).

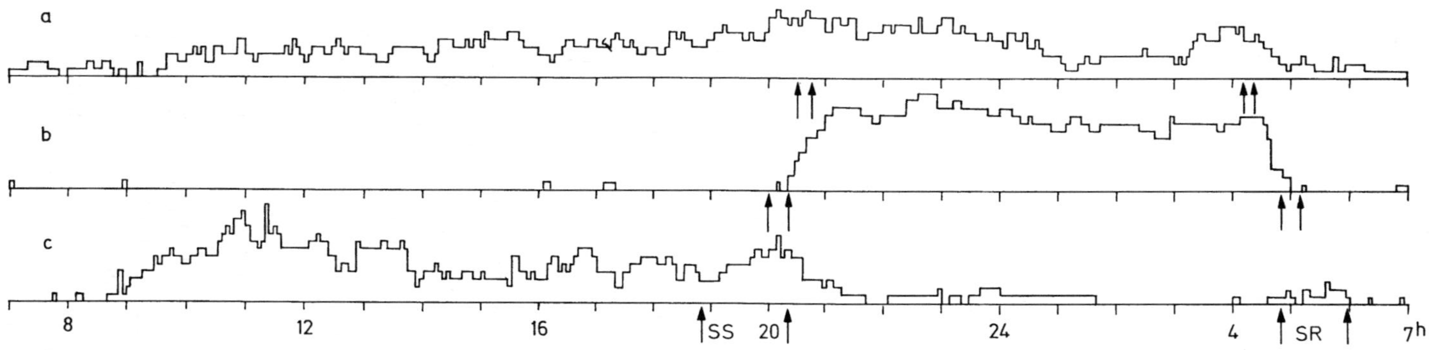

Fig. 103a–c. Activity diagram for three single individuals of *Carabus auratus*, recorded in the laboratory using natural light. (a) percent daytime activity 28 %, (b) 3 %, (c) 79 %. *SR:* sunrise; *SS:* sunset. *Arrows* indicate the region over which *SR* and *SS* altered during the course of the experiment. From Thiele and Weber, 1968

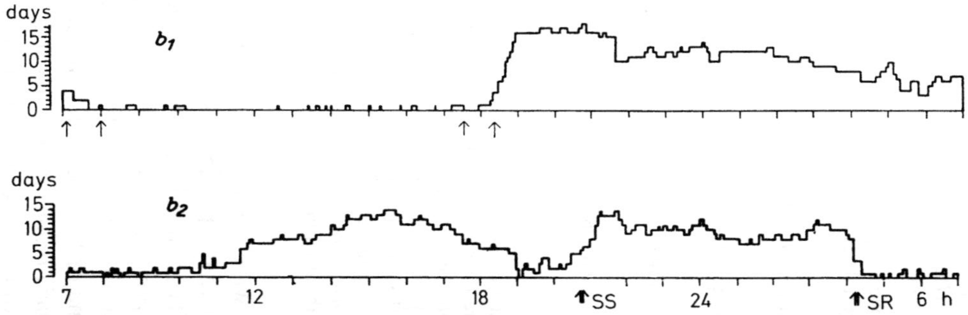

Fig. 104. Activity of *Carabus cancellatus* during the course of the day, registered in the laboratory under natural light conditions. *(b_1)* recording for 27 days from 12 Feb.–8 Mar.; *(b_2)* recording for 15 days from 24 Jun.–8 Jul.: additional daytime activity during reproductive period. *Arrows* indicate regions over which *SR* and *SS* altered during the course of the experiments. From Weber, 1966c

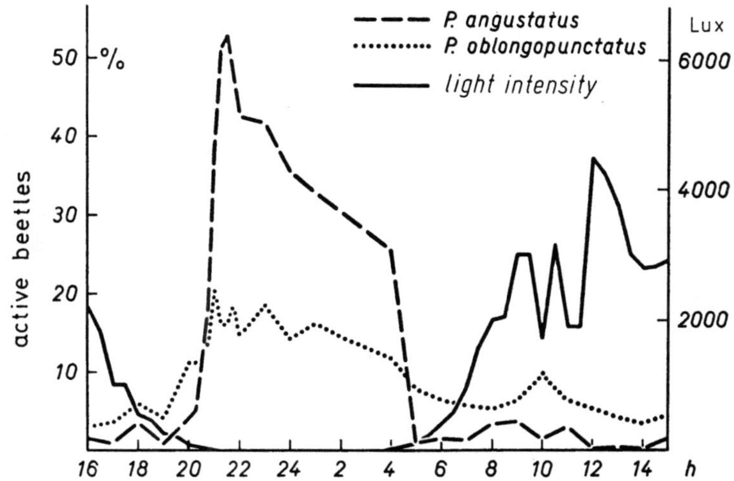

Fig. 105. Activity of *Pterostichus angustatus* (---) and *P. oblongopunctatus* (····) during the course of the day, both species kept under the same light conditions. (——): light intensity curve. From Paarmann, 1966

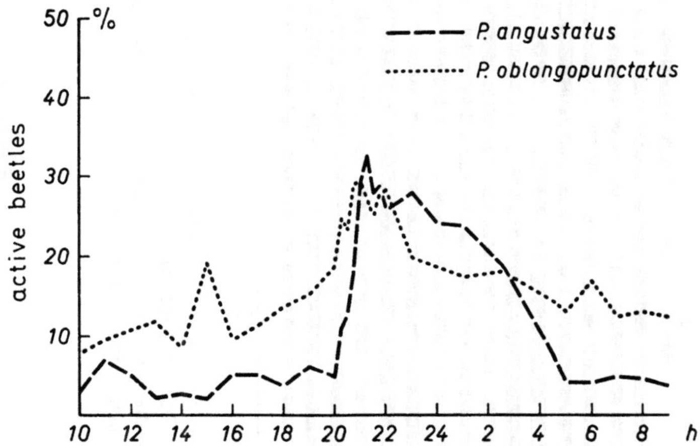

Fig. 106. Activity of *Pterostichus angustatus* and *P. oblongopunctatus* during the course of the day in experiments in which *P. oblongopunctatus* received only very weak light. From Paarmann, 1966

I. The Control of Daily Rhythmicity by Endogenous Factors

Like almost all other organisms in which it has been studied, carabids, too, exhibit circadian rhythmicity in their running activity. This means that they show a spontaneous frequency differing slightly from the 24-h rhythm in the rotation of the earth, for which the term "internal clock" has become accepted. By means of this internal clock, organisms are prepared for activity even before the external conditions essential for activity prevail (e. g. obligatory species-specific

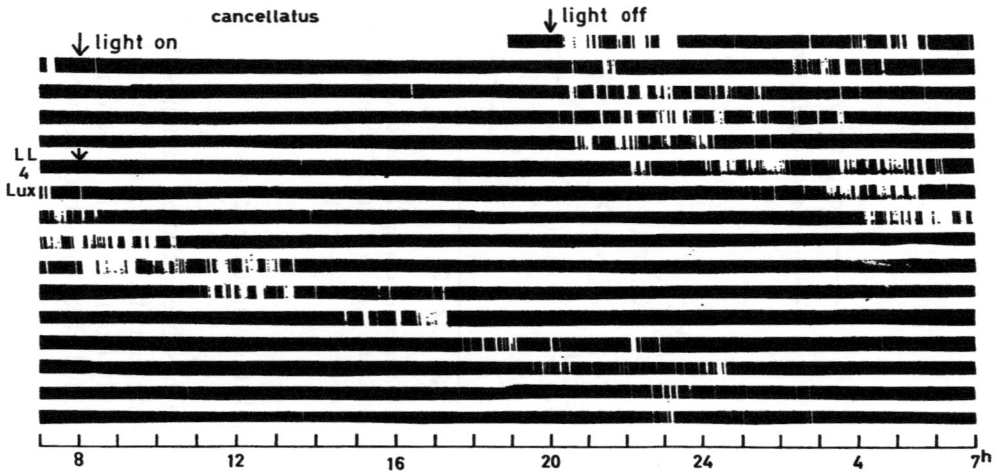

Fig. 107. Running periodicity of a *Carabus cancellatus* individual kept in alternating 12:12 h light–dark followed by continuous light (4 Lux=LL). Smoked paper recordings. *Light bands* represent the bursts of activity. In alternating light–dark the period of activity amounts to an average of 24 h whereas it is longer than 24 h in continuous light. From Weber, 1970

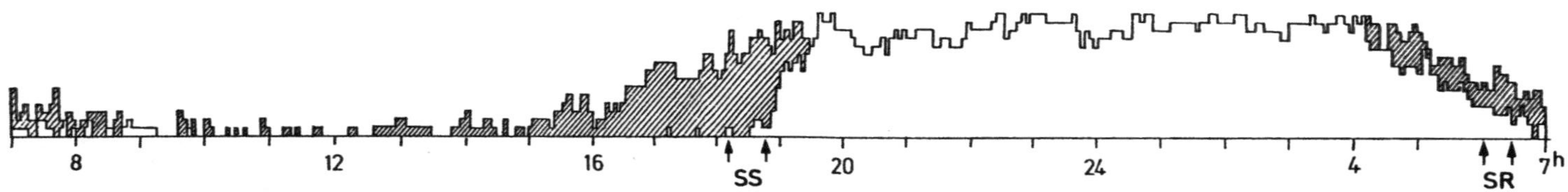

Fig. 108. "Preactivity" in the evening and "postactivity" in the morning of a group of *Carabus problematicus* (five females, five males) registered in the passage leading to a dark hide-out *(hatched)*. *White:* activity registered in the actograph. Recordings from 16 days. *Arrows:* displacement of sunrise and sunset during the course of the experiment. From Thiele and Weber, 1968

light conditions), and are thus in a position to exploit these conditions as soon as they obtain. This endogenous rhythm is termed "circadian", the length of its period being only approximately one day (lat. "circa diem"). It can be compared with a clock that gains or loses slightly each day and therefore has to be reset continually. This is achieved in the case of the organism by means of external stimuli which effect a daily resynchronisation. Such external stimuli are called zeitgebers, the most important being the daily change of light and darkness (LD).

Under constant conditions, particularly of light (constant darkness-DD, or constant light-LL), the spontaneous frequency becomes independent of the periodicity of the earth's rotation. The onset of activity is slightly advanced or retarded daily and a so-called "free-running periodicity" is established (Fig. 107)[13].

The first demonstration of a circadian rhythmicity in carabids was achieved by Kirchner (1964) for *Carabus purpurascens*. Quantitative measurements of the spontaneous frequencies for *Carabus problematicus*, *C. cancellatus* and *C. nitens* were made over a maximum of 36 days (Weber, 1967, 1970). The period length for the dark-active species *C. problematicus* and *C. cancellatus* is less than 24 hrs in DD. In keeping with this finding is the observation that *C. problematicus* under experimental conditions exhibits a measurable "preactivity" before onset of darkness. This is clearly a manifestation of the "preparedness for activity" mentioned above. Although at first suppressed by light (masked), activity shoots up rapidly after sunset (Thiele and Weber, 1968; Fig. 108). Circadian rhythmicity is not measurable in every individual of a carabid population, a question that has been studied quantitatively in some detail on *Pterostichus nigrita* (Leyk, 1975). Of 98 individually investigated specimens 50% showed a well-developed circadian rhythm in DD. The period length in this nocturnal species was almost invariably less than 24 h (22–24 h), and in only two of the individuals was it greater than 24 h. Fifteen per cent of the animals were so inactive in the actograph that no value could be obtained, and 35% were aperiodic.

Many arguments speak in favour of regarding biological rhythms in most cases as spontaneous oscillations. In the majority of organisms the rhythms cannot adapt to all zeitgeber frequencies, but are simply "entrained" by those that do not deviate too widely from the organism's own frequency.

Circadian rhythmicity in the carabids so far investigated, as compared with those of other organisms, is characterized by certain peculiarities that point to an extreme plasticity of the endogenous rhythms with respect to exogenous zeitgebers. One of these is the extraordinary breadth of the region over which spontaneous periodicity can be entrained by zeitgeber frequencies widely deviating from the 24-h rhythm. *Pterostichus vulgaris*, *P. niger* and *Harpalus rufipes* adapt well to such rhythms, *Pterostichus madidus* and *Carabus problematicus*, however,

[13] Older investigations of Brehm and Hempel (1952) on *Pterostichus vulgaris* and of Williams (1959) on *Pterostichus madidus* led to the statement that an exact 24-h rhythm persists under constant conditions. William's graphs, however, permit an interpretation of the data as being based upon a circadian periodicity deviating slightly from the 24-h rhythm. Brehm and Hempel gave neither values nor graphs. Aschoff (1960) comments: "In this case one cannot exclude that an overlooked or unknown periodic factor of the environment was effective as Zeitgeber".

adapt less accurately (Kirchner, 1964; Fig. 109). Recent studies have revealed that the periodicity in activity of *C. problematicus* can be entrained by LD 8/8 (Thiele and Weber, 1968) and that of *C. cancellatus* by LD 8/8 and 16/16 (Lamprecht and Weber, 1970, 1971; Fig. 110). Carabids also adapt rapidly to phase reversal of the zeitgeber, e. g. exchanging the light and dark periods in LD 12/12. *Carabus problematicus* exhibited rapid resynchronization after such a phase reversal, one or two transitional periods being necessary at 75–250 lux and three to four at 0.3–12 lux (Weber, 1966). Kirchner (1964) found that resynchronization following phase reversal of the zeitgeber was achieved after one or two transitional periods in *Carabus nemoralis*, *Pterostichus vulgaris*, *P. niger* and *Harpalus rufipes*, and after two or three in *Abax ater*.

Since it is impossible in cockroaches, for example, and still less in vertebrates, to bring about resynchronization by means of LD 8/8 the question arises as to whether the extreme plasticity in the endogenous periodicity of carabids

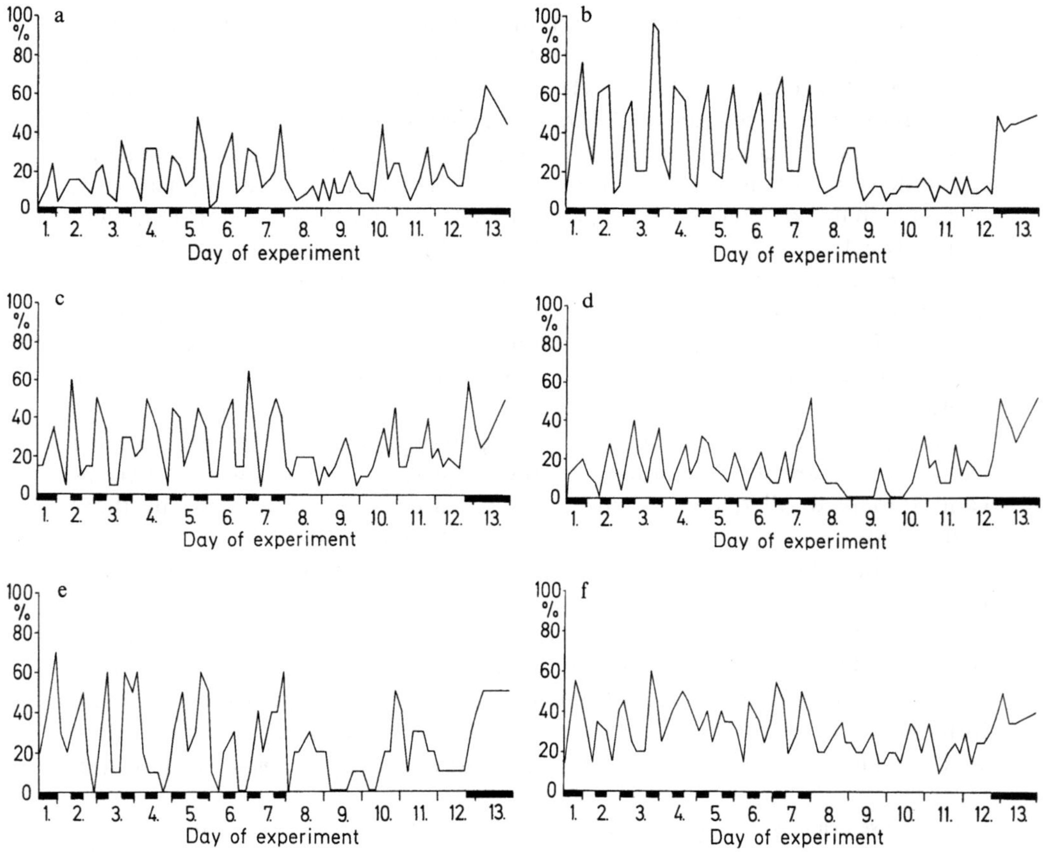

Fig. 109 a–f. Activity of carabids in an artificial 16-h day (LD 8/8) at constant temperature. *Black bands* indicate: darkness. (a) *Pterostichus vulgaris* (25 individuals); (b) *P. niger* (25 individuals); (c) *Abax ater* (20 individuals); (d) *Harpalus rufipes* (25 individuals); (e) *Carabus problematicus* (25 individuals); (f) *P. madidus* (20 individuals). *Ordinate:* percent active individuals. From Kirchner, 1964

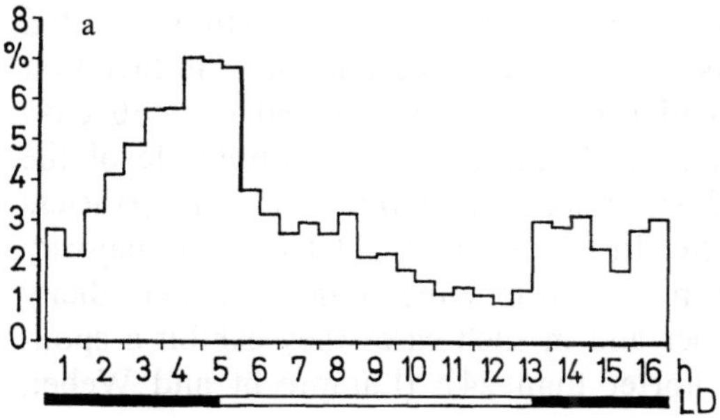

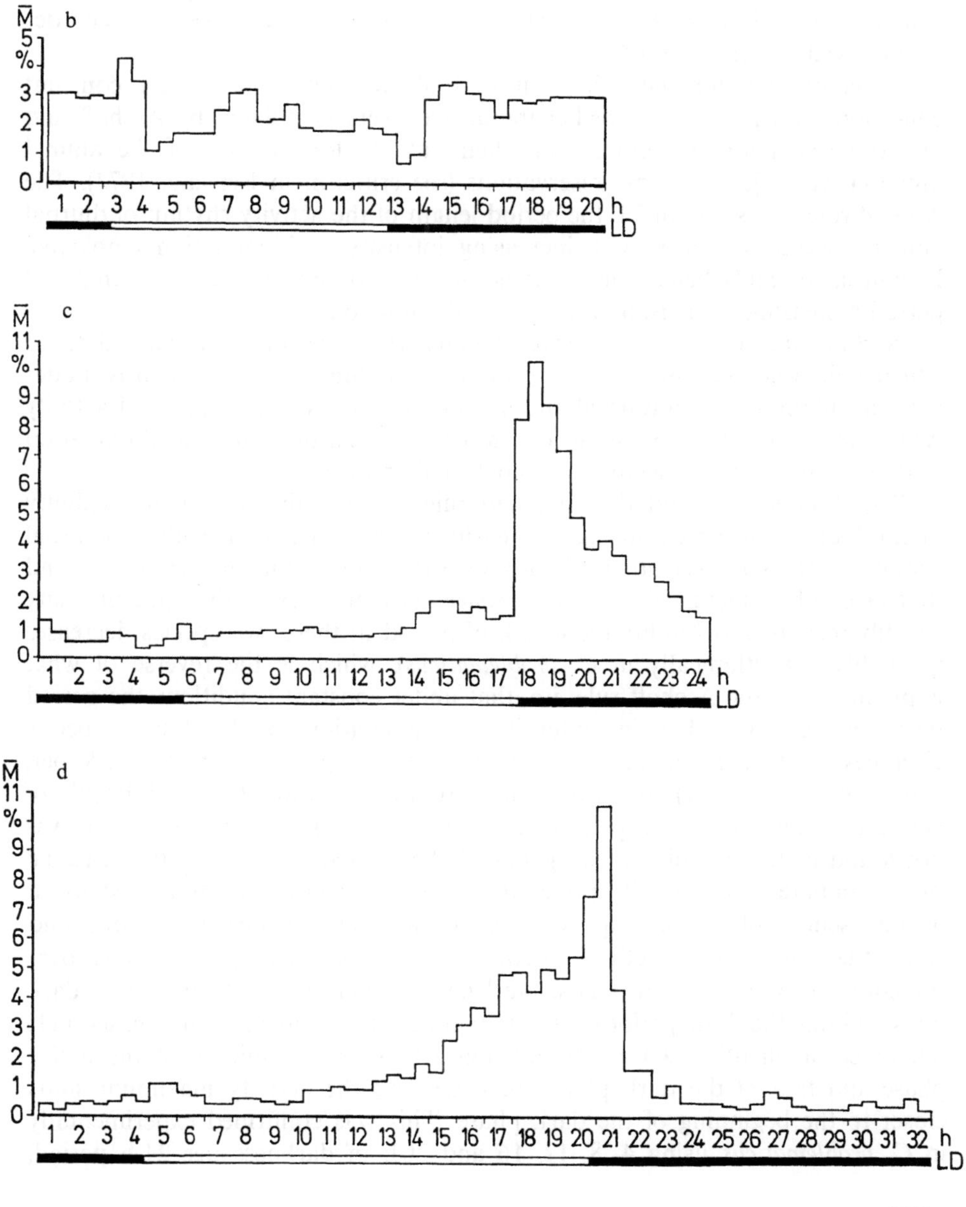

Fig. 110a–d. Averaged course of activity of *Carabus cancellatus* in different photoperiods. (a) LD 8/8; (b) LD 10/10; (c) LD 12/12; (d) LD 16/16. *Ordinate*: activity at the hour in question as percent of total activity. From Lamprecht and Weber, 1971

involves self-sustained oscillations or whether the circadian rhythm detected under constant conditions represents a free damped oscillation. The fact that no damping of the magnitude of oscillation could be observed over 36 days might also have been due to repeated reinforcement of the amplitude of the oscillations by aperiodic stimuli. Self-sustained stimulation could nevertheless be demonstrated in *Carabus cancellatus*. In LD 8/8 and LD 16/16 the majority of individuals investigated are synchronized, but in some animals the periodicity is dissociated from that of the zeitgeber and in such cases they exhibit a spontaneous periodicity that is slightly shorter than 24 h (Lamprecht and Weber, 1970).

During the main active phases shorter bursts of activity lasting about 30 min can be registered in *Carabus purpurascens*, if a sufficiently sensitive time recorder is employed (Kirchner, 1964).

A set of rules for the behaviour of biological oscillators under constant conditions and rhythmic zeitgeber frequencies were postulated by Aschoff and his co-workers for vertebrates, but their validity for the rest of the animal kingdom is subject to some reservations (see criticism in Rensing, 1973). The Aschoff rule states that in LL the period length of the activity rhythm of diurnal animals becomes shorter with increasing intensity of illumination employed. Nocturnal animals behave in a diametrically opposed manner, the length of period increasing with rising intensity of illumination.

Aschoff and Wever (1963) have formulated a circadian rule according to which still other parameters of the activity rhythm behave differently under constant illumination in diurnal and nocturnal animals: the quantity of activity A and the ratio of active to resting hours $\alpha : \varrho$ increases with strength of illumination in diurnal animals and decreases in nocturnal animals.

The Aschoff rule and the circadian rule are of only limited applicability in carabids. With increasing light intensity the period length of the nocturnal species *Carabus problematicus* already exhibits a maximum at 1.5 lux (!) and that of another nocturnal species, *C. cancellatus*, at 4 lux. If the light intensity steadily increases to 250 lux the length of period in these two species decreases (probably logarithmically in *C. problematicus*), which is the reverse of what is predicted by the Aschoff rule, i. e. that under constant conditions the period increases steadily with rising intensity of illumination. In the diurnal species *C. nitens*, no increase in period is observed when light intensity rises (Weber, 1967, 1968 a; Fig. 111). A correlation with the α/ϱ ratio (ratio of length of active to length of resting period) in accordance with the circadian rule was not found in the carabids investigated (Weber, 1968 a). However, the seasonal rule formulated by Aschoff applies to *C. problematicus*. This is an extension of the Aschoff rule to encompass the situation under zeitgeber conditions, and states that the period length of nocturnal species shortens progressively over the course of the year with increasing length of night (i. e. with increased dark phase within the 24-h period). This is manifest in an increase in phase-angle difference or, in other words, the difference between the middle of the active phase and that of the dark phase increases and the activity maximum shifts closer to the beginning of the dark phase. This was confirmed experimentally in *C. problematicus* using 4, 8, 12, 16 and 20 h of darkness per 24-h period,

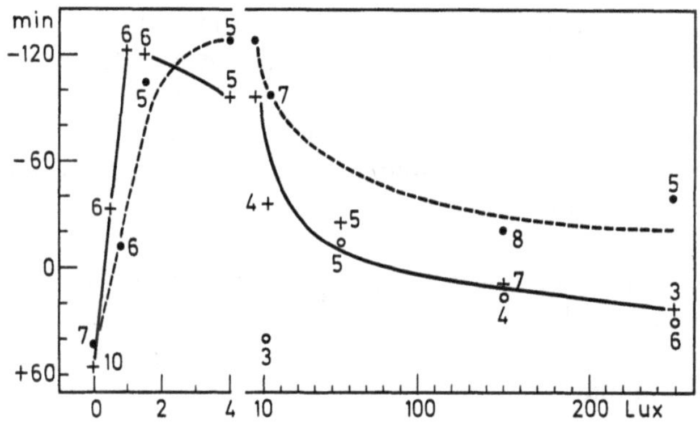

Fig. 111. Dependence of period length of daytime activity (ordinate given as difference $24-\tau$) of 3 *Carabus* species under constant light conditions, on light intensity in Lux (*abscissa:* note change of scale between 4 and 10 Lux): +——: *C. problematicus;* ●---: *C. cancellatus;* o: *C. nitens* (three values not linked up to a curve). From Weber, 1968a

although the individual values obtained showed a wide scatter (Thiele and Weber, 1968).

In experiments in the field an activity maximum could be observed in many nocturnal carabids soon after the onset of darkness (Vlijm *et al.*, 1961; Lauterbach,

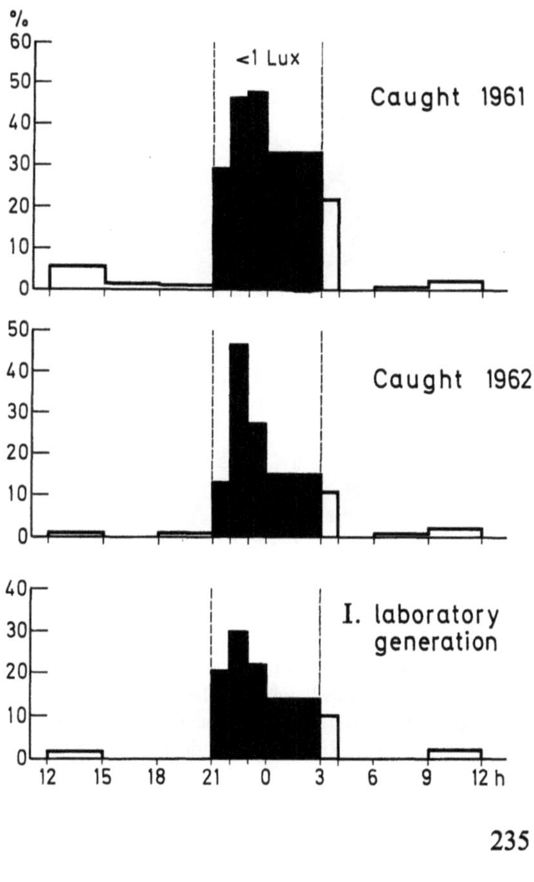

Fig. 112. Patterns of activity for three different populations of *Pterostichus nigrita* in field experiments. Length of experiment 23 May–7 Jun. *Ordinate:* percent active animals. From Thiele and Weber, 1968

1964; Paarmann, 1966; Thiele and Weber, 1968; Fig. 112). The maximum occurs 1–2 h after sunset and its displacement over the course of the year runs parallel with the time of sunset. Although in the open a decline in activity towards midnight parallels the decreasing temperature, a similar decline is also seen in *C. problematicus* under constant temperature conditions in the laboratory (Thiele and Weber, 1968), which suggests the involvement of an endogenous component.

Little is known about hormonal processes in controlling the activity rhythm of carabids. Klug (1958) found a daily periodicity in volume of the corpora allata in *Carabus nemoralis*, with a maximum at 5 a.m. and no well-defined minimum.

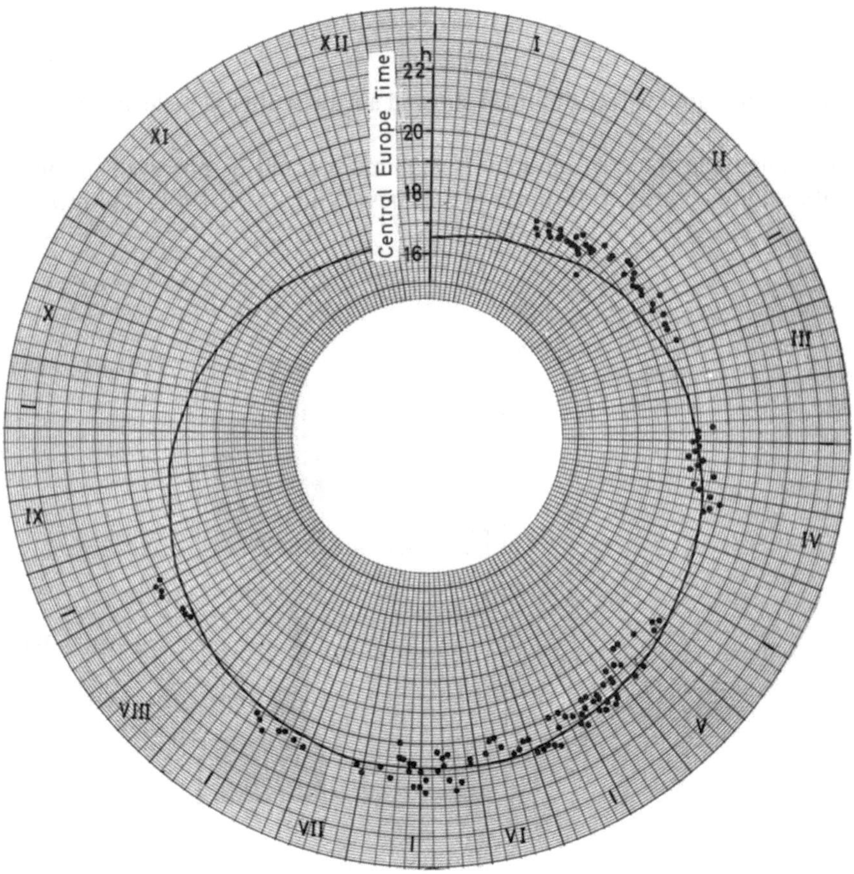

Fig. 113. Displacement of time of sunrise *(continuous line)* and onset of nocturnal phase of activity *(dots)* in *Carabus cancellatus* over the course of the year, represented in a polar-coordinate system. Above, hourly coordinate. *Roman numerals* indicate months. On the whole, onset of activity shifts with time of sunset. Recordings made in the laboratory under daylight conditions. From Thiele and Weber, 1968

II. The Role of Exogenous Factors as Zeitgebers

The most important zeitgeber for periodicity in activity is light. An indication of this can be seen in the correlation between light and dark activity on the one hand and light preference on the other, as well as in the rapid adaptation to phase shifts in the period of illumination and the entrainment of periodicity by an L:D ratio deviating widely from the 24-h day (see preceding section).

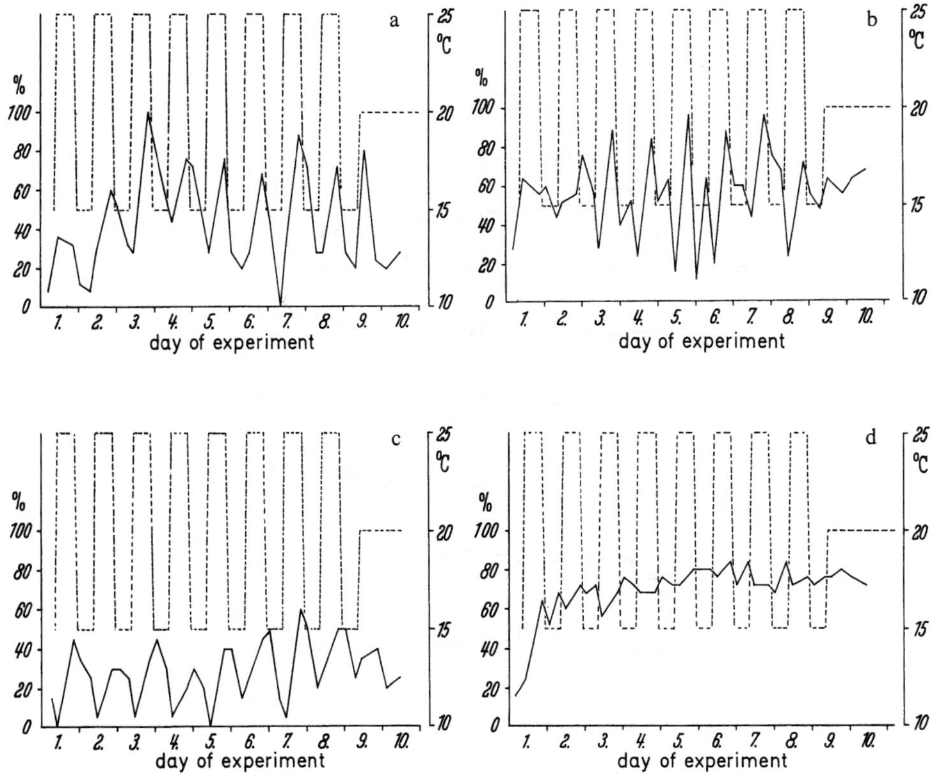

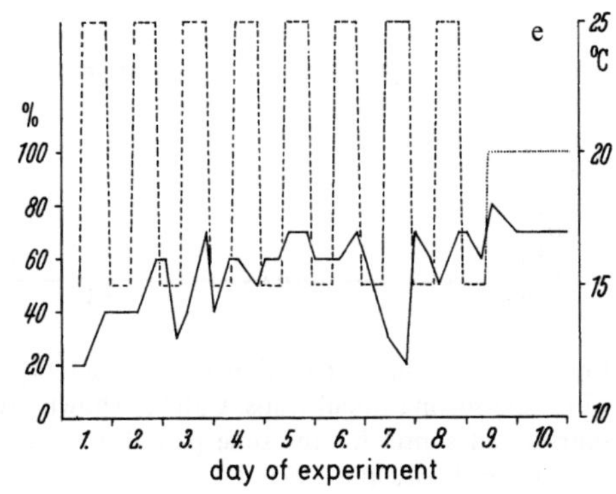

Fig. 114a–e. Activity of carabids in constant darkness with 24-hour temperature changeover. ——: Activity; -----: temperature. (a) *Harpalus rufipes*, 25 individuals; (b) *Pterostichus vulgaris*, 25 individuals; (c) *Abax ater*, 20 individuals; (d) *P. niger*, 25 individuals; (e) *Carabus problematicus*, 20 individuals. *Left ordinate:* percent active individuals. From Kirchner, 1964

As would be expected, the onset of activity of dark-active species changes with time of sunset, as Lauterbach (1964) was able to demonstrate for numerous species, especially *Abax ater*, in field experiments. The activity maximum invariably occurs 2h after sunset. In laboratory experiments under natural light the onset of nocturnal activity in *Carabus cancellatus* shifts in parallel with the time of sunset from January to August (Thiele and Weber, 1968; Fig. 113). The daily temperature course is of lesser importance as a zeitgeber but its role apparently varies according to habitat. A 24-h temperature rhythm under DD induced no rhythmicity in experiments with the forest species *Pterostichus niger* and *Carabus problematicus*, but a weak rhythmicity could be invoked

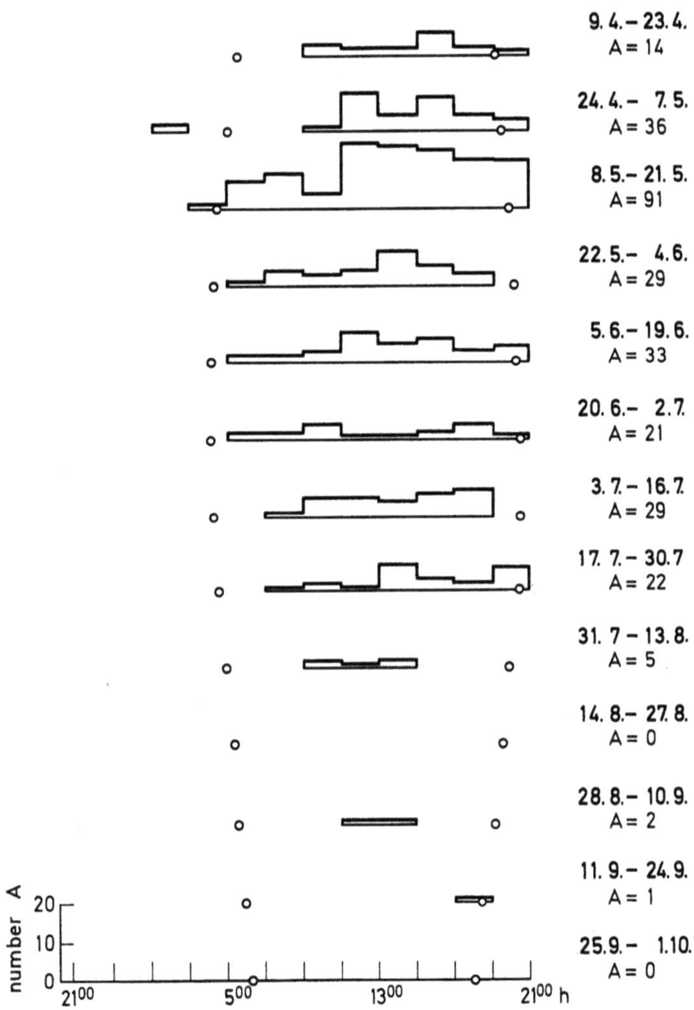

Fig. 115. Seasonal distribution of daily activity in *Notiophilus biguttatus*, based on catches made with time-sorting pitfall traps. Catches shown are for two weeks, at 2-h intervals. *Circles* indicate sunrise and sunset for the time period given on the right. *A:* number of animals caught in this time. From Kasischke, 1975

in the strongly eurytopic forest species *Abax ater* and a pronounced one in the field species *Pterostichus vulgaris* and *Harpalus rufipes* (Kirchner, 1964; Fig. 114). Desert carabids from Tunisia *(Anthia venator)* exhibit activity maxima at the temperature minima in DD under fluctuating temperature conditions but are nevertheless governed by light changes in LD and are then dark-active (Cloudsley-Thompson, 1956).

The purely diurnal species *Notiophilus biguttatus* exhibits a maximum in activity at midday and in the afternoon hours, more or less in keeping with the course of the daily temperature curve (Kasischke, 1975). From April until May, with generally increasing temperature, the time at which activity commences each day shifts from the morning towards sunrise, so that during the reproductive phase from May to June the period of activity is particularly long. Later in the year it shortens again by beginning later (Fig. 115). At a constant high temperature (20° C) in the laboratory, on the other hand, actograph experiments revealed that the animals reacted immediately to "light on" with enhanced activity, because they did not have to wait for the advancing day to bring a warming-up (Kasischke, 1975).

III. The Importance of Daily Rhythmicity in Habitat-Binding

The distribution of activity patterns in carabids differs considerably according to environment. Forest species are predominantly night active whereas a large proportion of the field species is diurnal (Greenslade, 1963a; Thiele, 1964a). A summing-up of the available data reveals the following (Thiele and Weber, 1968): 60 % of the forest species are nocturnal and only 5 % diurnal, whereas 42 % of the field species are distinctly diurnal and 50 % nocturnal. The preponderance of night-time activity in forests or in any denser type of vegetational cover is confirmed by a study of the numbers of individuals. In oak-hornbeam forests only 8 % diurnal individuals were found (n=2220), and 27 % (n=322, Lauterbach, 1964) in adjacent young plantations.

In a newly planted and therefore thin forest a relatively high proportion of predominantly diurnal animals (13 out of 38 species = 34 %) was found, particularly members of the genera *Amara*, *Notiophilus* and *Dyschirius* (Kasischke, 1975).

The predominance of night-time activity in forests is probably connected with, among other factors, the high moisture requirements of the majority of forest carabids (see Chap. 5.A.IV.). In coastal grazing land on the North Sea coast with sparse vegetation and strong nocturnal radiation daytime maxima in the activity of soil arthropods, based upon individual density, are found (with two peaks in the distribution of activity: in August/September maxima between 10 and 13 h and between 15 and 17 h). In ungrazed, more elevated salt meadows, on the other hand, a nocturnal maximum can be observed (22–24 h). In contrast to the findings for individual density, however, the largest activity biomass is also found at night on pastureland, since mainly the largest forms are nocturnal (e. g. *Pterostichus niger*, *P. vulgaris*, *Harpalus rufipes*, Heydemann, 1967).

Using time-sorting pitfall traps the existence of different daily patterns of activity for forest and field dwellers could be established, especially from the work of Novák (1971, 1972, 1973, 1974). His results confirm the predominance of daytime activity in field habitats and of nocturnal activity in forest habitats. In both cases the duration of daytime activity over the course of the year varies with the species spectrum. In both forest and field the daytime activity of carabids is higher in the spring, although the seasonal difference is far greater in fields. In the latter, too, nighttime activity dominates in the autumn. In the spring months activity in fields in the light phase totals 2.2 times that in the dark phase whereas in the autumn months it is 7.5 times larger at night than in the daytime (Novák, 1972). An attempt to interpret the differences in time of day at which activity occurs in field and forest biotopes will be made at a later stage (see Chap. 7).

The nocturnal pattern of activity of dark-active species may also exhibit connections with the type of habitat involved. In *Nebria brevicollis* activity maxima were found on the Dutch island of Schiermonnikoog in the early part of the night (20–22 h), although they were much smaller on sandy areas and sedgy regions than in a marshy area with taller vegetation. In more open countryside the activity of *P. niger* is rather uniformly distributed throughout the night. In *Calathus melanocephalus*, however, activity before midnight is much more pronounced on sandy areas than in sedge vegetation (Vlijm, 1960; Fig. 116). The maximum in the early part of the night, typical of forest carabids, was found by Kirchner (1960) to be less well developed in *Pterostichus vulgaris* and *Trechus quadristriatus* on cultivated fields. Instead, activity declines very gradually after midnight and only drops suddenly at sunrise (Fig. 117).

Differences in the time of day at which forest and field populations of one and the same species are active have been observed in the eurytopic species *Pterostichus madidus*. Whereas forest populations of this species are almost without exception night-active the field inhabitants are additionally active in the daytime (Williams, 1959). This is interpreted by Williams as resulting from differences in availability of prey, but it might also be due to an escape reaction in the face of excessive insolation, as is strongly suggested by experiments involving this species (Thiele and Weber, 1968). Inherited differences in behaviour between the two populations cannot, however, be excluded. "*P. madidus* from woodland were nocturnal in an insectary while grassland individuals showed initially more diurnal activity. Subsequently under the woodland type of conditions in the insectary the grassland specimens assumed the same rhythm as those from woodland. That in this case it was not a simple response to temperature or humidity, and that an inherent rhythm was present is shown by the fact that a week was required for the two habitat groups to achieve the same daily activity pattern" (Greenslade, 1963).

Annual and daily rhythms are apparently very closely connected. Ninety per cent of the species reproducing in autumn are nocturnal as compared with only 33 % of those reproducing in spring, and whereas diurnal species are completely lacking among the autumn breeders they account for 42 % of the spring animals (Thiele and Weber, 1968). In an investigation on spring breeders in a complex of moist woodland and open habitats an average of 70 % of the

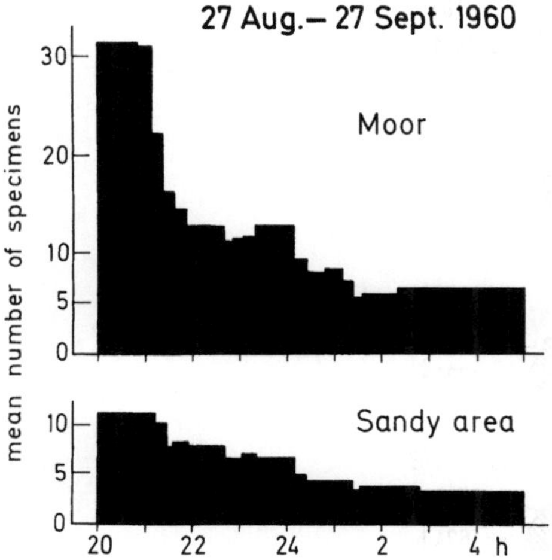

Fig. 116

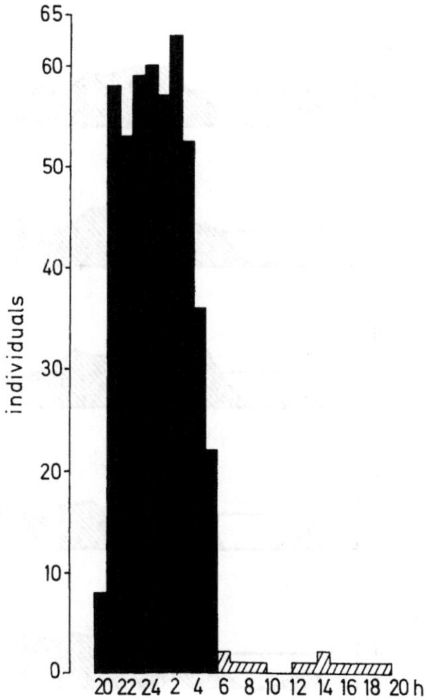

Fig. 117

Fig. 116. Activity of *Nebria brevicollis* during the night between 8 p.m. and 5 a.m., on a moor *(above)* and a sandy area *(below)*. *Ordinate:* mean number of animals caught per trapping interval. From Vlijm, 1960

Fig. 117. Number of *Pterostichus vulgaris* caught on a vegetable field in 24h. Night activity *black*, daytime activity *hatched*. From Kirchner, 1960

individuals of 16 species from open areas were found to be diurnal as compared with only 27.7% of the individuals of 13 forest species. Among the autumn breeders an average of 30.7% of the individuals of seven species from open land were diurnal as compared with only 9.6% of the forest species (Schiller and Weber, 1975). Apparently it is the warmth-loving spring breeders here that exhibit most daytime activity. An interpretation will be attempted on p. 281ff.

Little is known about the geographical distribution of different types of daily rhythms. Nocturnal activity seems to dominate in Mediterranean regions, even in species living in open country, although little material is so far available (Thiele and Weber, 1968). It is possible that the daytime hours are simply too warm for carabids to be active, since Rensch (1957) also found in the tropics (India, Egypt) that even the extremely thermophilic Cicindelinae are inactive for six or seven hours during the warmer part of the day and confine their activity to the morning and late afternoon. Under arctic conditions species known to be nocturnal in central Europe are diurnal, even at strong light intensities, in the face of the constant daylight of midsummer, an example being

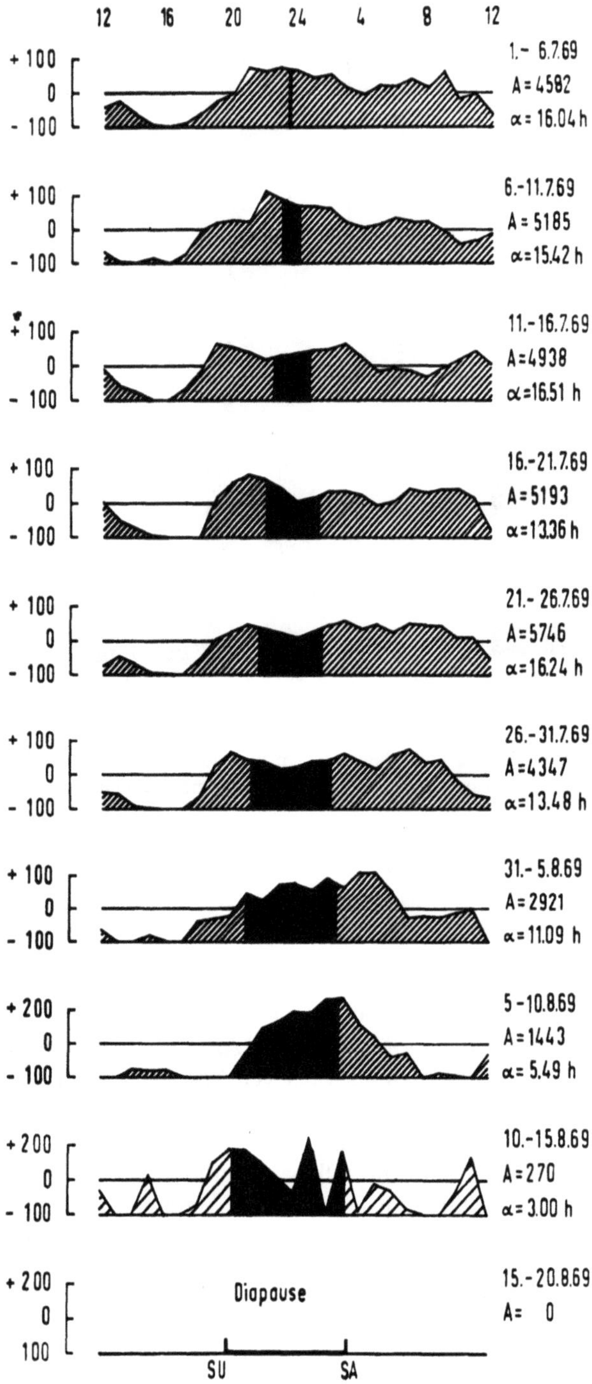

Fig. 118. Pattern of activity of a male *Carabus glabratus* in subarctic Sweden during the summer months, represented as a five-day total curve. *Ordinate:* percent deviation of the hourly value of activity from the 24-h mean. *A:* number of recordings in five days. α = mean duration of activity in 5 days. The interval between sunset and sunrise is shaded *black*. From Neudecker, 1971

provided by *Carabus glabratus* (Neudecker, 1971). A daily rhythmicity is nevertheless retained and in experiments a minimum is invariably observed when the light intensity is at its highest (Fig. 118). Arhythmic behaviour in individuals of *Carabus violaceus* originating far to the north was reported by Hempel and Hempel (1955) but could not be confirmed by Neudecker (1971). Although their rhythmicity is not so pronounced in the polar midsummer as in populations from temperate latitudes a distinct maximum in activity nevertheless is retained at the darkest hours of the day. Following midsummer, and with the lengthening nights, a type of rhythmicity, in which activity is confined to the actual nighttime, makes itself increasingly obvious even in the far north (Fig. 119).

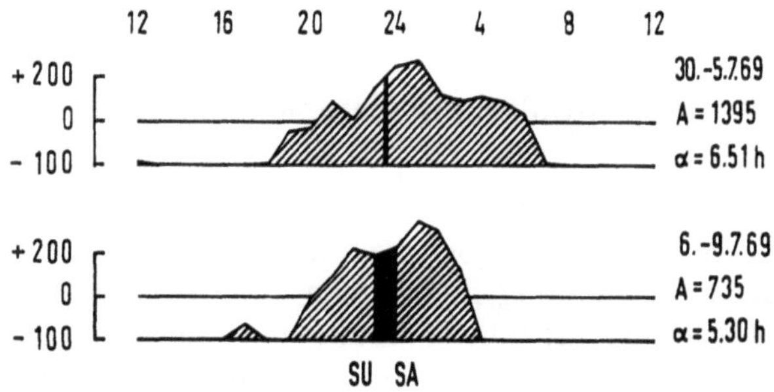

Fig. 119. Periodicity of activity of a male *Carabus violaceus* in subarctic Sweden in summer, given as five- and three-day total curves. Explanation see Figure 118. From Neudecker, 1971

A particularly interesting aspect is presented by the daily rhythms of cave-dwelling carabids. In other cave-dwellers such as diplopods, amphipods and fishes a highly disrupted circadian rhythm or aperiodic type of behaviour can be observed in experiments in DD. A comparison of the troglobiontic *Laemostenus navaricus* with the merely troglophilic species *L. terricola* and *oblongus* (Lamprecht and Weber, 1975) revealed that in DD the exclusively cave-dwelling *L. navaricus* exhibited a less well developed circadian rhythmicity than the other two species that are occasionally also encountered outside caves (Table 49). It is interesting

Table 49. Differences in daily rhythmicity in cave-dwelling carabids of the genus *Laemostenus*. (From Lamprecht and Weber, 1975)

Species	Experiment	Number of active animals	Percentage animals showing circadian periodicity
L. terricola	DD	26	100%
L. oblongus	DD	25	100%
L. navaricus	DD	42	71%
L. terricola	LL	59	75%
L. oblongus	LL	29	90%
L. navaricus	LL	28	100%

that in very weak DD, however, *L. navaricus* exhibited a more pronounced circadian rhythmicity than the other species. The authors regard this as indicating a special plasticity of the troglobiontic species by means of which it is not only well adapted to the interior of caves but also to the regions near their entrance. Well inside the cave an aperiodic pattern of activity, in which short bursts of activity alternate with periods of inactivity, is probably more effective than a circadian pattern of behaviour for catching prey that is equally aperiodic in its activity. Periodic activity is held to be of selective advantage in the neighbourhood of the cave entrance, in that it represents an adaptation to the daily pattern of abiotic factors and the periodic behaviour of prey and predators in this locality.

IV. Does Moulting in Carabids Exhibit a Daily Rhythmicity?

The daily rhythms in activity so far discussed can be demonstrated in single animals. Emergence of the adult from the pupa, on the other hand, is an individual event. A moulting rhythm can only be demonstrated therefore, if, as a result of recording the time of moulting in a large number of members of a population, the individuals are seen to moult mainly or without exception at definite, narrowly limited times of day (Remmert's "moulting at fixed time of day").

The adaptive significance of such a rhythm in short-lived species is that it favours encounters between the two sexes by temporally concentrating the moulting. Another possible advantage is that it ensures that the insects emerge at a time of day at which climatic conditions are most favourable.

Moulting rhythmicity is widespread in insects. According to a survey made by Remmert in 1962, 224 species from various orders exhibited a moulting pattern, whereas only 18 species showed aperiodic moulting. Up to that time studies on moulting rhythmicity in the insect order richest in species, the Coleoptera, had not been carried out. This prompted Thiele and Paarmann (1968) to investigate whether such a rhythm could be demonstrated in carabids.

The use of "moulting clocks" is not practicable in working with carabids since the young adults remain in the pupal cradle for at least a few days. The pupae were therefore placed in transparent plastic boxes divided into compartments and were photographed hourly with a recording camera. The distribution of moulting over the course of the day could then be read off from the film strip. LD light conditions were employed (18:6). Flash photography was used during the dark phase but in some experiments recording was interrupted in order to avoid disturbance due to the flash. In such cases only the total number of beetles that had emerged during the dark period could be counted from the first photograph taken following "light on" in the morning.

In four species rather more beetles emerged per hour in the light phase than in the dark phase whatever the conditions employed. Nevertheless, this difference could only be statistically confirmed in one species (*Pterostichus angustatus*) where large numbers of individuals were employed (Fig. 120). The characteristic moulting peak seen in so many insects at a particular time of day did

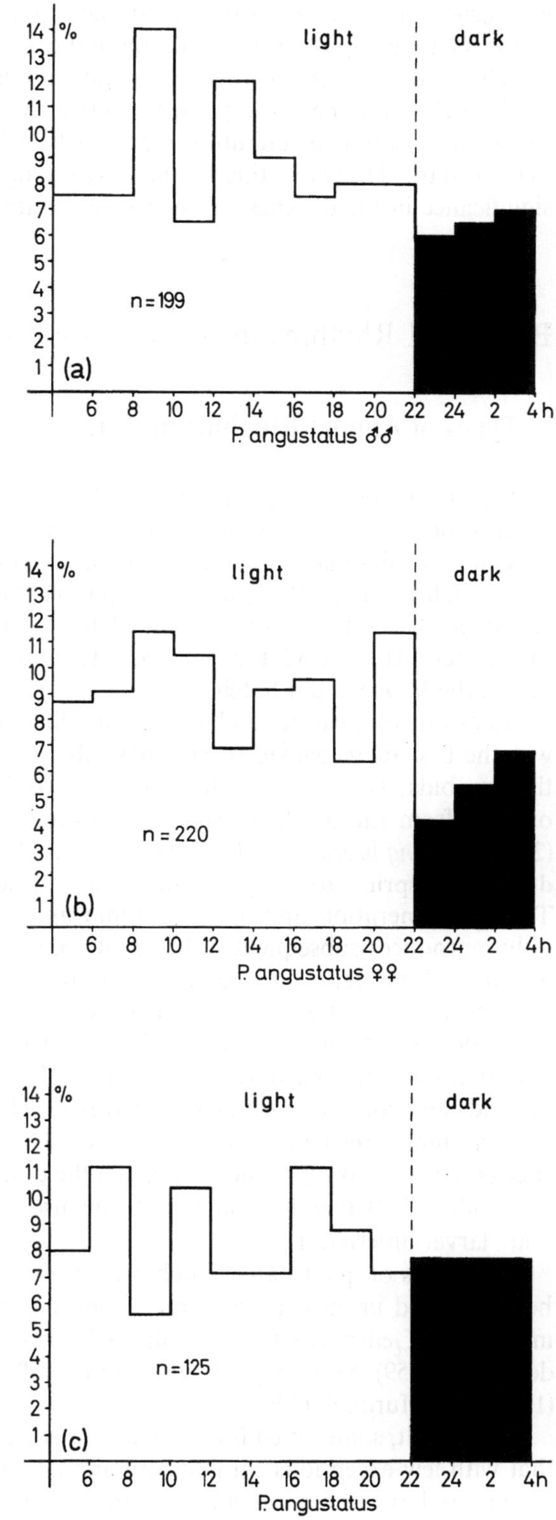

Fig. 120a–c. Distribution of emerging times of *Pterostichus angustatus* from pupae in long day LD 18/6. (a) and (b) males and females, using flash for automatic photographic recordings; (c) both sexes without flash. From Thiele and Paarmann, 1968

not occur in a single case. All four carabids studied have therefore to be allocated to the small group of insects with aperiodic moulting.

The different frequencies of moulting observed in *Pterostichus angustatus* in the light and the dark phases suggest the existence of a latent tendency to synchronisation of moulting that can be triggered off by the change from light to dark. However, this is only weakly developed and is of no ecological significance in the darkness of their subterranean habitats.

B. Annual Rhythms in Activity, Reproduction, and Development

I. Types of Annual Rhythms in Carabids

At least in the temperate zones all carabids are apparently univoltine inasmuch as only one of the series of generations reproduces each year (Thiele, 1971).

Carabids hibernate either as larvae or as adults. In the case of *Amara infima*, Schjøtz-Christensen (1965) is of the opinion that hibernation may occur in the egg stage. The reproductive period of this species stretches from autumn into the winter. At +10°C the embryos developed slowly for 82 days but died before the larvae could hatch.

Reproduction can take place at quite different times of year. Larsson (1939) was the first to recognize several basically different reproductive types among the carabids: (1) The so-called "*autumn-breeders*" that reproduce in autumn or even from the height of summer onwards and usually hibernate as larvae. (2) The "*spring breeders with autumn activity*" that hibernate as adults and reproduce from spring to early summer, after which most of the beetles die off. The new generation appears in autumn and becomes fully active although it only reproduces subsequent to hibernation. (3) "*Spring breeders without autumn activity*". Their reproductive rhythm is the same as that of group 2, except that the young beetles exhibit little activity following eclosion in autumn.

Although his conclusions were based solely on dated catches from museum material, Larsson clearly recognized the different types of reproduction among carabids and correctly assigned most species. His designations "spring breeders" and "autumn breeders" have become widely accepted, although for the many species that already reproduce in summer the term "autumn breeder" is inaccurate. Lindroth (1949) therefore suggested the use of the terms "adult hibernator" and "larval hibernator".

By means of quantitative catches with Barber traps, Larsson's findings have been followed up in various parts of Europe by, to mention some of the more important, Geiler (1956/57, 1960); Scherney (1955); den Boer (1958); van der Drift (1959); Skuhravý (1959); Grüm (1959); Lauterbach (1964); Greenslade (1965) and Murdoch (1967).

As a result, some amendments can be made to Larsson's scheme. He assumed that with few exceptions carabids are annual species. This holds true inasmuch as one and the same generation almost invariably reproduces only once in the same year (an exception to this rule is provided by *Abax parallelus*, of which Löser (1970) reported that the female produces two batches of eggs in one summer within the space of a few weeks. This is connected with the fact that it is a species that exhibits brood care (see p. 79). In the laboratory Paarmann (1966) was able to demonstrate that females of *Pterostichus angustatus* can reproduce three times in all, at approximately six-monthly intervals. Krehan (1970) observed (also in the laboratory) at least two laying periods in females of *P. cupreus* and *P. coerulescens*. In both of these spring breeders only a minute portion of a population hibernates twice in the field, so that in these species, too, extremely few animals go through two reproductive periods. It was only recently that Van Heerdt *et al.* (1976) reached the opinion that it might be possible to distinguish females of *Pterostichus oblongopunctatus* in the first, second, or third years of their lives from a study of the size of the corpora lutea. Females of these ages all take part in the reproduction of the population. From this, the authors concluded that in the course of four years about 40% of the reproducing females were two or three years old. The question of whether such a distinction between age classes is reliable demands further consideration in the opinion of the present author.

More recent investigations have shown, however, that not only the larvae but also a considerable proportion of adults of autumn breeders hibernate, after which they can enter upon a second reproductive period. Gilbert (1956) had already reported that in England *Calathus melanocephalus*, *C. fuscipes* and *C. erratus* hibernate after reproducing and are able to reproduce a second time. In Holland Vlijm *et al.* (1968) found that about 60% of *C. melanocephalus* hibernate, 54.5–76% of the hibernating females achieving a second period of reproduction. Schjøtz-Christensen (1968) found that 33.5% of a population of *Carabus hortensis* in Denmark hibernated, and Krehan (1970) reported that at the commencement of the reproductive period 30% of the females of a population of *Pterostichus vulgaris* were animals that had already reproduced in the previous year prior to hibernation.

So far the existence of two populations side by side, the one reproducing in spring and the other in autumn, has only been demonstrated in a few species. Detailed population analyses by Schjøtz-Christensen (1965) in Denmark showed this to be the case for *Harpalus smaragdinus*, *H. anxius*, *H. neglectus* and *H. aeneus*, and in a further paper (1966a) he mentioned five more species of *Harpalus* that probably exhibit a similar type of behaviour. Löser (1970) confirmed the existence of a similar situation in *Abax ater*. The best known species of this type so far are *Harpalus smaragdinus* and *Abax ater*. In both species it could be ruled out that allochronous populations could exist side by side without mixing with one another. For *Abax ater* Löser was able to show that, dependent upon microclimate and weather conditions, populations with two peaks of activity could change over to a one-peak type of activity and vice versa (see p. 262f.). By rearing exclusively in the laboratory Löser found that the larvae developed equally well under winter or summer conditions.

A further peculiarity of some species of autumn breeders is the occurrence in summer of a period of inactivity in the adults, connected with gonadal dormancy. In contrast to "normal" autumn breeders the beetles emerge in the spring instead of in summer, several months prior to the reproductive period. Following brief activity they enter upon an aestivation dormancy, as could also be shown by den Boer (1958) for *Leistus ferrugineus*. A more detailed analysis based on dissection of females of *Nebria brevicollis* was made by Gilbert (1958). A summer dormancy of this kind was found by van der Drift (1959) in *Carabus problematicus* and by Lehmann (1965) in *Nebria salina*. Thiele (1969 a) confirmed the occurrence of aestivation dormancy in *Nebria brevicollis* and *Patrobus atrorufus* in laboratory experiments and in the field.

The distinction made by Larsson between spring breeders with and without autumn activity seems to be unimportant, since the degree of activity of young beetles in autumn may vary from habitat to habitat and from one geographical region to another. The present state of our knowledge suggests that at least the following types of annual rhythms can be distinguished in carabids:

1. Spring breeders which have summer larvae and hibernate as adults.
2. Species which have winter larvae and reproduce from summer to autumn but exhibit no adult dormancy.
3. Species with winter larvae, the adults of which emerge in spring and undergo aestivation dormancy prior to reproduction.
4. Species with flexible reproductive periods. In such species spring and autumn reproduction can occur side by side in one and the same population and, what is more important, the larvae, in contrast to those of the above three types, can develop equally well under summer or winter conditions. Reproduction can take place at very different times of year according to climate and weather.
5. Species which require more than one year to develop.

II. The Adaptation of the Activity and Reproduction Rhythms to the Annual Cycle of Environmental Factors

The larvae of the majority of carabids are very specific in their demands as to climatic conditions. They are not able to survive unfavourable seasons, especially winter, at every stage of their development so that for most species of carabids it is of great adaptive value if maturation and reproduction can be so directed as to coincide with the season best suited to that species (see following section). This is well illustrated in the case of carabids living on river banks, where their existence has only been rendered possible by a synchronization of their reproductive rhythm with the annual flood cycle. Investigations carried out by Lehmann (1965) on the banks of the Rhine showed that only spring breeders can exist within the flood zone since they are able to complete their development between the winter and summer high water in good years, although they develop even better if no summer flood occurs. In northern Norway, on the other hand, larval hibernators are able to exist on river banks because the rivers freeze in winter (Andersen, 1968).

The existence of a correlation between reproductive type and distribution in different habitats was already observed by Larsson. This relationship became even clearer following an investigation using trap catches. According to Larsson spring breeders dominate in open country and autumn breeders in the forest. Greenslade (1965) found that only one-third of the forest species but six out of 11 grassland species were adult hibernators. The ratio of spring, summer and autumn breeders was 2:4:3 among forest species and 5:4:2 among grassland species. Thiele (1969) reported finding 50% spring breeders among the forest species and 62% among those living in open country. The per cent of spring breeders in both forest and open country rises in damp to wet habitats. A large proportion of the spring breeders amongst the forest-dwelling carabids was found by Thiele (1962) and Thiele and Kolbe (1962) in the damp Fagetalia forest communities rather than in the drier Quercetalia. In England Murdoch (1967) observed that 20 out of 21 species living in wet habitats were adult hibernators, but at least nine out of 12 "non-wet" species were autumn breeders with winter larvae. "It would be of interest to determine what selective forces produced and maintain this difference in life history patterns... Possibly the eggs and larvae are not so well adapted to withstand inundation. In addition, in winter the water often covers much of the litter, which is a very rich food source in these environments, and thus it would be disadvantageous to have larvae at this time. The larvae of *Leistus rufescens*, the only wet species in this study which has winter larvae, are well adapted to climb and were often found above the litter layer (at 6 ins to 3 ft) in shallow logs and other places where Collembola (their prey) hibernate". Kirchner (1960) reported finding spring and autumn breeders in a ratio of 6:1 (individuals) in fields with clay soil but only 2:1 on sandy soils.

A geographical difference can also be observed in the composition of the carabid fauna as regards developmental type. On the North Sea coast Heydemann (1964) found that larval hibernators accounted for 66% (individuals) of the population in fields with polders whereas near Leipzig Geiler (1956/57) reported a figure of only 47%.

An Atlantic climate with mild winters offers the larval hibernator apparent advantages as compared with a continental climate (see also Lindroth, 1949). This is also confirmed by Heydemann's (1962b) summary of data taken from the literature.

Proportion of species hibernating as larvae:

East Scandinavia	7%
North Scandinavia	23%
South Scandinavia	23%
West Scandinavia	54%
Shetland Isles	52%
Ireland	47%
Greenland	50%

Further, at least 10 out of 18 indigenous Icelandic species i.e. 56%, are, according to Larsson and Gigja (1959), larval hibernators. Heydemann (1962b) reported that in the maritime climate on the west coast of Schleswig-Holstein 38% of the species found were larval hibernators as compared with 18% further inland.

Investigations on the annual rhythm of carabids in North America indicate that the same types occur as in Europe, although so far only few publications are available (Gilbert, 1957, in Illinois, USA; Rivard, 1964a, in Ontario, Canada).

In North Africa autumn breeders are the numerically dominant species, whereas according to Larsson (1939) at least 70% of all species in central and northern Europe are spring breeders. However, in North Africa the humidity factor makes winter reproduction compulsory (see also p. 257). Gonadal development in *Harpalus litigiosus* began in September/October but in *Laemostenus picicornis* it was not yet complete by the end of December, and reproduction occurred in the months of January and February. Observations on *Calathus fuscipes* and *C. mollis* in Cyrenaika confirmed that it is indeed a case of a variant of the autumn breeding type. Both species are typical autumn breeders in the temperate zone. In N. Africa gonadal development commenced at the beginning of October, eggs were laid in November/December and the young beetles emerged from the beginning of March until mid-April (Paarmann, 1970).

Next to nothing is known so far about the annual rhythm of carabids in the tropics, where the temperature is more equable and the photoperiod changes only slightly if at all during the course of the year. Preliminary investigations covering almost a year have been made at different altitudes and in various kinds of vegetation on the western shores of Lake Kivu and on the eastern limits of the Congo Basin (Paarmann, to whom my thanks are due for permission to mention results as yet unpublished. See also Paarmann, 1976b). Unfortunately, the catches of carabids in a primary rain forest with almost constant temperature and soil humidity were so small that unequivocal conclusions could not be drawn. The carabids inhabiting the banks of streams and rivers in the rain forest exhibited a definite annual rhythm in gonadal development. In May, 85% of the beetles caught were sexually mature; this figure sank to 37% by August and then rose slowly. In the cultivated region west of Lake Kivu the carabids found between 1460 and 2000 m, just as those in the mountain forest above 2000 m, had a well developed reproductive rhythm coupled with an activity rhythm. This was correlated with the dry period of three months, from June to August, during which time the soil humidity sank drastically. The annual fluctuation of the monthly mean temperature was at most 2.5° C (in a bamboo forest). Catches made by hand in a mountain forest gave the following figures for sexually mature beetles: June 12%, July 9%, August 14%, a sudden jump to 74% in September, in October and November 95% and with the appearance of the young beetles in December the figure slowly dropped again. The preparation of beetles caught in traps gave similar results. In lower-lying regions the per cent of sexually mature animals rose more slowly than in the mountain forest: July 5%, August 6%, September 19%, October 45%, November 66%, December 72%. This is quite certainly connected with the fact that the dry period lasts longer in this region than in the higher forest. The proportion of sexually mature animals sank to a different extent from species to species. In extreme cases no sexually mature animals were caught over a period of four months. Scarcely any fluctuation at all was observed in only two species, caught in a damp habitat on the shore of Lake Kivu (*Abacetus* spec. and *Paramegalonychus brauneanus*). It appears that even in the tropics a well-developed annual periodicity in both activity and reproduction is to be expected in carabids.

Dissection of wild specimens as well as breeding under nearly natural conditions revealed that in cave-dwelling Trechini no annual rhythm in reproduction can be observed (Deleurance and Deleurance, 1964b). This finding was borne out by the observation that the neurosecretory cells in these species (e. g. *Aphaenops cerberus*) always exhibit the same degree of activity and show no signs of cyclical activity (Deleurance, 1967).

III. The Regulation of Annual Rhythms by External Factors

The factors influencing annual patterns of development have been investigated experimentally in about 10 species of carabids. It appears that in most species an obligatory dormancy occurs during the cycle of development. In agreement with H. J. Müller (1970) dormancy is defined as any facultative or obligatory inhibition or retardation in development occurring during the ontogenesis of an insect (Thiele, 1971a). "Dormancy" is used here in preference to the ambiguous term "diapause". Müller (1970) lists the following important types of dormancy which can be found in carabids (see Thiele, 1973b).

Quiescence is a facultative retardation or total stop of the development, which means an immediate answer to the lowering of optimal conditions of environment into suboptimal spheres which is reversible at any time.

Parapause is an obligatory dormancy, where a clear phase of induction cannot be recognized. It appears in a genetically fixed stage of development and is independent of external factors. Its termination demands at least one drastic alteration of the level of only *one* environmental factor. The termination of the parapause is also possible in two steps and then demands two alterations of the level of one factor.

Eudiapause is a facultative dormancy where an induction phase is clearly recognizable. If the environmental conditions during this phase differ from the optimum, the organism reacts with dormancy. If the dormancy is induced by *one* factor, it can only be terminated by changes of the level of *another* factor. Until now only such examples of eudiapause are known in which a certain photoperiod acts inductively and the termination of dormancy is produced by a change in the level of temperature.

In my opinion the most important step in Müller's system is that he clearly distinguishes between those dormancies which are only induced by suboptimal conditions (following the deterioration in the case of quiescence, anticipating it in the case of eudiapause) and the obligatory dormancy he calls parapause. Moreover Müller was the first to see that there is a fundamental difference between the two physiologically based types parapause and eudiapause, which were until that time confused under the term diapause: the parapause is under the control of only one factor, the eudiapause under the control of two factors, the terminating factor always being different from the inducing factor.

The number of types of dormancy observed in carabids is large; in species so far investigated in detail no two behave identically with respect to dormancy although certain basic types are clearly distinguishable. Dormancy, either obligatory or facultative, has been found to occur at the larval or adult stage (in this case during gonadal maturation) in all species studied.

The types known at present can be grouped as follows:

1. Spring Breeders With no Larval Dormancy but Obligatory Dormancy in the Adults (Parapause) Mainly Governed by Photoperiod

This type of development was first described by Thiele (1966) in females of *Pterostichus nigrita.* Under extreme long day (LD 18/6) at 20° C the polytrophic ovaries of *P. nigrita* remain almost entirely undifferentiated. Without a change of temperature changeover to short day (LD 8/16) leads to vigorous growth and to a rapid differentiation of the ovaries up to the stage of previtellogenesis. This step takes a minimum of two weeks in a few individuals but four to six weeks in the majority of females. The stage of previtellogenesis cannot be exceeded under continued short day. For its onset, however, it is immaterial whether the animals have been reared in short day, whether the adults have been exposed to short day after emerging or whether they have been transferred to short day following initial long day treatment.

The second stage in the development of the eggs, encompassing yolk formation (vitellogenesis), development of the chorion and growth of the oocytes up to their size at the time of laying, can only proceed if the animal is transferred from short day to long day. An abrupt changeover from short day to long day is just

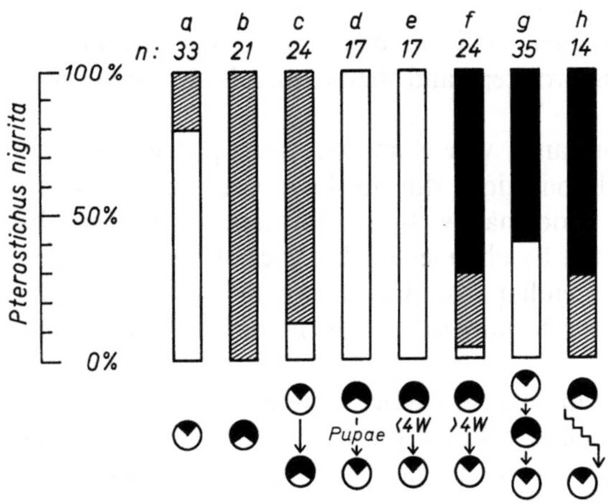

Fig. 121. State of development of the ovaries of *Pterostichus nigrita* under the influence of different photoperiods. Percent of the females with poorly differentiated ovaries *(white columns)*, at the stage of previtellogenesis with large oocytes *(hatched columns)* and with eggs ready for laying *(black columns)*. *a:* Long day LD 18/6; *b:* short day, LD 8/16; *c:* transfer of adults from long day to short day; *d:* transfer of pupae from short day to long day; *e:* transfer of adults younger than four weeks from short day into long day; *f:* transfer of adults older than four weeks from short day into long day; *g:* treatment of adults with long day/short day/long day; *h:* gradual increase of the length of day acting on adults from LD 8/16 to LD 18/6. The temperature was always 20° C. From Thiele, 1966

as effective as a slow increase from 8 up to 18 h of daylight over a period of five weeks (Thiele, 1966; Fig. 121).

As regards control of maturation of the ovaries by photoperiod *P. nigrita* is neither a short-day nor a long-day animal, but is one of the few species of insects in which a short-long-day type of egg maturation has been demonstrated. A control mechanism of this kind for the hibernation dormancy of the adults, involving the appropriate sequence of first short day then long day is a simple way of ensuring that univoltine insects reproduce in the spring. The larvae of spring breeders require high temperatures for their development (see Chap. 5.B.6.). It is of interest that each step in the development of the ovary has its own particular photoperiod.

Short-day development of *P. nigrita* is (within wide margins) only relatively little influenced by temperature. Lowering the temperature over a period of two months from +20° C to +5° C with a subsequent return to the higher temperature under constant long day, in contrast to the alternating photoperiod described above, results in ripening of eggs in only a small proportion of females (Thiele, 1968 a). From this it can be concluded that the maturation of the animals in the field in spring is also largely the result of a changeover from winter short day to summer long day. A figure of 13 hrs of light per 24 hrs has been found to be the critical photoperiod for the onset of vitellogenesis. This is borne out by the observation that in the field maturity is achieved shortly after the vernal equinox.

In the course of further investigations on carabids spring breeders were found in which the dependence of development on photoperiod was similar to that seen in *P. nigrita*. The development of *P. angustatus* under short day followed by long day is almost identical (Thiele, 1968a; Fig. 122). In both species gonadal development in the males can be suppressed by long day, but can be set off at any time by short day. Thus male development, too, is governed by photoperiod although in a somewhat simpler manner than in the females

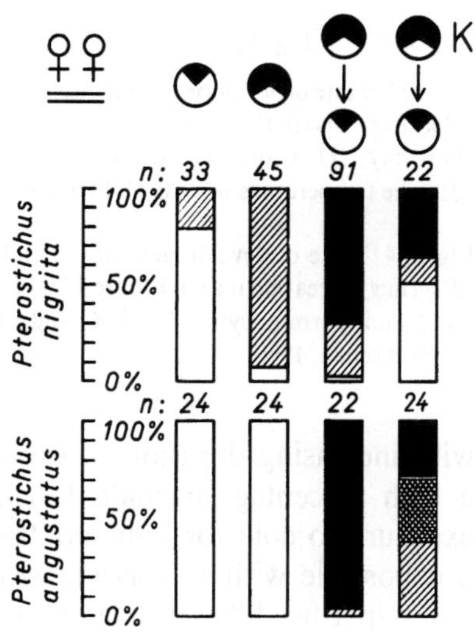

Fig. 122. State of development of the ovaries in females of two spring breeders under the influence of different photoperiods. *Column 1:* beetles under long day and 20° C; *Column 2:* beetles under short day and 20° C; *Column 3:* beetles after at least four weeks short day transferred to long day, 20° C; *Column 4:* beetles following cold/short day treatment transferred to long day and 20° C. Shading as in Figure 121. *Spotted:* females probably following egg deposition ("spent"). From Thiele, 1968 a, supplemented

(*P. nigrita:* see Ferenz, 1975a, *P. angustatus:* unpubl. data of Thiele). Dormancy in the males is a photoperiodic quiescence.

For the development of females of *P. cupreus* and *P. coerulescens*, likewise, a sequence of short day followed by long day is essential; under constant photoperiod maturation is hardly ever achieved. Temperature has only a slight influence on photoperiodism (Krehan, 1970; Fig. 123). The males of both species, however, behaved in a completely different way: photoperiod alone had scarcely any effect and only if short day was combined with low temperatures (+2° to +3°) did a large number of males achieve sexual maturity in the following long day (Fig. 124). In addition, the proportion of mature males increased

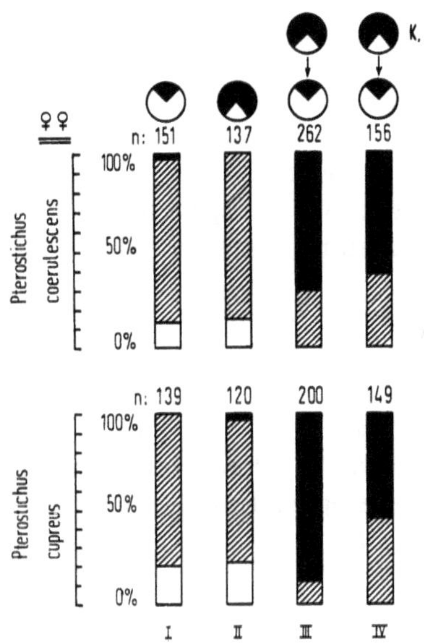

Fig. 123

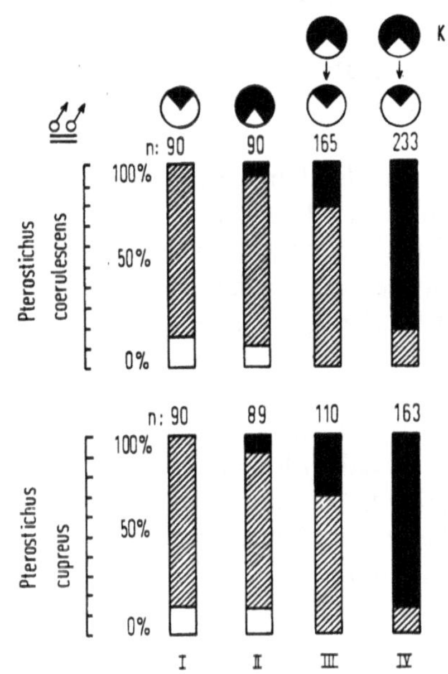

Fig. 124

Fig. 123. State of development of the ovaries in females of two spring breeders of the genus *Pterostichus (Poecilus)* under the influence of different photoperiods. *I:* Continuous long day; *II:* continuous short day; *III:* short day → long day; *IV:* cold/short day → warm/long day. Shading as in Figure 121. The temperature was 20° C. From Krehan, 1970

Fig. 124. State of development of the male gonads in two spring breeders of the genus *Pterostichus (Poecilus)*. Treatment as in Figure 123. Percent of males with juvenile gonads *(white)*, with spermatogonia and spermatocytes *(hatched)* and with spermiozeugmata *(black)*. The temperature was 20° C. From Krehan, 1960

with increasing duration of exposure to low temperatures (Table 50). In fact a high percentage of male *P. cupreus* could even be brought to maturity by exposure to cold for four months under constant short day although this was not possible with *P. coerulescens*. In these two *Pterostichus (Poecilus)* species a sex-specific difference in the control of gonadal development can therefore

Table 50. Percentage of mature males of *Pterostichus coerulescens* and *P. cupreus* in experiments with increasing duration of cold treatment. (From Krehan, 1970)

Experimental treatment	*P. coerulescens* n	*P. coerulescens* +	*P. cupreus* n	*P. cupreus* +
2M SD+2M SD (5–7° C)+1M LD	16	25%	10	40%
2M SD+2M SD (2–3° C)+1M LD	25	48%	19	53%
2M SD+3M SD (2–3° C)+1M LD	21	71%	17	82%
2M SD+4M SD (2–3° C)+1M LD	79	81%	64	87%
8M under natural conditions (mid-September until mid-April)	30	90%	—	—
2M LD+2M SD (5–7° C)+1M LD	12	17%	20	30%
2M LD+2M SD (2–3° C)+1M LD	26	42%	18	44%
2M LD+4M SD (2–3° C)+1M LD	68	74%	60	82%
9M under natural conditions (mid-August until mid-April)	35	94%	22	100%

+: mature males; n: no of experimental males; M: months; SD: short day; LD: long day. The temperature before and after the cold period was 20° C

be said to exist, the photoperiodicity of the males being highly dependent upon temperature.

Yet another type of behaviour is observable in females of the short day-long day animal *Pterostichus oblongopunctatus:* short day brings about previtellogenesis almost only within the very narrow temperature range of 10°–15° C. At temperatures above *and* below this range the short day effect is almost completely eliminated and as a result of the absence of previtellogenesis the gonads of the females fail to achieve maturity in the subsequent long day. The limits of the temperature effect are very sharply defined; at 15±1° C short day has a maximum effect but at 17.5±1° C almost none (Thiele, 1968, 1971a, 1975; Fig. 125). In the field, short day can therefore only play a part in maturation in autumn and spring but not in conjunction with the low temperatures of the winter months, as has been confirmed experimentally (Thiele and Könen, 1975).

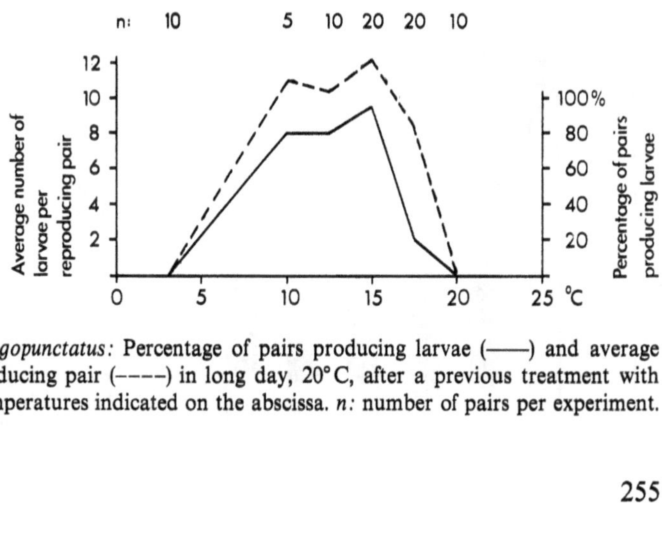

Fig. 125. *Pterostichus oblongopunctatus:* Percentage of pairs producing larvae (——) and average number of larvae per reproducing pair (----) in long day, 20° C, after a previous treatment with two months short day at temperatures indicated on the abscissa. *n:* number of pairs per experiment. From Thiele, 1975

All five species mentioned are the same type of spring breeders with respect to their annual patterns of activity and reproduction, although they exhibit very different modalities in the photoperiodic and thermic control of gonadal development. In all of these species dormancy is obligatory in at least one sex; complete maturation of adults cannot be achieved under constant long day or constant short day. The alteration of only one factor (the photoperiod) is, in all females and in the males of some species, largely responsible for the course taken by dormancy, which for this reason should be termed a parapause. Temperature influences the photoperiodicity to varying degrees from species to species.

2. Spring Breeders With no Larval Dormancy but a Facultative Dormancy in the Adults Governed by Photoperiod (Photoperiodic Quiescence)

This type of development has so far only been confirmed in *Agonum assimile*, also a spring breeder. In contrast to the above-mentioned species, dormancy in females of *A. assimile* was early recognized as being facultative; even under continuous long day some females produced ripe eggs (Thiele, 1966). Laboratory breeding, however, was unsuccessful. The factors controlling development were first elucidated by Neudecker and Thiele (1974; see also Thiele, 1971a). The males of *A. assimile* were shown to achieve maturity under continued long day. "The maturation processes in the two sexes are extremely dissynchronized at 20° C and long-day; the females mature more rapidly than the males, produce ripe eggs after only one or two months and in all probability lay them in an unfertilized state. The males mature one or two months later, but by the time they are capable of fertilization the females have laid all their eggs. This situation is prevented in nature by the onset of short day" (Fig. 126). Under short day neither sex of *A. assimile* achieves maturity (Thiele, 1968a). "The onset of autumnal short day puts an end to gonadal development in the field and so prevents the dissynchronization between the sexes observed under experimental conditions. A very slow gonadal development under short day permits

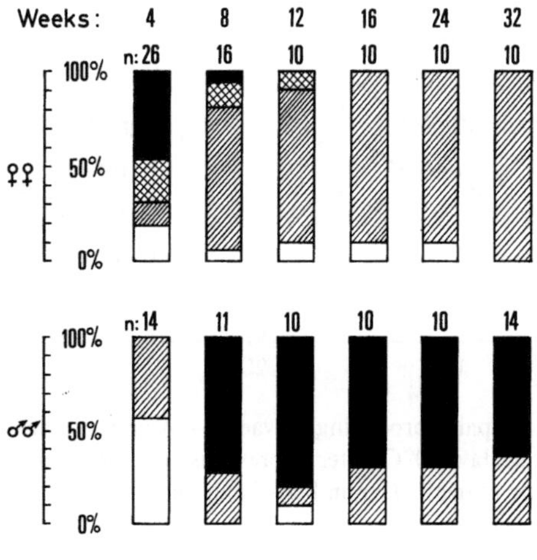

Fig. 126. State of development of the gonads of females and males of *Agonum assimile* in continuous long day (LD 16/8) and 20° C after different time intervals. Shading for females, as in Figure 121. *Cross-hatched:* females with commencing vitellogenesis. *Males: white:* immature; *hatched:* partially mature (accessory glands partly filled); *black:* completely mature (accessory glands filled; many spermiozeugmata). From data of Neudecker and Thiele, 1974

the males to catch up with the females, and with the increasing length of day in spring the two sexes are able to attain maturity at the same time... Since the development of *A. assimile* under long day can be completed without the intervention of dormancy, and in view of the fact that animals taken from their winter quarters can usually be brought to reproduce under artificial treatment with long day, the inhibition of gonadal development of *A. assimile* under short day can be regarded as a photoperiodic quiescence, capable of reversal at any time" (Thiele, 1971 a).

3. Autumn Breeders With a Thermic Hibernation Parapause at the Larval Stage and no Dormancy in the Course of Adult Development

This type of development has been investigated experimentally in *Pterostichus vulgaris*. The adults of both sexes achieve sexual maturity in about three weeks at approximately 20° C, irrespective of photoperiod and without the occurrence of dormancy. At lower temperatures only a thermic quiescence occurs. The larvae, too, develop independently of photoperiod (Krehan, 1970). At the third larval stage, however, cold is obligatory if the cycle of development is to be completed. At continued temperatures of 15° C and above, the animals only reach the third larval stage and never achieve metamorphosis. Dormancy can be overcome, however, if the third stage experiences temperatures below +8° C, whereby the proportion of animals undergoing metamorphosis is raised both by progressively lower temperatures and by increasing duration of cold (Table 51, see Thiele and Krehan, 1969; Krehan, 1970). "A rise in temperature is necessary for metamorphosis to take place; development cannot be completed under continued low temperature. Since the inhibition in development is obligatory and only a change in temperature at the appropriate time can bring about metamorphosis, the larval dormancy can be regarded as a thermic parapause" (Thiele and Krehan, 1969).

The control of gonad maturation in *P. niger* is quite similar to that in *P. vulgaris* (Witzke, 1976). Larvae of the third instar died at constant temperatures of +20° C. If, however, the larvae were exposed to natural winter temperatures (as low as −10° C) by digging into the soil, pupation became possible, provided that the beetles were maintained at +20° C in the laboratory afterwards.

Pterostichus madidus has a similar cycle of development to *P. vulgaris*, and could be reared up to pupation under constant temperatures of 10°, 15° and 20° C, although the mortality at the third instar was extremely high and only a very small proportion of the animals reached the pupal stage. "Most third instar larvae died when apparently full grown, irrespective of the temperature at which they were kept... It is possible, therefore, that *P. madidus* has a partial diapause in this instar, as in *P. melanarius (=P. vulgaris)* (Thiele and Krehan, 1969). Attempts to aid the development of *P. madidus* by chilling third instar larvae were inconclusive" (Luff, 1973).

In Cyrenaika in North Africa autumn-breeding species become "winter breeders", reproducing from November until March. It is impossible for these species to reproduce in late summer as they do in central Europe, on account

Table 51. Development of *Pterostichus vulgaris* with and without cold treatment in the 3rd larval instar under different photoperiods. (Each test involved 50 larvae; in series III and VI of 1964/65 only 40 larvae each.) (From Krehan (1970)

Year	Developmental stage achieved	*Short day* Temporary natural winter cold I.		Temporary artificial cooling II.		Continuous warmth III.		*Long day* Temporary natural winter cold IV.		Temporary artificial cooling V.		Continuous warmth VI.	
		No.	%	No.	%	No.	%	No.	%	No.	%	No.	%
1964/65	L III	38	76	34	68	6	15	35	70	29	58	7	18
	pupa	8	16	7	14	0	0	7	14	6	12	0	0
	adult	2	4	1	2	0	0	2	4	2	4	0	0
1965/66	L III	39	78	32	64	10	20	41	82	31	62	13	26
	pupa	16	32	14	28	0	0	18	36	11	22	0	0
	adult	12	24	9	18	0	0	13	26	7	14	0	0
1966/67	L III	44	88	30	60								
	pupa	21	42	16	32								
	adult	14	28	8	16								

of the humidity and low temperature required by the larvae. The time needed for development was reduced in *Broscus laevigatus* and *Pterostichus atlanticus* by exposing the first larval stages to 14° C (the mean winter temperature in Cyrenaika), which also reduced mortality in the latter species (Paarmann, 1973).

4. Autumn Breeders With a Thermic Hibernation Parapause in the Larvae and a Photoperiodic Aestivation Parapause in the Adults

In this group of autumn breeders the process of development is complicated by the fact that two variously governed obligatory dormancies succeed each other at two stages in the development of the individual. Here, too, cold is obligatory for the hibernating larvae and again at the third larval stage for

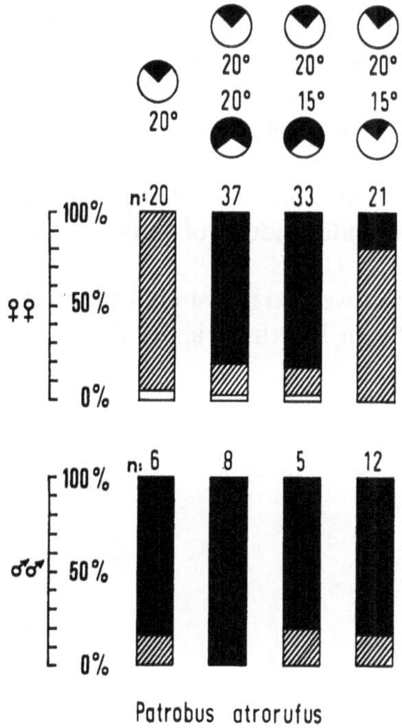

Fig. 127

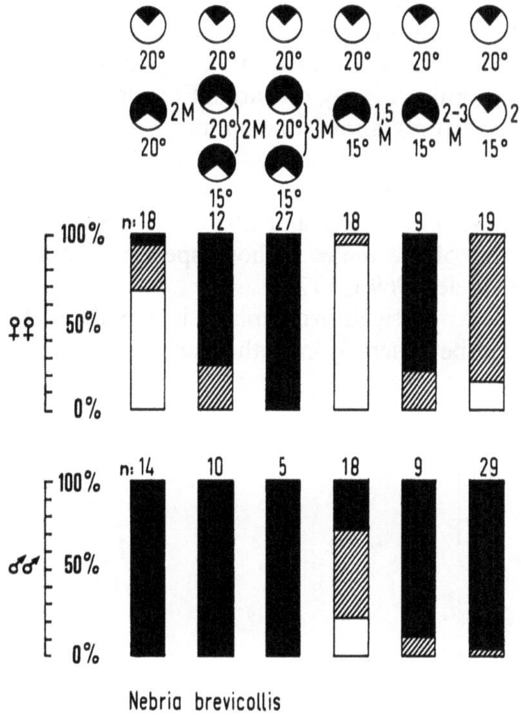

Fig. 128

Fig. 127. State of development of the gonads in males and females of *Patrobus atrorufus* under the influence of different photoperiods. *Left to right:* continuous long day at 20° C, long day → short day with and without lowering of temperature, continuous long day with lowering of temperature. For state of development see Figures 121 and 126. From Thiele, 1971 a

Fig. 128. State of development of the gonads of females and males of *Nebria brevicollis* after treatment with short day of different durations and different temperatures. All experiments following treatment with long day and 20° C. Extreme right: treatment with lowered temperature in continuous long day. *M:* months. State of development: females see Figure 121. *Males:* with immature gonads *(white)*, with small accessory glands *(hatched)*, with large accessory glands and sperms *(black)*. From Thiele, 1971 a

Patrobus atrorufus, but from the first stage onwards for *Nebria brevicollis*, the larvae of which hatch later in the year. In contrast to *Pterostichus vulgaris* in which development takes about nine and one-half months (Krehan, 1970), that of the species under consideration here is much quicker, taking five months in *N. brevicollis* and six and one-half months in *Patrobus atrorufus*. The beetles emerge already in spring and the cold-requiring larvae would be confronted with, for them, intolerable summer temperatures if there were to be no delay in the reproduction of the adults. "A mechanism preventing reproduction at this time, when temperatures are too high for the development of their larvae is indispensible. This mechanism is the arrest of ovarian maturation of the females resulting in an aestivation parapause" (Thiele, 1969a). It has been shown experimentally that extreme long day, such as prevails in summer, suppresses ovarian maturation. Changeover to short day provides the signal for egg ripening (Figs. 127, 128), whereby the critical photoperiod for short day is extremely high in both species (16–17h light per day, Figs. 129, 130). This means that, as perceived by these animals, short day conditions already prevail from the beginning of July onwards. Females of *P. atrorufus* matured under experimental conditions about one month after changeover from long day to short day, those of *N. brevicollis* two to three months after. This is in agreement with the findings that, in the field, *P. atrorufus* and *N. brevicollis* begin to reproduce about one month and 2.5 months respectively after the summer solstice. Maturation of the males of both species is more or less independent of photoperiod (Thiele, 1969a, 1971a).

An aestivation dormancy in gonadal maturation, governed in a similar manner, can be observed in both sexes of *Broscus laevigatus* in North Africa (Paarmann,

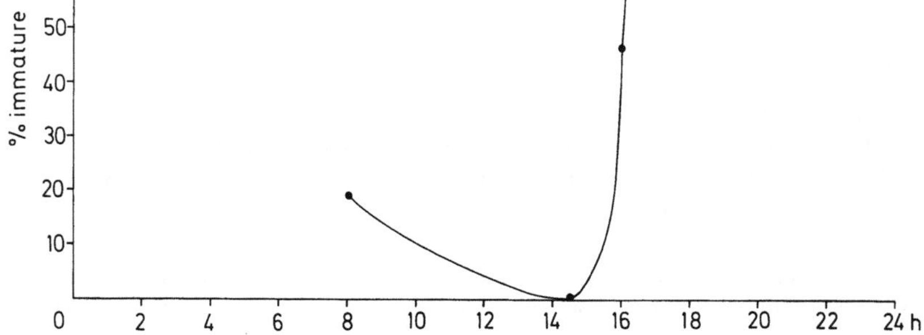

Fig. 129. Critical photoperiod of the females of *Patrobus atrorufus*. Percent immature animals in continuous long day (LD 18/6) and after transfer from this to LD 16/8, LD 14.5/9.5, and LD 8/16. LD 8/16 at 15° C, otherwise 20° C. From Thiele, 1971a

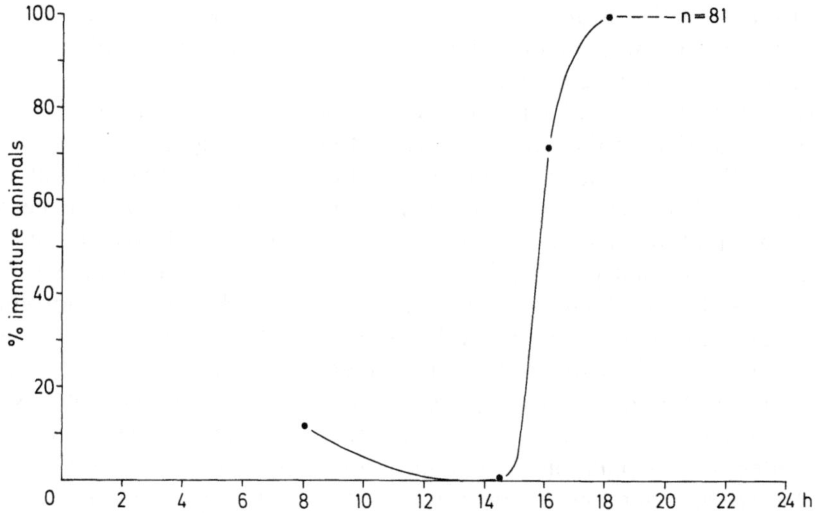

Fig. 130. Critical photoperiod of females of *Nebria brevicollis*. Experimental conditions as in Figure 129. From Thiele, 1971 a

1974). Short day of LD 10/14 following on long day of LD 14/10 (photoperiods corresponding to the seasonal changes in length of day in Cyrenaika) led to maturation of 79 %, and a change from LD 18/6 to 8/16 to maturation of 90 % of the animals. The animals also achieved maturity in DD at 20° C, although the process took seven months. A temperature of 30° C and DD totally suppressed maturation, just as did constant long day (LD 14/10) at 20° C. Maturation in this species therefore requires the joint action of an optimum temperature and a changeover from long day to short day. Further, in females of *N. brevicollis* maturation in short day proceeds much better at 15° than at 20° C (Thiele, 1971 a).

A phase of locomotory inactivity running parallel to the summer interruption in ovarian development has been shown to last about one and one-half months in *P. atrorufus* and three and one-half months in *N. brevicollis* (Thiele, 1969 a, 1971 a). Penney (1969) found that the summer inactivity of *N. brevicollis* depends upon nutrition and not on the photoperiod; following fat storage the animals become inactive in summer whereas starved animals remain active. In the field the large quantities of food available in spring ensure storage of reserve materials and thus provide the basis for reduced activity in summer.

Phases of locomotor inactivity also occurred in summer in field experiments with *Carabus problematicus* (summer sleep, Thiele, unpublished).

5. Species With Unstable Conditions of Hibernation and Potentially Lacking Dormancy

It was already clear to Lindroth (1949) that not all species could be fitted into Larsson's (1939) scheme of developmental types, for which reason he spoke of species with unstable patterns of hibernation, in which the latter could occur just as well in the larval or in the adult stage. The factors controlling development have only been investigated experimentally in one such species, i.e. *Abax ater* (Löser, 1970). An influence of photoperiod on the ripening of the eggs could be excluded since the females developed and laid eggs equally well at LD 8/16 and LD 16/8. Under favourable conditions they matured without dormancy. A thermal quiescence may occur at lower temperatures, and it has also been shown that the larvae can be brought to metamorphosis in the laboratory at 18°C and that photoperiod plays no role in larval development. The larvae thus exhibit no obligatory dormancy. They also achieved metamorphosis under simulated winter conditions in the laboratory, development being slowed down during the cold treatment (thermal quiescence). This was the first experimental demonstration in a carabid species of such a degree of eurypotency with respect to developmental requirements that the larvae can exist both in summer and winter. At no point in development does an obligatory dormancy occur in this species, but low temperature at the third stage reduces mortality. This is Löser's explanation of the greater frequency of *A. ater* in highland regions and cool habitats.

Investigations in the field have revealed that *A. ater* emerges in spring and in autumn. Animals that have emerged in spring reproduce in summer whereas those emerging in autumn reproduce only after hibernation. Under favourable temperature conditions the latter mature soon after hibernation so that two peaks of activity can be observed: "While in higher regions, on slopes with northern exposition or in cold years, only one period of activity (in summer) is found, in lowlands, on slopes with southern exposition or in warm years two peaks of activity (the first in spring) appear" (Löser, 1970). The occurrence of only one period of activity is due to the delay in maturation of hibernated females until the summer, or to the hibernation of the vast majority of a population in the larval stage under constant cool conditions. *A. ater* exhibits an extremely flexible type of annual rhythm. A shift from two peaks to one peak of activity

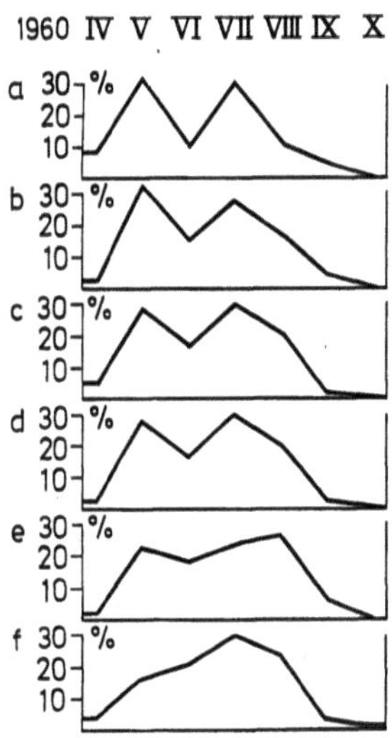

Fig. 131a–f. Transition from a two-peak to a one-peak annual rhythmicity curve in *Abax ater* from warmer to cooler forest habitats in western German hills. (a) young spruce plantation; (b) spruce stand; (c) oak-hornbeam forest; (d) oak-birch forest; (e) beech forest on sandstone; (f) beech forest on limestone, particularly cool. *Ordinates:* percent values of annual catch. From Lauterbach, 1964

has been demonstrated in this species, paralleling the temperature conditions of the microclimate in forest plant communities (Lauterbach, 1964; Fig. 131).

In the summer-wet lowland regions of North Africa (Morocco), where even in winter the temperatures are high enough for their development, potentially non-dormant species can develop throughout the entire year. In central Europe *Pogonus chalceus* and *Bembidion andreae* are univoltine spring breeders and presumably undergo a period of quiescence as adults at the cold time of year. In North Africa, on the other hand, as a result of the favourable conditions, several generations per year are possible (Paarmann, 1975, 1976c).

6. Species Requiring Two Years to Develop

Although it appears likely that this is the case in a number of species (*Carabus silvestris*, Hůrka, 1973, some Cicindelinae, Shelford, 1908; Hamilton, 1925; *Pterostichus metallicus*, Weidemann, 1971) experimental investigation of control of development by environmental factors both in the field and in the laboratory has been reported solely for *Abax ovalis* (Lampe, 1975).

A. ovalis provides an interesting example of the uncertainties connected with drawing conclusions as to the type of development solely on the basis of annual periodicity of activity. In this species one peak of activity is observed regularly in spring and a second peak in late autumn. According to Larsson this would indicate a spring breeder, and *A. ovalis* was in fact at first taken for such (Thiele, 1962). However, for many years the observation was made that in natural light in the laboratory *A. ovalis* that had been caught wild laid eggs in August or September (Thiele, 1961).

This is, in fact, the normal time for *A. ovalis*, which passes the winter in the larval stage, to lay its eggs. Increasing temperatures in the succeeding summer effect a synchronization of pupation and emerging, as a result of which a maximum activity, mainly of young beetles, can be observed with great regularity in September/October. Following hibernation the males are mature and copulate with, as yet, immature females. Maturation of the ovaries only begins with the onset of summer long day. Almost two years elapse, therefore, between hatching of the larvae from the eggs and egg deposition by the resulting females. Whereas the females of *A. ovalis* live for several years and probably reproduce more than once, the males are shorter-lived, a considerable part of the population dying after copulation (Lampe, 1975).

A development lasting two years, as investigated in detail by Lampe (1975) in *A. ovalis*, can probably be interpreted as an adaptation to extreme climatic conditions (cold winter and/or cool summer). Several observations add weight to this argument. Three carabid species live on the subantarctic Crozet Islands (Possession Island, approx. latitude 46° south, 3000 km SE of the coast of S. Africa and equally far from Antarctica). For the two species *Amblystogenium minimum* and *A. pacificum* field investigations pointed to "an unsynchronized larval life lasting a year or more", and for the adults "an unsynchronized emergence in late summer and autumn of adults that do not breed until the following late summer, so that the total life cycle takes two years or more" (Davies, 1972).

Recent investigations also emphasize the possibility (already pointed out by Lindroth, 1949) that species with larval hibernation become two-year animals in regions with severe winters. Concerning *Calathus melanocephalus* van Dijk (1972) wrote: "The hypothesis may be forwarded that going from an Atlantic to a continental climate the larval overwintering species reproduce more and more in the second year or even in later years... the earlier in the year the season conditions become unfavourable (for instance because of very low temperatures e.g. continental climate), the more beetles will reproduce only for a very short period and may be expected to continue reproduction in the next year after overwintering". *Pterostichus madidus*, primarily an autumn breeder, has also proved to be a two-year spring breeder under subarctic conditions (Luff, 1973, according to unpublished data of Houston).

7. *Summary of Results Concerning the Control of Annual Rhythms by External Factors*

The most important factors influencing annual rhythms are photoperiod and temperature. Both factors can exert a greater or lesser degree of mutual influence. This influence is exerted at different stages in development either at the larval stage or during gonadal development in the adult stage. In many cases the control mechanisms in females are more complicated than those in the males (with the exception of the *Poecilus* species investigated).

"All mechanisms known to control dormancy in carabids effect an adaptation of the reproductive rhythm to the annual cycle of environmental factors. Investigations on the requirements of the different stages of development have shown that all types of dormancy are a necessary adaptation of the species concerned, aimed at ensuring that maturation and reproduction occur at such a point in the annual cycle of events that the larvae can hatch under temperatures conductive to their further development. Adult dormancy and the requirements of the larvae are mutually dependent. There can be no doubt that such complex processes of development are the result of a long process of phylogenetic adaptation" (Thiele, 1971a; see summarizing Table 52).

Pterostichus nigrita is an example in which the photoperiodicities of geographically different populations vary to such an extent that it is permissible to speak of physiological subspecies (see also p. 318f.). An intraspecific differentiation of this nature, in addition to the differences in photoperiodic control of dormancy between closely related species, confirms the statement made by Danilevski (1965): "There is no doubt ... that photoperiodic adaptations may serve as one of the important early steps in the process of divergence".

IV. The Hormonal Regulation of Annual Rhythmicity

The first exact description of the endocrine glands in carabids was that of Klug (1958/59) for *Carabus nemoralis*. He reported periodic fluctuations in the volume of the corpora allata in this species over the course of the year, with

Table 52. Types of dormancy and dormancy control in carabids

Adults Season of dormancy (if any)	Type of dormancy	Terminated by	*Larvae* Season of dormancy (if any)	Type of dormancy	Terminated by	Examples
Hibernation	Photoperiodic Parapause	short day → long day	No dormancy			*Pterostichus nigrita* ♀ *Pterostichus angustatus* ♀ *Pterostichus cupreus* ♀ *Pterostichus coerulescens* ♀ *Abax ovalis* ♀
	Photoperiodic Parapause	short day (with temperature optimum at 15° C) → long day	No dormancy			*Pterostichus oblongopunctatus* ♀
	Photoperiodic Parapause	short day (at low temperatures) → long day	No dormancy			*Pterostichus cupreus* ♂ *Pterostichus coerulescens* ♂
	Photoperiodic Quiescence	short day	No dormancy			*Pterostichus nigrita* ♂ *Pterostichus angustatus* ♂ *Pterostichus oblongopunctatus* ♂
	Photoperiodic Quiescence	long day	No dormancy			*Agonum assimile*
No dormancy			Hibernation	Thermic Parapause	Cold	*Pterostichus vulgaris* *Pterostichus niger* *Pterostichus madidus*
Aestivation	Photoperiodic Parapause	long day → short day		Thermic Parapause	Cold	*Patrobus atrorufus* *Nebria brevicollis*
No obligatory dormancy (thermic quiescence possible)			No obligatory dormancy (temporary cold lowers mortality however)			*Abax ater* *Abax ovalis* ♂

maxima in May and September/October, the May maximum coinciding with the peak of reproductive activity. A detailed description of the endocrine glands of *Pterostichus nigrita* was provided by Hoffmann (1969). The number of medial neurosecretory cells of the A type found in the pars intercerebralis (approx. 30) is remarkably high. Ganagarajah (1965) reported the constant number of 16 for *Nebria brevicollis*, and Klug (1958/59) found in *C. nemoralis* that the number varied over the course of the year from about 15 in the inactive periods to about 30 during the active phases. A steep rise in the volume of the corpora allata during the reproductive period as compared with early dormancy was found by Ganagarajah (1965) in *N. brevicollis*. At the end of the reproductive period the volume of these glands decreases again. The amount of stainable substance stored in the neurosecretory cells of the pars intercerebralis is variable in early dormancy, but is always high during the reproductive period. During this stage transport of neurosecretions along the nervi corporis cardiaci is very active whereas it is almost non-existent in early dormancy. In females of *P. nigrita* Hoffmann (1969) found only a slight increase in axonal transport of neurosecretion at the stage of previtellogenesis as compared with that seen at the beginning of dormancy, but reported a large increase at the vitellogenesis stage. The volume of the corpora allata, however, increases steadily throughout all three stages of development. Since, in *P. nigrita*, previtellogenesis takes place in short day and vitellogenesis in long day a close correlation between governing external factors and hormonal activity in dormancy can be assumed. Emmerich and Thiele (1969) were able to elicit previtellogenesis but not vitellogenesis in *P. nigrita* by injection of farnesyl methyl ether. This, in conjunction with the findings of Hoffmann (1969), would suggest that the short-day process (resulting in previtellogenesis) involves mainly secretion of the hormone of the corpora allata and the long-day process (resulting in vitellogenesis) consists essentially in activation of the pars intercerebralis. Although Ganagarajah (1965) observed no significant storage of secretion in the corpora cardiaca in *Nebria brevicollis*, Hoffmann (1969) reported considerable storage during egg maturation in *P. nigrita*.

It has been shown recently that the implantation of corpora allata from sexually mature females (or males) into totally immature females that had been kept under extreme long day leads to previtellogenesis, in contrast to the situation in control animals (Ferenz, 1977). Since injection of synthetic juvenile hormone has an identical effect it seems to be almost certain that in *P. nigrita* the juvenile hormone and the "previtellogenesis hormone" are identical. Ferenz (1977) implanted brains and elicited normal egg formation in this way, thus providing further evidence that neurosecretions are responsible for vitellogenesis.

In many insects the corpora allata alone suffice to trigger off the entire process of egg maturation. This does not apply to *P. nigrita* nor, in all probability, to some other carabids, although the corpora allata hormone alone evokes maturation in the males (injection experiments with juvenile hormone, Ferenz, 1977, 1975a). One and the same hormone, therefore, governs both short-day processes, i. e. previtellogenesis in the females and complete maturation involving the formation of spermatozeugmata and growth of the accessory glands in males.

(Note at correction: Recently Hölters and Könen (pers. communication) detected that in *Pterostichus angustatus*, *oblongopunctatus* and *nigrita* the application of higher doses of juvenile hormone or an application of lower doses in frequent intervals could achieve not only previtellogenesis but also vitellogenesis in the females, so that rather a more intensive stimulation of the corpora allata than a direct influence on oogenesis is the pricipal effect of the activation of the neurosecretory cells following the change from short day to long day.)

The critical photoperiod for both of these short-day effects was found by Ferenz to be similar, i.e. approx. LD 15/9 for previtellogenesis and approx. LD 17/7 for male maturation. The photoperiod response curves for previtellogenesis and vitellogenesis overlap in the region between LD 12/12 and LD 16/8. Ferenz supports the hypothesis that these response curves reflect the quantity of previtellogenesis (=juvenile) hormone and vitellogenesis hormone formed under different photoperiods. In the region where the curves overlap the two hormones are therefore thought to be produced simultaneously in amounts adequate for bringing about complete maturation of the eggs under constant photoperiod. This hypothesis has been confirmed experimentally. For *P. nigrita* from Lapland the critical photoperiod for previtellogenesis is LD 19/5, that for vitellogenesis the same as in central European animals, i.e. LD 13/11. In these animals maturation was possible under constant conditions of 16–18 h of light per day, which is in good agreement with the above hypothesis.

Thus the following picture of the humoral control of maturation in *P. nigrita* emerges: a single hormone, produced under short day, suffices to bring about maturity in the males, whereas for the two steps involved in maturation of the eggs the females require two hormones, the formation and release of which depend upon two different photoperiods.

Under short-day conditions at 20° C *P. nigrita* eats no more than under long-day at the same temperature (Kreckwitz, 1974). Although the animals are night-active, so that short day offers them a potentially longer period of activity, they consume the same amount of food as under long-day conditions.

This fact, together with the experiments described above, points to the conclusion that photoperiod does not simply influence the metabolic balance via the potential duration of activity, but rather exerts a direct influence on the hormone system involved in maturation.

V. Time Measurement in Photoperiodism of Carabids

The control of maturation processes by photoperiod in numerous carabids indicates that the animals are able to distinguish between short and long day. So far only in *Pterostichus nigrita* have preliminary investigations been made into the nature of the processes involved in time measurement (Thiele, 1976).

If, following treatment with one month long day, one month short day or one month continuous darkness (from emerging of adults) the females are exposed to short day interrupted at different times by light of 2-h duration in the dark phase, complete maturation of the eggs takes place in up to 100% of the females, depending upon the timing of the light interruption. If the animals are kept with correspondingly treated males, the normal quantity of fertile eggs are laid. Optimum curves for per cent maturation and number of offspring

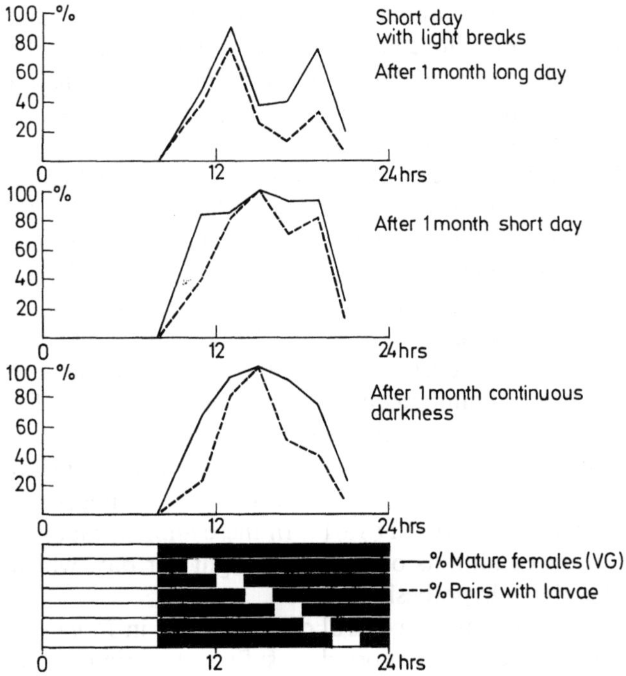

Fig. 132. *Pterostichus nigrita:* Effect on maturation of short day (LD 8/16) with 2-h light interruptions following treatment with one month long-day, short-day or continuous darkness. The ordinate values are given above the temporal positions of the light interruptions, as indicated in the diagram below. VG = vitellogenesis

show a maximum effect of the light disturbances if occurring from about the 12th to the 18th h following the onset of the main light phase (Figs. 132, 133).

This undoubtedly supports the theory that the long-day effect that leads to vitellogenesis is not the result of a summation effect of the total number of light hours per 24-h cycle. The total light in all experiments with light interruptions shown in Figures 132 and 133 is only 10 h, which is well below the critical photoperiod of LD 13/11. Corresponding experiments with a main light phase of only four h led to the same results, although the daily total of light was only six h.

These data provide reasons for assuming that a periodically recurring light-sensitive phase is important for the long-day effect. Vitellogenesis occurs if light falls within this phase. If, on the other hand, it is consistently dark during this phase, then vitellogenesis does not occur.

VI. The Behaviour of Carabids in Winter

Even in the coldest season overwintering carabid larvae can be very active, but since they are mostly hypogeic the degree of activity cannot be judged

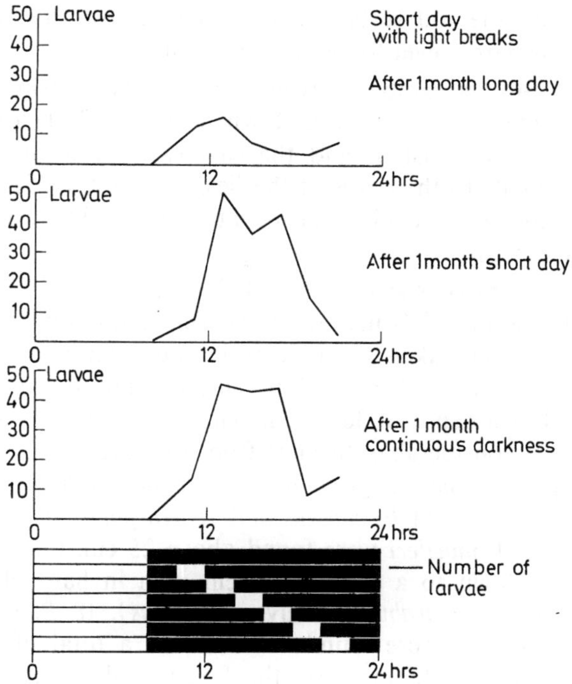

Fig. 133. *Pterostichus nigrita:* number of progeny under different light-interruption programmes. Explanations see Figure 132.

from trap catches. An exception in this respect is provided by those species in which the adults undergo summer dormancy, in connection with which the winter larvae develop particularly rapidly (Thiele, 1969). "Catches of larvae when expressed as percentages of adults trapped in the same area during a year reached figures of 498 % and 42 % respectively at two locations for *Nebria brevicollis* and 20 % for *Patrobus atrorufus* (Thiele, 1964a). In *Leistus ferrugineus* the larvae comprised 1340 % of the adults. Lehmann (1965) in a locality with many *N. salina* found 938 larvae and 1162 adults (larvae = 81 %). Heydemann (1962b) caught 2862 larvae of *N. brevicollis* and *N. salina* besides 6852 adults (larvae = 40 %). Such high percentages of larvae active on the soil surface were not found in other carabid species with winter larvae (where they mostly remain far below 10 %) and seldom in species with summer larvae. This form of activity shows once more that the larvae of these species are extremely adapted to activity in winter climate. On the soil surface they are much more exposed to severe climatic conditions than larvae inhabiting deeper layers of soil. This fact is in agreement with their fast development at low temperatures and their emerging early in the year". The physiological explanation of this extraordinary tolerance to cold is unknown (Thiele, 1969).

The numbers of larvae of *Pterostichus niger* and *P. vulgaris* caught on cultivated fields was only loosely correlated to temperature (Weber, 1965). At −1°C the larvae were still active. In autumn the larvae of *Harpalus rufipes* disappeared from the soil surface, and unlike *Pterostichus* the third stage of this species was hardly ever found above ground.

Adults exhibit scarcely any activity in winter. Adults of *Bembidion* were caught at temperatures down to almost 2° below zero. *Trechus quadristriatus* was less resistant and the catches sank drastically due to migration of the animals to the edges of the fields (Weber, 1965). The present author, however, trapped active adults of the latter species in each month of the year in western Germany.

Carabid species that inhabit cultivated fields usually spend the winter in the soil itself (Scherney, 1961). Ten samples of $0.5\,m^3$ ($1\,m^2$ dug out to a depth of 50 cm) taken from fields in the vicinity of forests yielded 14 *Carabus* of four species, 14 *Pterostichus vulgaris*, nine *P. cupreus* and eight *Harpalus rufipes*, all at a depth of 30–35 cm. The same number and type of samples from fields at a considerable distance from the forest yielded 18 *Carabus* of five different species (*auratus*, *granulatus*, *cancellatus*, *ullrichi*, *purpurascens*), 12 *Pterostichus vulgaris*, eight *P. cupreus* and 15 *Harpalus rufipes*. Only a few *Agonum dorsale* and *A. muelleri* were found above 25 cm, the rest further down. *Carabus* can dig down to a depth of 45 cm even in hard clay soil. On the edges of the fields 35 *Carabus* (mainly *cancellatus*), 18 *P. vulgaris*, 11 *P. cupreus* and 13 *H. rufipes* were found. This makes a total of 5.3 carabids/m^2 remote from the forest, 4.5/m^2 near the forest, and 7.7/m^2 in field borders, probably as a result of migration. These figures agree well with catches of carabids emerging from their winter quarters at the end of the winter on demarcated plots of $4\,m^2$, i.e. 3.5–5.7/m^2.

Carabids from fields near forests migrate into the forest itself during winter and can be found "on the sheltered side of older and partially decayed tree stumps ... where the rotten layer is at least 8–10 cm thick. *Carabus granulatus*, *Pterostichus vulgaris* and *Harpalus rufipes* bury themselves 5–10 cm in the ground beneath this rotten layer. Winter quarters are mainly tree stumps about 5–8 m from the edge of the forest" (Scherney, 1955). Apparently the migration from field to forest in winter is obligatory for only a few species, such as *Agonum dorsale* (see p. 154f.). Seasonal migration of *Pterostichus nigrita* between a bog (in spring and summer) and a forest (in winter) has been described (den Boer, 1965).

Species that spend the winter in the tree stumps themselves are of a quite distinct nature, and usually prefer dry conditions (see p. 195). Such species can withstand very low temperatures and according to Asahina and Ohyama (1969) are also frost resistant. In Japan, adult carabids e.g. *Pterostichus orientalis*, were found overwintering in the deeper xylem layers. "These could survive freezing at −10° C for a few days. However their supercooling ability was very low. Ice formed spontaneously within these insects generally when they were cooled to about −7° C. All of these insects in the xylem of decayed wood were considered to be rarely exposed to temperatures lower than −10° C".

Some species such as *Agonum assimile* and *Agonum dorsale* (see Chap. 3.B.I.) and, in N. America, *A. decentis*, *A. gratiosum* and *A. puncticeps*, tend to aggregate in their winter quarters (Larochelle, 1972b). At the edge of the forest, *Carabus granulatus* was found in groups of 14–63 between the larger roots of tree trunks (Scherney, 1961).

Beetles inhabiting river banks can only overwinter in ice under certain conditions. In a field experiment, 70 carabids of four species were submerged 15 cm in brackish water that froze within two days and then became covered with 10 cm of snow. This experiment was carried out in southern Finland from 9.1.–22.2.1945 (Palmén, 1949); only three *Bembidion doris* and one *Agonum thoreyi* survived. The chances of survival are much greater for carabids overwintering below the ice in hollow stems of reeds, as was reported by Palmén for *Dyschirius globosus*, *Bembidion quadrimaculatum*, *Agonum thoreyi* and especially for *Dromius sigma* which lives mainly in the vegetation layer. In a field experiment involving in each case 50 *Agonum thoreyi*, 64 % survived one month below the ice in hollow reed stems, 98 % in Petri dishes beneath the snow, but only 18 % in Petri dishes above the snow. The insulating effect of the snow and of the air in the reed stems raised the proportion of animals surviving the winter to a remarkable degree.

Chapter 7

Choice of Habitat: The Influence of Connections Between Demands Upon Environmental Factors and Activity Rhythms

The influence of biotic factors on habitat selection was dealt with in detail in Chapter 3 and found to be indecisive. We shall therefore now consider the part played by abiotic factors, the influence of which has already been illustrated by numerous examples. Which factors are decisive, and in what manner can they combine to influence choice of habitat? Let us take some of the species of the genus *Abax* as examples (the following is in part translated from Thiele, 1968b, 1974).

In central Europe *Abax ater* is one of the most common forest-dwelling carabids. It is markedly eurytopic and occurs in forests of every plant-ecological type. *Abax ovalis*, on the other hand, is a creature of the mountain forests of the hill and lower montane regions. In lowlands its occurrence is sporadic, its centre of distribution obviously lying in the tall deciduous forest type known as Fagetalia, which is characterized by moderate coolness and a uniform humidity of the forest floor together with the air layer immediately above it (Thiele and Kolbe, 1962; see p. 22f.). The two species are so much alike that non-specialists are unable to distinguish them unless their individual characteristics are pointed out. Experimental data have helped to explain the differences in their distribution: In a moist, ring-shaped temperature gradient apparatus ("Temperaturorgel", temperature organ) *Abax ater* is eurythermic, the temperature of preference exhibiting a broad maximum between 15° and 25° C. *Abax ovalis*, however, is stenothermic, with a readily reproducible steep maximum around 15° C (Thiele, 1964a). If the behaviour of the two species is tested over a ring-shaped gradient in relative humidity rising from 40% to 100% ("humidity organ"), it could repeatedly be shown (at different temperatures and at various seasons) that both species prefer moisture but that the preference is stronger in *Abax ovalis* (Thiele, 1964a), see p. 61f. In this case the stenotopic species *A. ovalis*, in contrast to *Abax ater*, is also stenopotent with respect to important climatic factors, and its behaviour under experimental conditions is congruent with its distribution in nature among habitats of different climatic types.

However, it would be wrong to regard the ecological distribution of carabids as being merely orientation responses to gradients in the microclimate. *Abax parallelus*, also, is of limited ecological distribution as compared with *A. ater*. Externally, it is even more similar to *A. ater* than is *A. ovalis*, and although it occurs in low country too, it is also obviously concentrated in the tall deciduous forests with an equable microclimate. In temperature and humidity gradients, however, the reaction of *A. parallelus* is identical with that of *A. ater*. Differences in their distribution can be regarded as reflecting very different types of behaviour with respect to provision for and care of their broods (Thiele, 1974). Females of *A. parallelus* dig nests in the ground during the summer, lay about 16 eggs and guard them until the larvae hatch in approx. two to three weeks (Löser, 1970, 1972; see Chap. 3.B.IV.). In order for the brood and the female watching over it to thrive, a reliably uniform soil humidity is essential. *A. ater* behaves quite differently, its females laying the eggs in self-moulded clay cocoons which prevent drying-out, but beyond which there is no further brood care. This means that the brood is much less dependent upon the microclimate than that of *A. parallelus*.

These three species show us that genetically anchored behaviour with regard to microclimatic factors can play an important role in habitat binding *(A. ovalis)*. In other cases, an analysis of microclimatic requirements has not served to elucidate the distributional patterns. It seems that complicated and only recently recognized forms of behaviour play a part in habitat affinity *(A. parallelus)*. Furthermore, certain environmental "requisites" must be available. For example, *Abax ater* can only lay eggs if clay is available for the egg case: if kept in peat in the laboratory it never lays eggs.

What then are the chances of arriving at general conclusions within the family of carabids? In any case investigations will have to be carried out on the behaviour of a large number of ecologically differing species with respect to several factors. Conclusions drawn on the basis of a statistical material of this nature would possess validity for the majority of species of any one ecological group. This can best be illustrated by considering, in a greatly abridged form, the situation of the central European forest and field carabids. The distribution of a species over an experimental gradient can be represented diagrammatically. Given identical methods of investigation, curves can be drawn from mean values for the constituent species of an ecological group, reflecting the average behaviour of the group as a whole (field or forest species, for example, Fig. 134).

The average curve for temperature preference of forest animals lies slightly more to the left, or in a colder region, than that of the field species. The differences in the other factors investigated are much more dramatic. Whereas the field carabids, on an average, are fairly evenly distributed over a gradient in relative air humidity (R. H.), the forest species exhibit a steep maximum in humidity preference at about 100% R.H. Even more pronounced is the strong preference of the forest carabids for the lower light intensities in an intensity gradient apparatus as compared with the indifferent field carabids. To sum up, forest carabids are much more strongly attracted to moisture and darkness than the field species but show only a slightly greater preference for cold. The conclusion to be drawn is that, for central European carabids, on a statistical average, the factors of humidity and light are of greater importance in determining habitat affinity than the temperature factor.

Such investigations, in contrast to those reported in Chapter 5.A.IV. are based upon an analysis of the carabid fauna of a small area and the habitat affinity of its carabid species. The conclusions agree, however, in principle, with the analysis, mentioned earlier, of a larger number of species taken from a greater area.

These conclusions on forest and field species are illustrated in more detail by the data in Table 43 (p. 208). Of the forest animals investigated 87% preferred darkness whereas 47% of the field animals preferred, or were indifferent to,

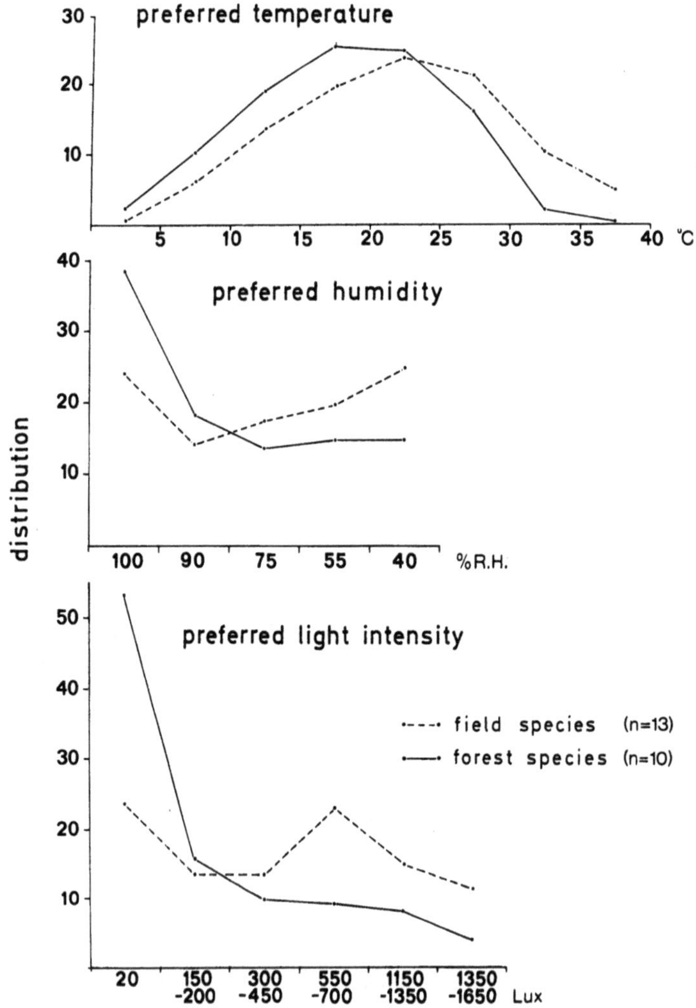

Fig. 134. Average curves from choice experiments of forest and field-dwelling carabids in temperature, humidity and light intensity gradients. *Ordinate values:* percent choosing the different levels of the gradient. The curve for preferred light intensity of the field animals is constructed from observations on only seven species. From Thiele's data, 1964a

light. Only 40% of the forest animals exhibited the preference for cold that their choice of habitat would suggest, but 94% of the field animals chose warmth or were eurythermic, as would be expected from the temperature conditions prevailing in their habitat. Thus the majority of forest animals appear to be more dependent upon the light factor whereas the temperature factor is of greater importance for field species.

Clearly defined reactions are obtained if carabids are tested over temperature gradients into which light gradients have been so incorporated that in one experiment a compound gradient of dark-cold → light-warm exists and in another a compound gradient of light-cold → dark-warm. Distribution in such a "combined gradient" apparatus of this type can be recorded either with reference to the temperature gradient or to the light gradient, and can be compared with the distribution over an isolated gradient. Two types of reaction can be distinguished. One group, whatever the combination offered, seeks the dark zone and even accepts temperature regions widely divergent from its optimum in an effort to attain darkness (Fig. 135). A second group, on the other hand,

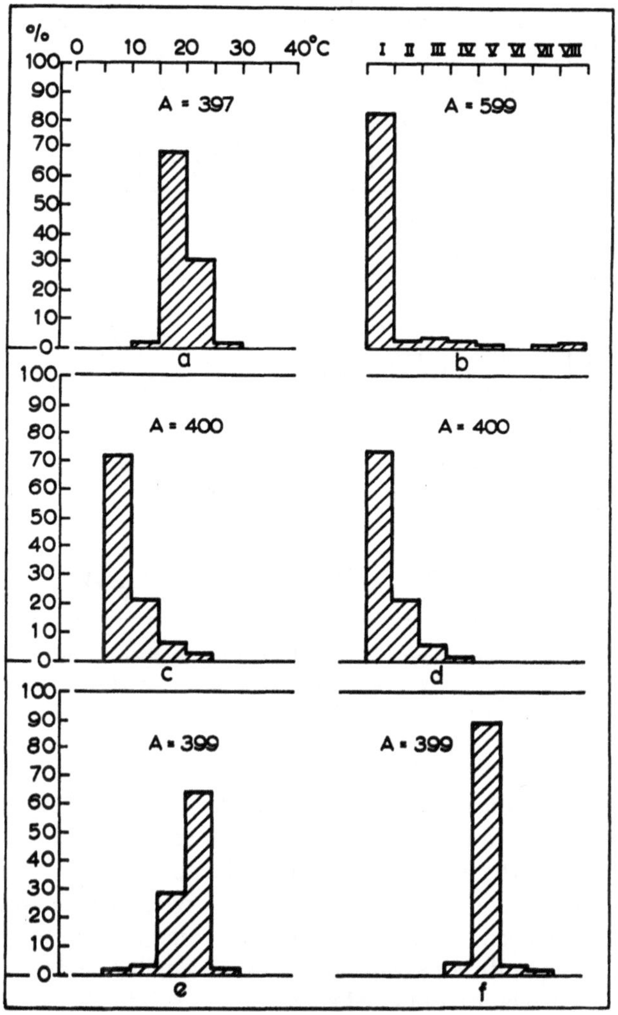

Fig. 135a–f. Behaviour of the forest-dwelling *Nebria brevicollis* as an example of a species in which orientation is largely based upon light intensity preference. (a) Distribution in temperature gradients; (b) distribution in light intensity gradients; (c) distribution in combination experiment with dark-cold; temperature level chosen is recorded; (d) same experiment; choice of light intensity level is recorded; (e) distribution in combination experiment with dark-warm; temperature level chosen is recorded; (f) same experiment; choice of light intensity level is recorded. Light intensity levels range from I = 0–20 Lux to VIII = 2300–2650 Lux. *Ordinates:* Choice in percent of all experimental data. A = number of registrations (usually 10 per animal). From Thiele and Lehmann, 1967

attempts to remain at its preferred temperature, turning up at widely differing light intensities in a variety of combinations (Fig. 136). Forest animals are mainly of the first type, field animals of the second. Again, the former are apparently more dependent upon light conditions and the latter upon temperature (Thiele and Lehmann, 1967).

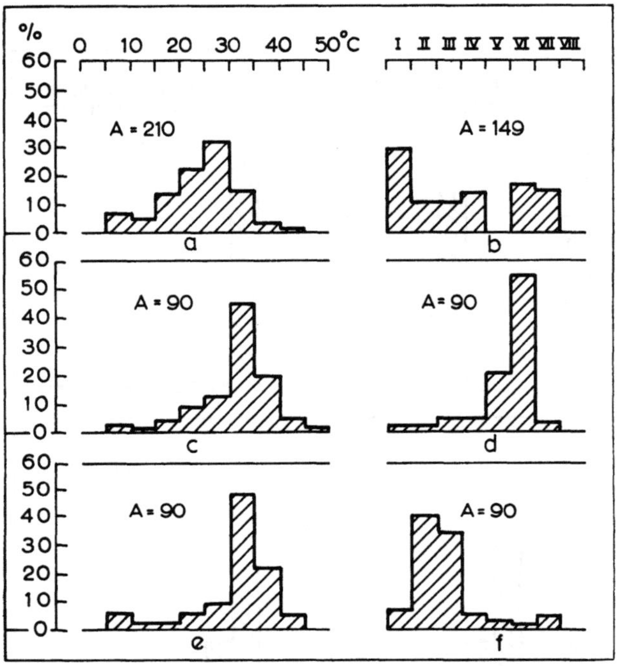

Fig. 136a–f. Behaviour of the field-dweller *Pterostichus cupreus*, an example of a species in which orientation is largely based upon temperature preference. Further explanation, see Figure 135. From Thiele and Lehmann, 1967

Experimentally determined preferences with regard to temperature, humidity and light can be expressed for each species by a value termed the "preference index", whereby lower values indicate a preference for cold, humidity and/or darkness and high values denote a preference for warmth, dryness and/or light (Lauterbach, 1964). If the preference indices are compared with a series based upon quantitative trap catches expressing the degree of affinity to a forest habitat, the correlation obtained substantiates still further the significance of the "preferences" in habitat affinity (Fig. 137). Even in a small vegetational mosaic consisting of deciduous forest, conifer forest and nursery areas, the fauna of an individual vegetational complex is distinguishable by its share in the preference indices (Fig. 138). In stands of mature trees the proportion of carabid individuals with a low preference index is always larger than in the less shady nurseries and is at its highest in the cool, moist beech forests into which a minimum of sunshine can penetrate.

Fig. 137. Connection between preference index (according to Lauterbach, 1964) and degree of binding to the forest. *Each point* indicates a carabid species. *Low index values* indicate preference for cold, moisture and darkness; high values indicate preference for warmth, dryness and light. Degree of binding to the forest is indicated by the percent individuals caught in the forest in an investigation involving quantitative catches in forests and neighbouring clearings. Redrawn from Lauterbach, 1964

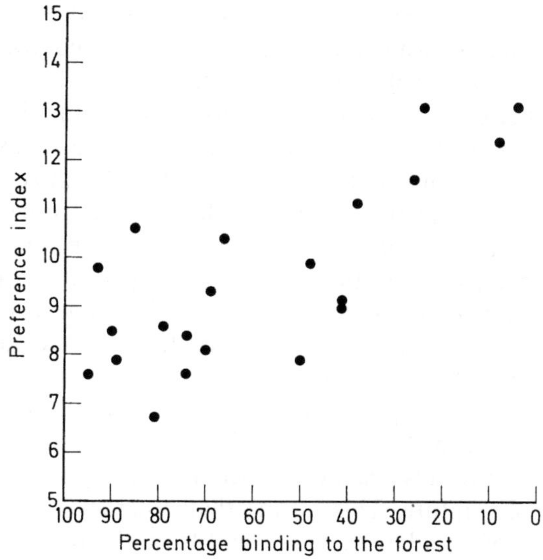

Apparently, some carabid species are predominantly dependent upon their reaction to one particular factor, some upon two reactions and only a few upon equidirectional reactions to gradients in all three factors. The latter type is obviously highly specialized, an example being provided by the inhabitants of cool, moist mountain forests (*Molops elatus*, *M. piceus*, *Abax ovalis:* positive reaction to coolness, moisture and darkness), or of particularly dry, warm field areas (*Carabus auratus*, *Pterostichus cupreus:* positive reaction to warmth, dryness, light). The majority of species found in our cultivated regions exhibit less extreme adaptations and react with greater flexibility to at least some of these factors (Thiele, 1964a; see also Table 42).

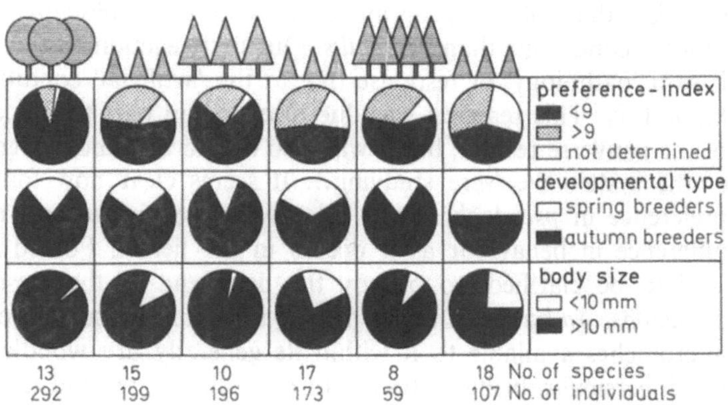

Fig. 138. Comparison of the carabid fauna of neighbouring habitats. *From left to right:* beech forest, young spruce plantation, mature spruce stand, young spruce plantation, spruce copsewood, young spruce plantation. From Lauterbach, 1964

One species, *Agonum assimile*, an inhabitant of cool, moist forests, exhibits reactions which, to all appearances, are contradictory. In a temperature gradient over which the humidity is largely constant (uniformly high air humidity) the species chooses the cool region, but in humidity gradients with constant temperature it chooses the driest zone. It is also extremely resistant to desiccation. The conclusion to be drawn is that the choice of habitat outside the laboratory is determined by a strongly dominating temperature preference. This assumption has been confirmed experimentally by observing *A. assimile* in a dry temperature gradient apparatus in which the temperature gradient is combined with a humidity gradient running in the reverse direction (cool-damp → warm-dry). Here, too, *A. assimile* seeks its preferred temperature rather than its optimum humidity (Fig. 69; p. 195; Thiele, 1964a, 1967).

It seems permissible to draw the conclusion that the species is predominantly guided in its choice of habitat by its preference for low temperatures and that this prevails despite a preference for dryness revealed in experiments at constant temperature. This means that preference for dryness is in itself considered to be of no adaptive value, and from the phylogenetic point of view has probably only been retained because it is not detrimental to the species. Perhaps this interpretation is too superficial, as the following observation shows. The coolest habitats in central Europe, such as the marshy alder woodlands or ash-alder woods on the banks of swiftly flowing mountain streams are also the wettest. *A. assimile*, however, is rarely or never found in such places, in conformity with its preference for dryness. Thus we can also interpret this as a protective mechanism preventing the species from drowning as a result of its extreme preference for cold.

A comparable example was reported by Pittendrigh (1958) for *Drosophila*. "We had expected in comparing the responses of *pseudoobscura* and *persimilis* to find adaptive differences that would explain the mesodistributions observed in the Sierra Nevada. Thus we expected to find *persimilis* selecting moister and darker alternatives than *pseudoobscura*, because, on the average, *persimilis* habitats (on mesoscale) are wetter and darker. We found, however, to our complete surprise, that all the geographic races of *pseudoobscura* we studied selected moister conditions than *persimilis*. Our first reaction to this result was to view it as a 'pathology' of response due to the 'artificial' conditions utilized in the laboratory. This reaction was due of course in part to our strong anticipation of the converse results; but it was also due to our knowledge of the complexity of the system we were studying... It seems clear now that the stronger wet preference in *pseudoobscura* is adaptively meaningful and reflects the normal difference in behaviour as it occurs in the field... Forced by other variables (still to be clarified) to occupy the drier end of the mesodistribution scale, *pseudoobscura* has been under heavier selection pressure to choose the moistest microniches available to it within its generally dry woodlands. On the other hand, *persimilis*, being restricted to generally wetter and darker woods has not been so severely obliged to seek the moistest niches."

Of the factors investigated here light seems to be of particular interest. Preference for darkness would appear to be an advantageous adaptation for animals requiring moisture, like the forest carabids, since dark refuges usually offer adequate moisture. The mechanism of dark preference can thus function equally well by night (if not absolutely dark) and by day. Lauterbach (1964) was able to show that *A. ater* always moves towards the darkest point on its horizon (see p. 201f.): forest carabids erring into open country can thus find their way back to an optimum kind of habitat (Neumann, 1971).

Light is also the most important zeitgeber for the daily rhythmicity of carabids (see p. 237ff.). This means a further adaptive advantage for the forest carabids as moisture-requiring animals, since they are active at night and thus avoid the time of day associated with strong sunlight and the attendant danger of drying-out. Quantitative data from laboratory and field experiments concerning the daily rhythms of species from various habitats illustrate this well (Fig. 139). The different types of behaviour are even better revealed in catches made in

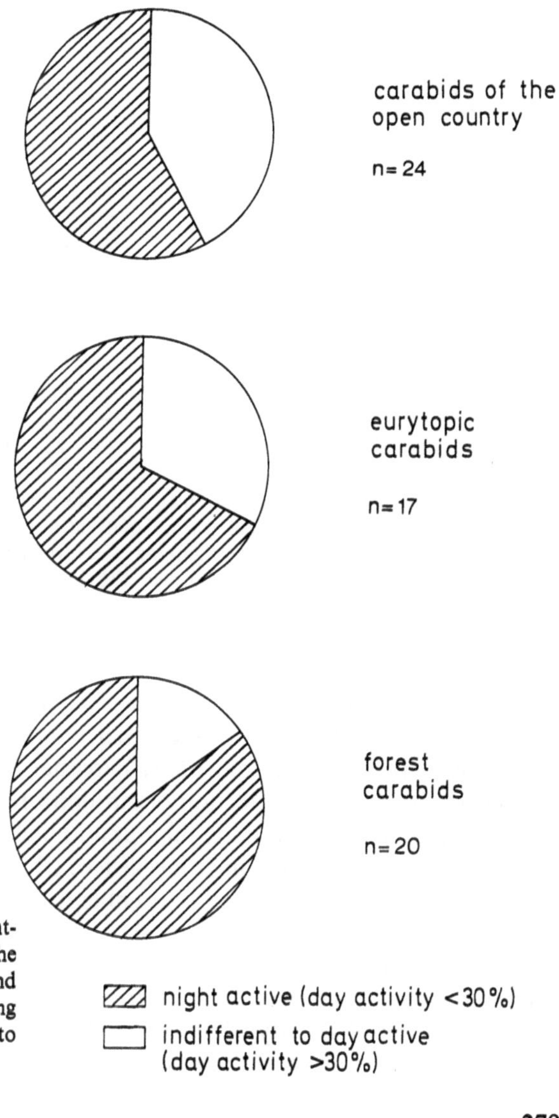

Fig. 139. Proportion of day- and night-activity among the carabid species of the open country and of the forests, and among the eurytopic species occurring in forest and field habitats. According to data of Thiele and Weber, 1968

time-sorting pitfall traps in the forest and immediately adjacent open habitats: the different times of activity of forest- and field-carabids are clearly demonstrated in the habitats (Fig. 140).

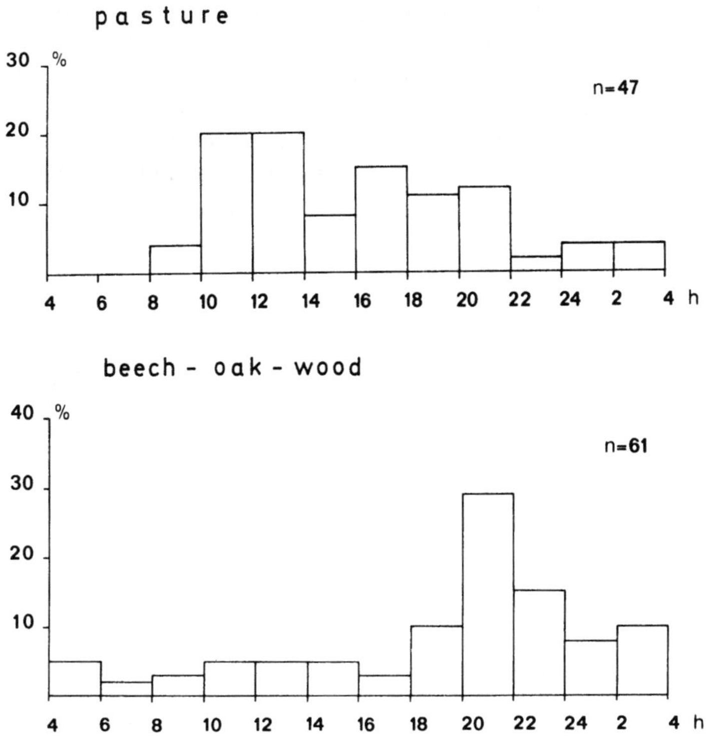

Fig. 140. Daily activity patterns of carabids of a forest habitat and a neighbouring pasture, from catches made in time-sorting pitfall traps. Total catches from mid-April until the beginning of June and mid-September until the beginning of November 1972. Investigations in cooperation with S. Löser at the Wildenrath Station near Erkelenz in the Schwalm-Nette nature park. From Thiele, 1974

Connections have also been established between habitat affinity and type of annual rhythm in carabids. Statistical relationships of this nature have long been known to exist (Larsson, 1939; Kirchner, 1960; Greenslade, 1965; Thiele and Weber, 1968). The proportion of autumn breeders with larval overwintering is higher among forest carabids than among field species, of which the majority are spring breeders with summer larvae (see Chap. 6.B.II.). This was well demonstrated by trap catches on the Bausenberg, a mountain of volcanic origin in the Eifel. It is an almost perfect truncated cone, offering habitats of dry grassland, scrub and, to the NE, tall, cool, moist beech forests (see Thiele, 1974). Trap catches from two years revealed that from spring to autumn a uniform displacement of the peak of activity of carabids takes place from the warm, dry grassland towards the tall, cool beech forests (Fig. 141).

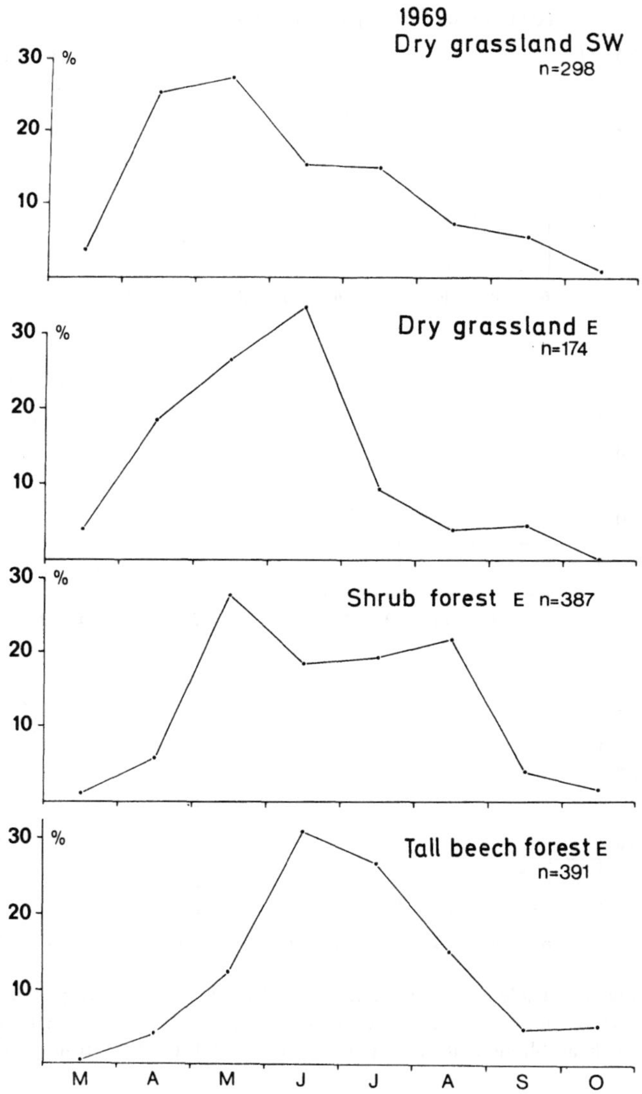

Fig. 141. Yearly periodicity in activity of carabids in plant societies with different exposures on the Bausenberg in the Eifel. *Ordinate:* Percentage of annual catch in the individual months. *Abscissa:* March until October. Compiled from data of Becker, 1975

Figure 142 shows the distribution of carabids at different times of day in one and the same habitat at various seasons. A young plantation was chosen, i. e. a habitat occupying an intermediate position between forest and field. Daytime activity predominates in spring, night-time activity in autumn, and in summer activity was recorded both during the day and at night. This is a reflection of a seasonal displacement in the species spectrum.

How does this fit into what has previously been said? Forest carabids, as we have seen, prefer a relatively high temperature, only slightly lower, in

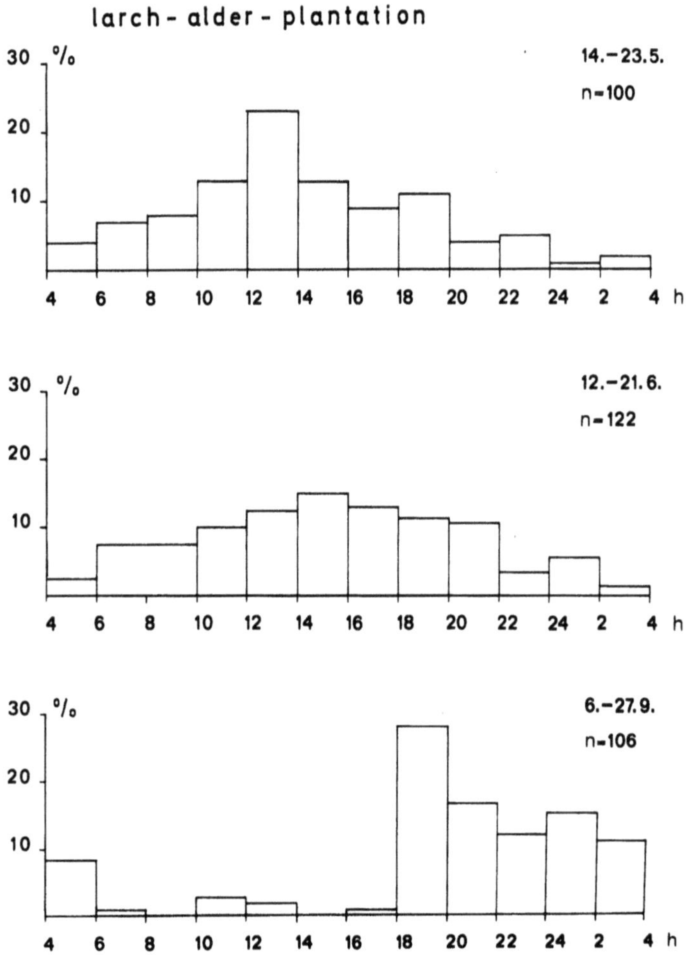

Fig. 142. Daily activity patterns of carabids in a larch-alder plantation near Cologne, using time-sorting pitfall traps, at three different seasons. *Ordinate:* animals caught as percentage of total catch, at 2-h intervals. Incorporating data of Bittner, 1971, from Thiele, 1974

fact, than that preferred by field animals. Since most field species are active during the day, early spring already offers temperature conditions adequate for full activity and reproduction. Nocturnal temperatures at this time are still too low, however, for the night-active animals and not until July–September are they sufficiently high for the activity necessary for reproduction (Fig. 143). Night-active forest carabids therefore reproduce almost exclusively in the autumn (Thiele and Weber, 1968). Close connections thus exist between microclimatic requirements and the daily and annual rhythms in activity. Such complex patterns of behaviour cannot have evolved independently but rather in close interdependence upon one another (Thiele, 1969b). With this statement we corroborate conclusions drawn by Pittendrigh (1958) from ecological studies on *Drosophila*

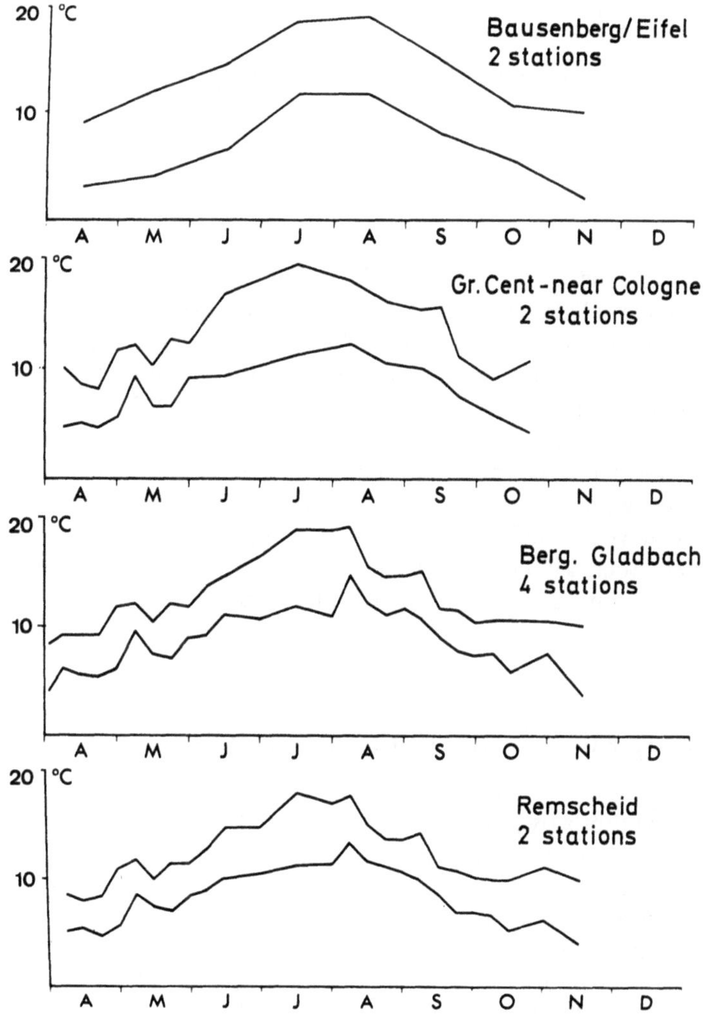

Fig. 143. Annual course of the maxima *(upper curves)* and minima *(lower curves)* in temperature of the soil surface in four habitats in 1972. *Abscissa:* April until December. The maxima in the months April/May correspond to the minima in the months of July to September. Drawn from recordings of K. H. Lampe, from Thiele, 1974

and *Anopheles:* "The organism is not just a system some features of which may or may not be adaptive; the living system is all adaptation insofar as it is organized".

Chapter 8

Dispersal and Dispersal Power of Carabid Beetles

The means of dispersal employed by carabids and the speed with which it is achieved are important aspects in the ecology and evolution of this group of animals. Many studies have therefore been concerned with direct observations on the dispersal power of carabids and discussions of its role in their geographical distribution and evolution, past and present. A basis for this line of research was provided by Lindroth (1945–49), who devoted a large part of his studies on the fennoscandian carabids to these aspects. He later carried out similar investigations in North America (Lindroth, 1961–69, 1969). In 1969 a symposium was devoted entirely to the question that also provides the title for this section (den Boer, 1971 b). However, a study of the publications cited is essential for a closer insight into the subject matter touched upon here[14].

A. Speed of Locomotion of Carabids

The author's own measurements of speeds of species taken from widely differing habitats revealed experimental values of the magnitude of about 10 cm/s at 20° C (between 3.9 and 16.1 cm/s[15]). These speeds were attained over short distances (30 cm) on a smooth surface (Table 53) and would correspond to several metres per minute. In fact, Heydemann (1957) did find comparable speeds under favourable conditions in the field (see Chap. 5.C.II.). With increasing resistance due to lack of space in the habitat, however, the speeds drop rapidly.

Extrapolation from such data to distances covered by carabids over longer periods of time in the wild is clearly not permissible. Nonetheless, trapping of marked animals revealed some surprisingly high values. Traps were set up in concentric circles at varying distances from the point at which the animals were released. The highest figures were recorded for two individuals of *Carabus problematicus*, which covered 70 m and 77 m respectively in the night following their release (Neumann, 1971). *C. problematicus* is a forest species but these individuals had been transported to open country for the experiment. Upon being set loose, they immediately set off in the direction of the forest silhouette

[14] Man himself has in some cases played an important role in dispersal of carabids, as, for example in the introduction of many European species into N. America. Lindroth describes how they were transported in the ballast material of sailing ships. Since, on the return journey to Europe, goods and not ballast were carried, American species were not introduced into Europe by this route. For details see Lindroth's book (1957).

[15] Values of this order were also found by Scherney (1960) for three *Carabus* species.

Table 53. Running speeds of some carabid species

	Speed in cm/s	Animals	n
Stenotopic forest species			
Molops piceus	3.9	4	40
Abax ovalis	9.7	10	100
Abax parallelus	11.0	8	80
Agonum assimile	9.9	10	100
Eurytopic forest species			
Abax ater	12.2	10	100
Pterostichus cristatus	13.0	10	100
Pterostichus niger	16.1	10	100
Pterostichus oblongopunctatus	9.4	10	100
Patrobus atrorufus	5.5	10	100
Nebria brevicollis	12.4	10	100
Eurytopic field species			
Pterostichus vulgaris	8.9	10	100
Pterostichus madidus	9.1	10	100
Agonum dorsale	8.1	10	100
Stenotopic field species			
Pterostichus cupreus	10.6	10	100

(Chap. 5.A.III.3.). "Such extraordinary feats can only be explained by the fact that the beetle is aiming at a certain goal, moreover a goal that is almost certainly cut off from its sight repeatedly by vegetation and irregularities of terrain. Furthermore, it has to traverse or circumvent considerable obstacles blocking its path so that the distances actually covered before its goal is reached are in fact greater than those measured".

In the case of *Pterostichus vulgaris* Kirchner (1960) observed an average speed of dispersal of 3 m in 24 h on a field. One animal even covered 15 m in the first night. Similar achievements were recorded by Skuhravý (1957) for several species on fields. For example, two of 22 *Pterostichus cupreus* trapped after their first night of freedom had covered 30 m, but the majority of the animals were caught at the site of release, and of 50 caught within two months only three were found at a distance of 45 m or more. The maximum dispersal measured amounted to 250 m in one month. In the case of *P. vulgaris* it took four days before one individual was trapped at a distance of 15 m and one at 30 m from the site of release, and only after one month was a single individual caught at a distance of 60 m. *Carabus cancellatus* wandered at the most 15 m in 24 h. In similar experiments with *Carabus cancellatus*, *granulatus* and *auratus* Scherney (1960c) found dispersal values of up to 120 m within the first 10 days (i.e. an average of 12 m per 24 h), and a maximum of 230 m after 50 days. All of these results are in good agreement and indicate average speeds of dispersal of a few metres per 24 h (as the crow flies) for larger carabids. Even for very small carabid species the values obtained were scarcely lower. *Bembidion lampros* wandered an average of 1.6 m/day (maximum 10.0, minimum

0.15 m/day), and *Trechus quadristriatus* an average of 0.4 m (maximum 3.0, minimum 0.05 m/day), according to Mitchell (1963).

In 1911 Burgess reported an experiment in which he placed a larva of *Calosoma sycophanta*, immediately after hatching, on a continually moving strip of paper and let it trace its path with ink. With no food uptake whatever the animal covered 2.7 km in 72 h!

Achievements of this magnitude suggest that dispersal over larger areas is possible even without flight. *Nebria salina* was found in 1941 in large areas of central Europe even though, at least outside Scandinavia, the species was previously strictly Atlantic in distribution. There are no reports of individuals having been caught in central Europe in the 19th century, and the first record in Germany is from 1903; in 1911 it was caught in Bohemia and in 1916 in Saxony (Horion, 1941). It seems possible that this extensive area was colonized within the space of a few decades. The species possesses well-developed wings although it has never been observed in flight, and its close relation *N. brevicollis*, of similar build, only rarely flies (Lindroth, 1945).

B. Concerning the Flight of Carabids

Many carabids are capable of flight. In all probability the well-developed alae and the ability to fly can be considered to be primal (den Boer, 1971 b, contribution to discussion made by van den Aart). Nevertheless, many species have lost their alae in the course of evolution or else they have become polymorphic or dimorphic. This means that within one species and even within a single population individuals with well-developed alae (macropterous) can occur side by side with individuals with reduced hind wings (brachypterous). This phenomenon is of considerable significance for the dispersal of carabids and probably for their evolution as well (Chap. 8.F. and 10.A.). Even animals with well-developed wings are not necessarily capable of flight. *Pterostichus oblongopunctatus* is invariably macropterous although it never flies, even under conditions which induce its close relative *P. angustatus* to take to the wing. In this case it might be that the wings of *P. oblongopunctatus* are simply too small, the ratio of body length to wing length being 1:0.68 whereas for *P. angustatus* it is 1:0.94 (Paarmann, 1966). Another possibility is that the musculature of the alae has become rudimentary (Tietze, 1963). Apart from direct observation, the ability to fly can be detected with so-called window traps. These are vertical pieces of glass supported at various heights above the ground and presenting an obstacle to insects in flight. After collision the animals drop into vessels of preserving fluid. In the recently reclaimed polder region of the Netherlands 31 species were caught in this manner. At the same time 67 species of carabids were caught in ground traps (Haeck, 1971). This is a fairly high percentage of flying carabids, but on newly won land the figure is much higher than in older, stable habitats (see below).

The so-called light-trapping method usually only practiced by lepidopterists also brings large catches of beetles, including carabids, thus confirming their ability to fly. However, the method has been little used so far (Kerstens, 1961; Scherf and Drechsel, 1973).

It should be pointed out that some carabids only fly at certain stages of their life. Paarmann (1966) showed that the young beetles of *Pterostichus angustatus* experience a swarming period in autumn, beginning four to five weeks after emerging and lasting about two months, after which they never fly again.

In the course of trapping with coloured dishes on lightships situated between 6 and 30 km from the nearest coast it could be shown that carabids (*Trechus quadristriatus*, *Bradycellus collaris*) and innumerable other insects can fly large distances (Heydemann, 1967b).

C. Anemohydrochoric Dispersal

This term was introduced by Palmén in 1944. Large numbers of flying insects are carried out to sea by air currents. If they fall onto the water surface they may be carried by ocean currents to foreign shores and thus colonize new land. Palmén demonstrated the significance of this process of dispersal from the Baltic via the Gulf of Finland to the southwest coast of Finland. He found 112 carabid species that had been washed ashore on these coasts. The following species were particularly numerous (more than 50 individuals in each case): *Clivina fossor*, *Bembidion obliquum*, *B. gilvipes*, *B. quadrimaculatum*, *B. doris*, *Acupalpus dorsalis*, *Amara plebeja*, *A. familiaris*, *Pterostichus minor*, *Agonum fuliginosum* and *A. thoreyi*.

This form of dispersal is perhaps nowhere else so common as in the Baltic and it is probably the low salt content of its waters that accounts for the fact that so many carabids survive drifting. In contrast, Lindroth (1963) found only dead or dying individuals among the innumerable carabids washed ashore on the coast of Newfoundland.

D. The Role of Aerial Dispersal in the Post-Glacial Expansion of Carabids

In his classical investigations on the fennoscandian carabids Lindroth (1945–49) proved the value of analysing wing polymorphism for the elucidation of the post- (and probably also inter-)glacial expansion of carabids in this region. In many cases of wing-dimorphic species that are distributed across central Europe but reach their northern limits of distribution in Fennoscandia, it could be shown that the proportion of macropterous individuals increases as the limits of its area of distribution are approached. This also holds for species that in central Europe are almost exclusively brachypterous (Fig. 144). Lindroth regards this macropterous vanguard as a kind of "parachute force", capable

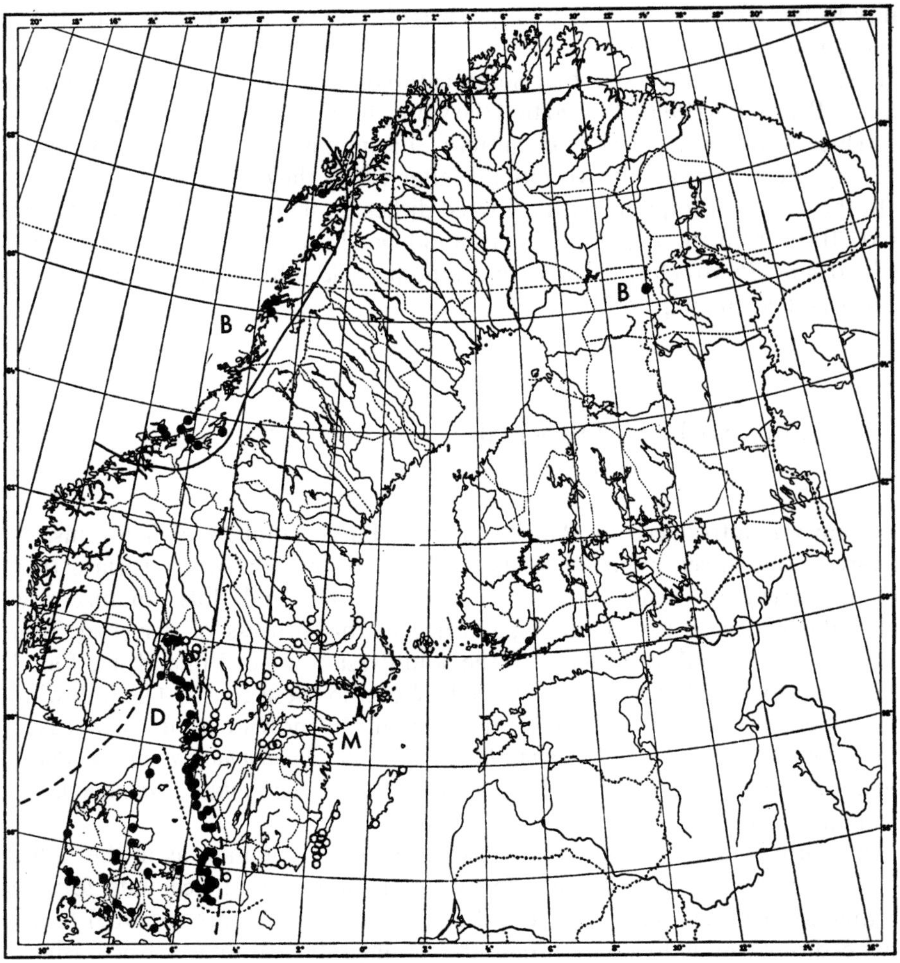

Fig. 144. Distribution of *Bembidion aeneum* in Fennoscandia, showing wing condition. *M:* macropterous; *B:* brachypterous; *D:* dimorphic populations. From Lindroth, 1949

of spreading out more rapidly at the extremity of the area of distribution than the pedestrians. This would at least leave open the possibility that the process of expansion is still incomplete, and in many individual cases Lindroth was able to reconstruct the expansion routes on the basis of his observations. The following detail is of particular interest. In western Norway a number of carabids are found in small areas that are far removed from the main southern Scandinavian–central European area. They are species that are usually dimorphic but in western Norway are exclusively or predominantly brachypterous (see below, as to the selective advantages of brachyptery in small isolated areas). *Bembidion aeneum* is a typical example (Fig. 144). Polymorphic populations are found in southern Scandinavia whereas on the northern boundaries of its area of distribution only macropterous, and in western Norway exclusively brachypterous individuals occur. A likely interpretation is that *B. aeneum* immigrated into

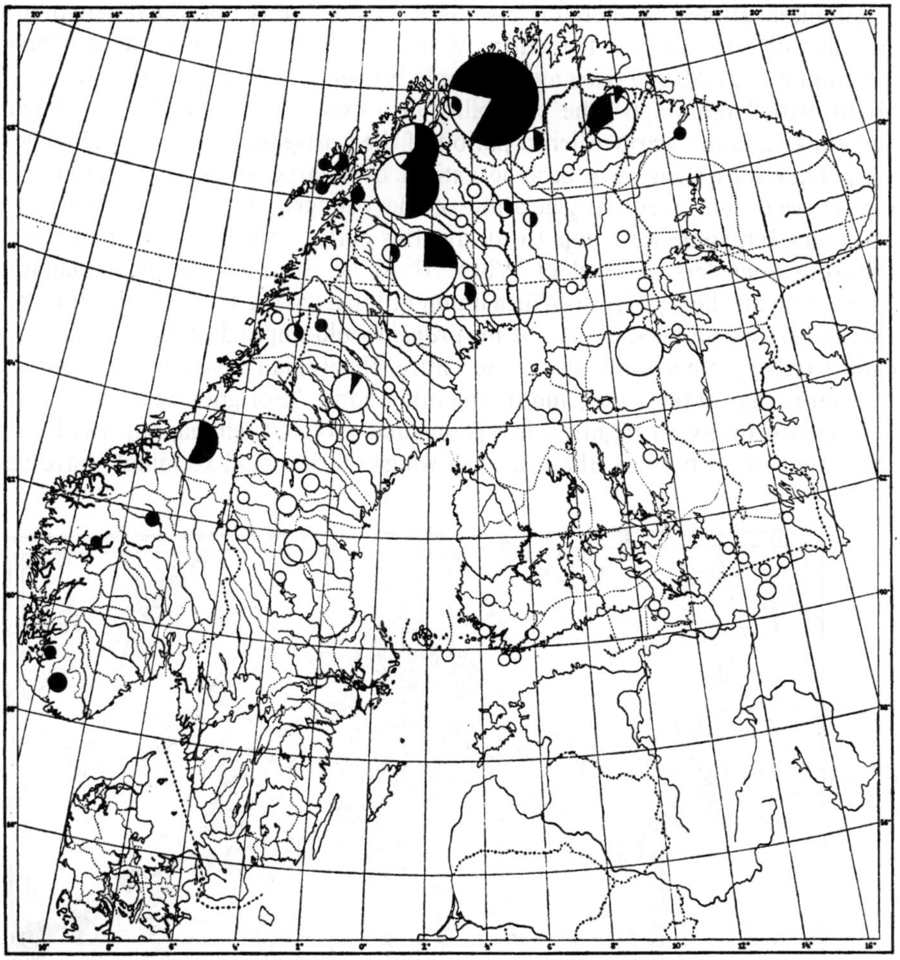

Fig. 145. Fennoscandian distribution of long-winged *(white)* and short-winged *(black)* individuals of *Bembidion grapei. Area of the circles* is in proportion to the number of specimens investigated (1–80). From Lindroth, 1949

Scandinavia from the south during the post-glacial period (which is still in process) and that the west Norwegian animals represent relics of populations that survived in ice-free glacial refugia. In western Norway only brachypterous individuals of populations of *Bembidion grapei* are found, but in moving southeast from this point, the proportion of macropterous forms gradually increases towards the boundaries of the distributional area, where only macropterous populations occur. Since *B. grapei* is not found south of Scandinavia expansion probably took place from glacial refuges (Fig. 145).

Botanists and lepidopterists have also offered arguments in favour of ice-free refugia in western Norway during the whole of the last glacial period (= Weichsel- or Würm-glacial, in North America = Wisconsin Glacial). However, the latest geological evidence speaks against this interpretation. Lindroth (1970a), in more recent publications, has therefore discussed the possibility that the brachypterous carabids did not survive the entire Würm glacial period in western Norway (beginning about 80,000 years ago) but only Würm II, following on the interglacial about 40,000 years ago. The possibility also exists that carabids first reached Scandinavia in the warm Alleröd period of the post-glacial (12,000–10,900 years ago). They may then have become extinct over large areas of Scandinavia in the renewed and extensive glaciation of the younger Dryas period that lasted 500 years, but some were able to survive this period in west Norwegian refuges. Lindroth considers survival of Würm II to be the more likely explanation. "Accepting the theory of post-Würm immigration from the south for the numerous species here exemplified by *Bembidion grapei* would imply dispersal of the genetically fixed flightless form along the entire Norwegian coast in the course of little more than 1000 years", and on the other hand: "For good reasons, selection is supposed to favour flightless forms on small isolated refugia, but it is beyond the realms of probability, that the long-winged form could seemingly have been

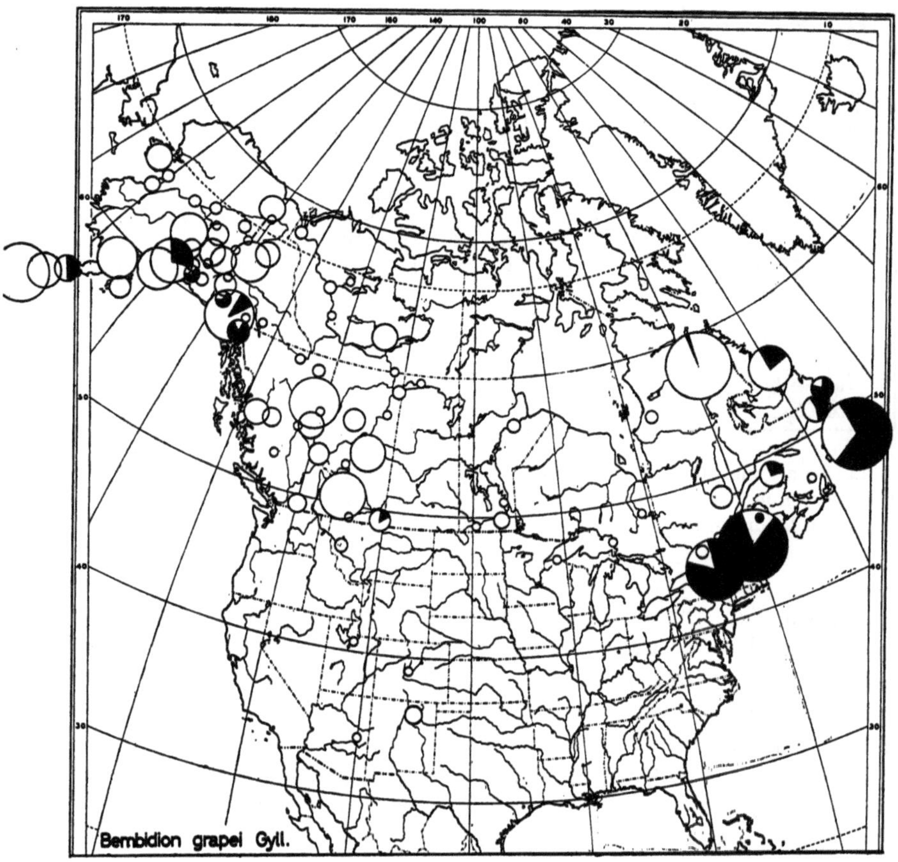

Fig. 146. North American distribution of long-winged *(white)* and short-winged *(black)* individuals of *Bembidion grapei*. The beetles spread from two glacial refugia on the coasts of Alaska and eastern Canada. *Area of the circles* is in proportion to the number of individuals investigated (1–71; the circle for the sample from Newfoundland, 150 specimens, is undersized). From Lindroth, 1969

totally exterminated over a period of 500 years at most, that is during the younger Dryas period". (See also Chap. 9.C.)

The principle of analysing wing polymorphism has also proved useful in studying the evolution of the carabid fauna of North America (Lindroth, 1969; Darlington, 1971; Fig. 146). In North America (Newfoundland) *Notiophilus biguttatus*, imported from Europe, provides an example of a species expanding its range of distribution by means of macropterous individuals (Haeck, 1971; Lindroth's contribution to discussion).

Further examples of the value of such analyses were offered by den Boer (1962). In western Europe *Calathus piceus* is brachypterous, but in the western and central parts of the Netherlands 45 % and 33 % of the populations, respectively, were found to be macropterous, and in Drenthe, a northeastern province of the Netherlands, even 93 % (of 235 individuals). Expansion of the species to central Europe is very recent: until 1860 it had never been caught in western Germany and is still in the process of expanding to the east. So far, only macropterous individuals have been found near Hamburg, in Denmark and in southern Sweden.

In Drenthe the boreo-montane species *Agonum ericeti* is at its southernmost limits. Since all individuals are brachypterous at this point, as in Scandinavia, den Boer assumes that the boundary is not recent but rather one of long-standing (macropterous individuals of this species do exist and have been encountered in the central European mountains). A different situation prevails with regard to the sub-boreal species *Agonum fuliginosum.* Whereas exceedingly few of the hundreds of Scandinavian individuals examined were macropterous, 14 were found among 55 individuals from the moor regions of Drenthe. This may indicate that the population in Drenthe is much more recent than that in Scandinavia and it can be assumed that the species has only recently spread to central Europe from Scandinavia.

E. Contemporary Processes Involved in the Expansion of Carabids to Land Freshly Available for Colonization

The above assumptions concerning the importance of wings for dispersal can in some cases be put to the test. One example is provided by newly won land, for the first time available for colonization by carabids and other organisms.

A case of this type is offered by the deep open lignite workings in western Germany. The spoil banks are replanted with trees and very soon a typical pioneer carabid community, rich in both species and individuals, is established (U. Neumann, 1971). In three-year-old poplar plantations the number of species and of individuals reached an absolute maximum. Many familiar species from cultivated fields were encountered (genera *Bembidion*, *Harpalus*, *Calathus*, *Amara*—the most frequent was *Pterostichus cupreus*, followed by about half as many *P. coerulescens*). Drastic differences in the carabid community were observed in the 11-year-old plantations as a result of the denser crown canopy and the approximation to a forest-like microclimate. Carabids typical of a "pioneer forest community" were found, headed by *Agonum obscurum*, *Pterostichus oblongopunctatus*, *Calathus piceus* and *Stomis pumicatus*. Up to this point winged species were quite obviously in the majority, accounting for 80% of the total. On the other hand, the older plantations of 25 years' standing or more had been reached by pedestrian forms, the dominant type being typical forest animals such as *Abax ater*, *Carabus nemoralis*, *Abax parallelus* and *Pterostichus madidus*. The proportion of species capable of flight sank to below 20% and in the mature forests in the same area, investigated for purposes of comparison, the figure was less than 10% (Fig. 147).

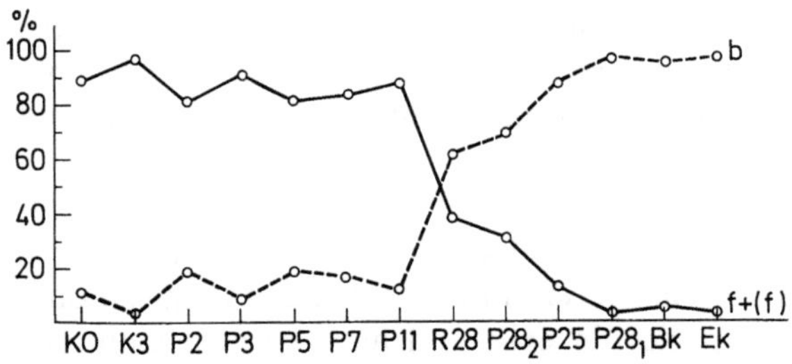

Fig. 147. Proportion of the carabid species demonstrably capable of flight, *f*, and probably capable of flight *(f)* in a carabid population in a succession series in reforested spoil banks of the open lignite workings near Cologne. *b*: the reciprocal proportion of brachypterous species. *K0*: fresh; *K3*: three-year unplanted spoil banks; *P2* to *P28*: two to 28 year-old poplar reforestations. *R28*: 28-year red-oak reforestation; *Bk*: natural beech forest; *Ek*: natural oak forest for comparison. From Neumann, 1971

A gigantic natural experiment is provided by the Ijssel sea polders, newly reclaimed from the sea in the Netherlands. Their colonization by soil arthropods forms the subject of thorough and long-term investigations (den Boer, 1968b, 1970; Haeck, 1971). On the polders of east Flevoland it could be shown that a large percentage of the individuals of wing-dimorphic species were macropterous, whereas on old land (Province of Drenthe) none, or only a very small percent, were observed in populations of the same species (Fig. 148): the per cent ratio for *Trechus obtusus* was 95:4 and for *Dyschirius globosus* 48:0. The "parachute force" hypothesis (p. 287) seems to win direct confirmation from these figures. It is odd, nevertheless, that Haeck (1971) caught none of these species in window traps in the polders of south Flevoland. Even in Drenthe, only on one occasion were two individuals of *P. strenuus* found in traps of this kind. None of the species in Figure 148 was ever directly observed in flight by Lindroth (1945), although he emphasized that the occurrence of 18 of the otherwise rare macropterous individuals of *Calathus melanocephalus* in sea drift from southwest Finland points to an ability to fly.

Some light may be thrown upon this apparently contradictory situation by an observation made by Neudecker on *Agonum assimile* (unpubl. verbal commun.). The species is consistently macropterous, but its wings are relatively

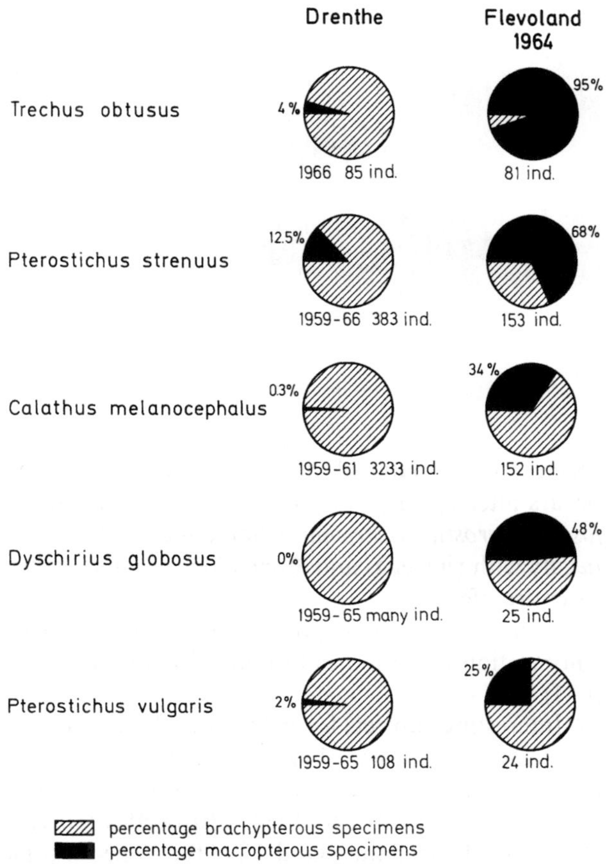

Fig. 148. Winged specimens *(black sectors)* of dimorphic carabid species in old (Drenthe) and young (Eastern Flevoland Polder) populations of carabids in the Netherlands. From den Boer, 1968b

small and Lindroth (1945) is of the opinion that the species is "certainly neither a good nor habitual flier"; he knows of only one direct observation of flight, from 2.7.44. The species has been kept and bred in our own laboratory for many years but has never been seen to fly. Neudecker, however, observed that young laboratory-bred animals spread their elytra in autumn in strong sunlight, unfolded their alae and undertook short hopping flights over a few centimetres (Fig. 149). It appears possible, therefore, that in open country exposed to very strong winds, like the polders, macropterous carabids can achieve dispersal much more rapidly than the brachypterous individuals, since their unfolded alae offer a considerable area of resistance to the wind. Experiments have been reported by Meijer (1971) from the dyked-in Lauwerszee in the northern Netherlands, where even wingless individuals of *Dichirotrichus pubescens* and *Pogonus chalceus* were carried by the wind close to the ground. "So it seems very likely that carabids are blown into the Lauwerszee once they have reached a barren area." In a further investigation in 1974 Meijer found 50 of the freshly immigrated

Fig. 149. *Agonum assimile* spreads its elytra and unfolds ist alae for a "hopping flight". Actual size ca. 1 cm. Photo, Neudecker

species to be capable of flight. In numerous species autolysis of the flight muscles occurs after egg ripening (*Dichirotrichus pubescens*, *Bembidion iricolor*, *Clivina fossor*, *Pterostichus strenuus*). Some species, however, such as *Trechus quadristriatus*, fly with ripe eggs and *Bembidion varium* remains capable of flight throughout its entire life.

It has already been pointed out that the ability to fly is invaluable in the colonization of periodically flooded *river banks* and small islands in rivers (see p. 41).

Interesting subjects for field research are *recently emerged islands*. Since 1959 an island has formed 14 km off the North Sea coast on the "Knechtsand" by flying sand having collected on wattle fences that were erected to protect sea birds. From 1972 to 1974, 210 beetles (50 carabids) were encountered on the $500 \times 250\,m^2$ of the island. 32 of these species (nine carabids) can be considered to be well established (continuous development on the island). Eurytopic species dominated among the carabids. Anemohydrochoric dispersal is assumed for *Dyschirius globosus*, which was only found in the brachypterous form. Introduction by man seems possible for a few species (*Broscus cephalotes*, *Calathus fuscipes*, Topp, 1975).

Extensive investigations are in progress on Surtsey, the already famous volcano that appeared off the coast of Iceland in 1963 (since 1965; Lindroth, 1970b, 1971a). In the four years up to 1968, 70 species of arthropods, mainly Diptera (43 species), had already been caught there. "*Amara quenseli* Schnh. is the only carabid beetle found alive on Surtsey... The specimen was a female with fully developed hind-wings and there is little doubt that it had arrived by air. The species is widely distributed in Iceland, also on Heimaey, the main island of the Westman group. However, it seems to occur on the last-named place exclusively in its brachypterous form (more than 300 individuals investigated). On the mainland of southern Iceland, in a distance of at least 30 km, populations are mixed, though with strong preponderance of the brachypterous form. This apparently was the region of departure for the Surtsey female. Due to lack of food, no carabid beetle is able to colonize Surtsey permanently at present time" (Lindroth, 1971a). It is thus surprising that as early as 1967 a macropterous carabid female of a predominantly brachypterous species landed on Surtsey from a distance of at least 30 km, underlining the importance of the role of rarely occurring macropterous individuals in dispersal.

In 1946 Lindroth showed that in the carabid *Pterostichus anthracinus* macroptery is recessive (p. 297). If this is of general validity it would mean that all macropterous carabids are homozygous. Considering the rarity of macropterous carabids in the original population, their high rate of occurrence in invaded regions can best be explained by relatively frequent meeting-up of homozygous migrants of the two sexes. It is unlikely that already fertilized "pioneer females" land in the newly-settled regions. Den Boer (1971) found only two individuals of two species containing ripe eggs among 89 carabid females caught in window-traps. Nevertheless, carabid females are often inseminated before their eggs have ripened. This would at first suggest the likelihood that insemination could be performed by the more common brachypterous males, and the resulting progeny would thus be largely heterozygous and phenotypically brachypterous. However, Meijer observed (in discussion with Haeck, 1971) that: "Among high numbers of brachypterous individuals a few macropterous males and females preferred to copulate with each other and not with brachypterous individuals". Doubts as to the general applicability of this statement have been cast by Lindroth (personal communication).

F. The Adaptive Significance of Dispersal Processes

The question as to whether the possession of a high dispersal potential is of adaptive value for a species has been put by den Boer (1971 b). A species with a strong tendency to dispersal suffers high losses, since the majority of the emigrating individuals land in unsuitable habitats or simply die without progeny for lack of a sexual partner at the right moment. At a rough estimate, about 4.5 billion insects per day are lost by drifting at the height of summer on the North Sea coast of Germany. This is equivalent to 270,000 kg for the months of July and August, merely from a coastal strip 30 km wide. "This biomass must arise largely from biocenoses in the vicinity of the coast and represents the tribute that has to be paid for the dispersal activity of numerous species" (Heydemann, 1967 b).

Two theses, one termed the "overflow hypothesis" and the other the "foundation hypothesis" have been put forward to explain the significance of powers of dispersal. The former hypothesis regards migrating individuals as a chance side effect of the increased numbers of a species, or, in the final analysis, a result of intraspecific competition by which part of a population is crowded out of the optimum environment. Primarily at least, the hypothesis does not connect any adaptive advantage with the possession of certain organs or types of behaviour facilitating rapid dispersal. Den Boer supports the foundation hypothesis and regards high powers of dispersal as being of selective value in diminishing the risk of extinction of a population. This seems to be particularly plausible in all cases where unstable habitats are colonized (Darlington, 1943).

For example, river bank zones within reach of high water are only exploited as habitats by species capable of flight. A particularly good example is provided by *Pterostichus angustatus* (Paarmann, 1966). The peak in distribution of this species lies in continental Europe and it is extremely stenotopic in Atlantic western Europe. Being a species with a preference for warmth and dryness, the larvae of which, especially the 1st instar, require plenty of warmth (p. 191), it is found scattered in western Europe in forest clearings, clear-felled and burned areas. Species which have a strong affinity for these unstable habitats can only colonize them if they have a highly dynamic ability to disperse. This was seen by Paarmann in *P. angustatus*, with its "ver sacrum"-like swarming phase in the young beetles. In the first year of his investigation (1963) Paarmann caught 64 individuals in a forest clearing as compared with six in the following year (1964) and none whatever in 1965 (see Paarmann, 1966). The decrease in numbers paralleled the disappearance of open spaces and the growth of *Molinia coerulea* grasses.

From the large numbers of individuals of some species caught in window-traps as compared with pitfall-traps *(Bradycellus harpalinus, Amara plebeja, A. familiaris, bifrons* and *apricaria)* den Boer (1971b) concluded "that in the life-cycle of the latter species dispersal must play a relatively important role. It may even be supposed that the small numbers of individuals caught in pitfalls result from a very great 'investment' in dispersal by flight... In other words such species would live more or less 'nomadic'".

Den Boer, in keeping with his theory of "spreading of risk" (p. 71) regards high powers of dispersal as being advantageous not only for species from "unstable" habitats but, as he says: "apparently flight is not only 'adaptive' in species living in temporary habitats, but more generally in species living in temporary (sub)populations".

A question requiring particular explanation concerns the loss of the ability to fly due to reduction of the hind wings. It is clear that in stable and smaller habitats this would present an advantage (in analogy to wing reduction in other insects on storm-swept islands, e.g. the Kerguelen flies, *Aptenus, Calycopteryx, Amalopteryx, Anatalanta*). This explanation has already been put forward by Lindroth for the wingless carabids in west Norwegian refuges, and Darlington (1970) has dealt with the question at some length for the tropical fauna. Most carabids of tropical lowlands are small and winged, good examples being found in South and Central America as well as in the Antilles and New Guinea. Near Santa Marta in Columbia Darlington found only two wingless species among 134, and on Barro Colorado (Panama) all 76 species encountered were small and winged. "Larger Carabidae and flightless ones do occur in Central America, but most are at considerable altitudes on the plateaus or mountains... This is apparently not a result of exposure on mountains, for the Carabidae concerned live in dense wind-free cloud forest... Lowland Carabidae tend to exist as widely dispersed populations in patchy and often unstable habitats, and the populations themselves may be patchy even when habitats are continuous... Under these circumstances populations are maintained by continuous redispersal, often by flight. Mountain Carabidae, however, exist in much smaller areas, their ranges on West Indian mountains being often of the order of one or a very few square miles against many thousands of square miles for most lowland species. Mountain habitats are likely to be continuous, within their narrow geographic limits. And mountain habitats are likely also to be relatively stable, because of increase of amount and regularity of rainfall with increasing altitude... It seems likely that, under these conditions, mountain Carabidae do (and must, if they are to survive) form fewer but denser, less patchy, and more stable populations than lowland Carabidae; that redispersal by flight ceases; *and that possession of wings becomes selectively disadvantageous...*" (Present author's italics).

Unfortunately, we know very little about the inheritance of wing dimorphism. Lindroth (1946, 1949) stands alone in having proved in breeding experiments on *Pterostichus anthracinus* that brachyptery is simple dominant to macroptery. It is of course questionable whether generalizations are permissible on the basis of this one finding. It would not necessarily alter the explanation of the rapid appearance of macropterous forms in recent Dutch polders: it is the macropterous forms that emigrate, are thus seldom seen in stable habitats, and perish if they cannot find a suitable (new) habitat in which to multiply rapidly. In the initially macropterous founder populations dominance of brachyptery would account for the relatively rapid elimination of long-winged forms under adequate selective pressure. Assuming a selective advantage of 0.1 % for brachyptery, Lindroth (1949) lists the following time intervals required for the proportion of brachypterous individuals in a population to increase as shown (one generation = one year):

From 0.001 % to	1 % brachyptery:	6920 years
1 % to	50 % brachyptery:	4819 years
50 % to	99 % brachyptery:	11,664 years
99 % to	99.999 % brachyptery:	309,780 years

At first the percentage of brachypterous forms increases relatively rapidly, but it takes a very long time before the macropterous forms completely disappear from the population and, even so, it can never be proved that the *gene* for macroptery is completely eliminated. This is in good agreement with observations. In addition the figures suggest a solution to the problem of the west Norwegian refugia, since elimination of macropterous forms would seem to be improbable in the mere 500 years of the younger Dryas period, even in the face of great selective pressure.

Chapter 9

Ecological Aspects of the Evolution of Carabids

A. Geological Age of the Adephaga and Carabids: Fossil Evidence

Compared with the large numbers of recent species of carabids, fossil evidence, even in the later geological deposits, is not as a rule particularly abundant. In an upper Pliocene site near Willershausen, West Germany, for example, Gersdorf (1971) found only four species of carabids among 56 classifiable fossils of beetles. Of 32 further specimens that could not be classified with certainty, one at the most may have been a further species of carabid.

The oldest fossilized carabid, *Umkoomasia depressa*, was described by Zeuner (1961) from middle Triassic deposits in Natal, South Africa, although Hennig (1969) considers this classification to be questionable. According to his view, *Umkoomasia* could be an earlier nonspecialized representative of the Adephaga, which may not have split up into families before the Jurassic period. But, on the other hand, even the attribution of Jurassic Adephaga to different families (e.g. that of *Grahamelytron crofti* Zeuner to the carabids) is open to criticism. "Unfortunately, we possess next to no fossilized beetles from the Cretaceous period" (Rüschkamp, 1932).

It seems unlikely, therefore, that carabid evolution dates back to the Paleozoic, but that the development of the family commenced, at the earliest, in the Triassic. Not until the Tertiary does the fossil documentation of the carabids become more abundant. About 30 genera had been identified by 1949 from Baltic amber (Oligocene) (Bachofen-Echt, 1949). The cicindelinid *Tetracha carolina* L. from amber is a species still in existence today, and perhaps the only amber species of insect that has survived up to the present time. Some doubts exist, nevertheless, as to the validity of this statement (see Lindroth, 1957). Although other recent genera have been found in amber, no species belonging to these genera is in existence today, a fact which would suggest a rapid evolution of the carabids in the Tertiary. Upper Pliocene fossils, on the other hand, conform well with recent carabids. The five Upper Pliocene carabids mentioned from Willershausen in northern Germany (three to five million years old) are all identical with recent species or genera *(Calosoma sycophanta, C. inquisitor)* (Gersdorf, 1969). No fossil beetle from Willershausen could be identified with a Miocene species (Gersdorf, 1969). In investigations on the phylogeny of North American Cicindelinae Freitag (1965) points out that the early Tertiary "was a time of extensive mountain building", an epoch, therefore, that favoured geographical isolation as a factor in speciation. The Pliocene, in contrast, was largely a period of geological quiescence.

A rapid evolution during the early epochs of the Tertiary seems to have taken place in the family Paussidae, which belong to the Isochaeta, a sister group of the carabids (see Chap. 1.A.). The recent Paussidae comprise about 375 tropical and subtropical species, which are all myrmecophilous. Nearly all the recent species show a striking reduction of the number of antennae segments from originally 11 down to only two. 20 species out of seven genera are known from the Baltic amber in the early Tertiary, all of them belonging to the most primitive Paussid stock. All these *species* have died out, only one *genus* from the early Tertiary (*Arthropterus* in Australia) surviving (Bachofen-Echt, 1949).

B. Centres of Development and the Routes of Dispersal of Carabids

Numerous and diverse adaptations exhibited by carabids to particular climates (e. g. in dormancy control), bring up the interesting question as to which climatic region of the Earth saw the development of the ancestral forms of our present-day carabids. The Migadopini (Darlington, 1969), a tribe confined to the southern regions of the Southern Hemisphere, is particularly interesting for zoogeographers. Four genera of this tribe live in Australia and Tasmania, four in New Zealand and seven in South America. The genera differ enormously from one another, and each is confined to one of the three regions mentioned. This does not support the idea, put forward in discussion, of a common ancestral form of the Migadopini on a hypothetical ancient southern continent. The fact that the nearest relations of the tribe are the phylogenetically primal Elaphrini, which are confined to the Northern Hemisphere, makes it thus more likely that the ancestral form of the Migadopini lived in the Northern Hemisphere, that the Migadopini themselves migrated across the tropics and became extinct in the north. The only relics in existence today are in the Southern Hemisphere. The assumption of a similar process in the case of the Broscini tribe is justified, since representatives are to be found in the northern and southern regions of the Earth but none whatever in the tropics. Various forms of the tribe of the Bembidiini occur in all parts of the world, although few species inhabit the tropics, and these mainly in the high mountains.

Findings of this nature suggest that the oldest and main centres of differentiation of carabids were situated in the Northern Hemisphere, but the question is where? Until recently it was thought that the tropics possessed a much less varied carabid fauna than the temperate zones and that only Cicindelinae exhibited any large degree of diversity of species in the former regions. More exact systematic investigations in tropical latitudes, however, have led to a revision of this idea. For North and Central America Willis (1972) was able to show a peak in species density for Cicindelinae in the southwestern U.S.A. (30 species), whence it decreases sharply both to the north and towards the tropical regions of Mexico (northern Canada and Yucatan, five species each, Fig. 150).

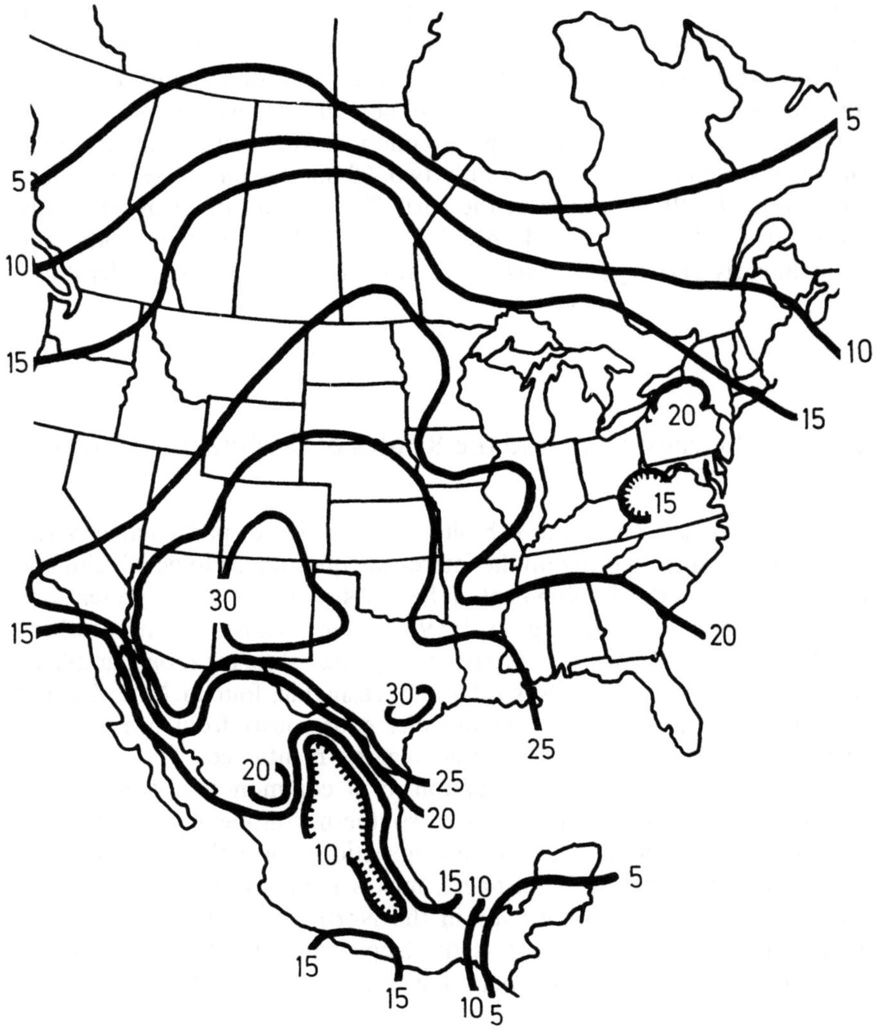

Fig. 150. Lines showing occurrence of equal numbers of species of Cicindelinae in North America and Mexico. From Willis, 1972

In the course of his studies on the carabid fauna of New Guinea Darlington (1971) devoted particular attention to the question of its evolution and to the history of its distribution. He had previously held the opinion that the number of carabid species decreased from temperate North America to the American tropics, although exact data were not available. In New Guinea, however, the carabid fauna proved to be of a diversity exceeding all expectations for tropical lowlands. It thus appears that in tropical lowlands of certain regions at least, carabids may be represented by a large number of species even though hard to find: perhaps the populations tend to be low in numbers of individuals.

On the basis of his investigations in New Guinea Darlington came to the conclusion that the tropics of the Old World were the largest centres of evolution and dispersal of carabids[16]. Australia, although less important, was another such centre.

In view of the many recent physiological and ecological studies on Pterostichini and Agonini it seems appropriate to quote Darlington's conclusions concerning their distribution: "Pterostichini and Agonini tend, as dominant tribes, to be complementary over the world as a whole... Both tribes are cosmopolitan, but unevenly so. In some parts of the world they occur in nearly equal numbers; in others one tribe or the other is overwhelmingly dominant. The tribes tend to be complementary within the Australian region, ... In (the whole of) Australia ... (with Tasmania) Pterostichini are dominant, with more than 350 known species against probably less than 20 species of Agonini, a ratio of nearly 20/1. But in New Guinea Agonini are dominant, with considerably more than 100 known full species ... against about 40 species of Pterostichini ... a *reversed* ratio of about 3/1...

"Over the world as a whole, there is a tendency for Agonini to be better represented in the tropics and Pterostichini in the temperate zones. Also it is probable that Agonini, which are phylogenetically less diverse, are more recent in origin than Pterostichini and that they have dispersed more recently. It is therefore likely that Pterostichini are dominant in Australia partly because Australia is more temperate than tropical in climate and partly because Pterostichini reached Australia before Agonini, and it is likely that Agonini are dominant in New Guinea partly because the climate there is fully tropical and partly because the carabid fauna of New Guinea is more recent in its origins than that of Australia...".

Darlington words his interpretations with extreme caution, and he emphasizes that his conclusions are of a tentative nature. Undoubtedly, the Earth's history has involved extremely complex currents and counter-currents in distributory movements. In any case is seems feasible that in the southern Asiatic area, over the course of a long period of time, large groups of carabids spread to New Guinea and Australia from Asia and the southern Asiatic islands, whereas the spread of carabid groups from Australia and New Guinea to Asia was obviously much smaller.

Goulet (1974) has arrived at well-founded conclusions as to the chronological sequence of evolution and dispersal of *Pterostichus* species of the subgenus *Bothriopterus* in North America. North American species of the subgenus form two groups, a boreal one to which *P. adstrictus* belongs, and a temperate one comprising all other species. *P. adstrictus* alone is widespread in Eurasia and is closely related to Eurasian species.

"The most diversified branches of *Bothriopterus* are found in Eurasia... These three points suggest that the area of origin of the Nearctic *Bothriopterus* is Eurasia, and that there were at least two widely separated periods of introduction

[16] This assumption is not incompatible with an amphitropic distribution of tribes thought to originate in the Northern Hemisphere. The tropical and subtropical climatic zones extended much further to the north in the Tertiary and waning Mesozoic, in which the differentiation and splitting up of carabids into their main branches must chiefly have taken place.

of *Bothriopterus* into the Nearctic Region, one rather recent, and one or two very old. Eurasia and the Nearctic Region were united in three main periods: in the Miocene until 12 million years ago, in the late Miocene from 10 million to 3.5 million years ago, and repeatedly in the Pleistocene with the advance and retreat of the ice sheet...

"Because *P. adstrictus* did not speciate in the Nearctic Region, I believe that it arrived during the Pleistocene (probably invading more than once, as Lindroth (1966) shows evidence of subspeciation in the Aleutians and California). Because the other Nearctic species are very different from Eurasian species, I believe they evolved here a long time ago and may have invaded the Nearctic Region at one of the two periods of land connections during the Miocene. Because *P. oregonus* and *P. pennsylvanicus* are adapted to more northerly climates (though mostly south of the boreal forest), I believe they came less than 10 million years ago when mixed and boreal forests were developing over Beringia... Also, because the remaining species are restricted to warmer climates (usually the deciduous forest biome), I believe that their ancestors came more than 12 million years ago...

"I believe that the first Nearctic invader, the *P. tropicalis* and *P. mutus* group ancestor, spread widely over North America and northern Middle America. Later the originally continuous range was separated into northern and southern populations by the development of grasslands in the southern U.S.A. The southern population became what we know today as *P. tropicalis* which is restricted to high altitudes in southern Mexico. The northern population was separated into eastern and western populations with the northward development of the southern grassland. In time the western population became *P. lustrans* and the eastern one became *P. mutus* and *P. ohionis* (a more southern species than *P. mutus*). These last three species probably evolved in the Nearctic Region as no close relatives are known in Asia...

"From the present distributions, three main groups may be seen: a boreal group, a temperate group, and a subtropical group. As the subgenus probably originated in Eurasia, a species with a southern distribution is probably older than a northern one. As climatic conditions cooled, first the warm-adapted, then the cool-adapted, and finally the cold-adapted species arrived. There was probably little or no displacement of already existing species as the invaders established themselves in the climatic zone for which they were already adapted".

This is a good illustration of the possibilities provided by an intensive study of the geographical and ecological distribution of a species for reconstructing the age, evolutionary connections and history of distribution of a carabid taxon.

An attempt of this nature has been made for still remoter epochs in the case of the Anisodactylina, a sub-tribe of the Harpalini (Noonan, 1973). Primitive taxa of this group live in the Australian region, whereas more highly evolved forms are found in the Northern Hemisphere. Sister species, in Hennig's sense of the term, inhabit two widely separated continents (Africa and South America). Noonan deduces that the Anisodactylina evolved towards the end of the Jurassic or at the beginning of the Cretaceous period on what, at that time, was the united Australian and Antarctic continent, and then spread northwards. This conforms with Darlington's theory, touched upon above, that Australia was one of the starting-points for the evolution of carabids. The North American species of the genus *Anisotarsus* can probably be attributed to two distinct invasions by South American ancestral forms. The more primitive European genus *Pseudodichirus* is a very close relative of the genus *Gynandrotarsus*, and it is therefore assumed that the ancestral forms of *Gynandrotarsus* reached North America in the late Cretaceous or early Tertiary across the European-North-American land bridge.

C. The Fossil History of Carabids in the Central European Pleistocene

Investigations on this subject have been carried out by Lindroth (1948, 1960) and, among others, by Coope (numerous recent publications on periglacial deposits in England, review 1970). An analysis of their carabid fauna and the accompanying flora and fauna reveals an amazing degree of constancy in morphological character and ecological demands of the species throughout the entire Pleistocene. This means that the Ice Age had little influence on the speciation and subspeciation of carabids. Only one case is known in Europe of an east-west disjunction of two Carabus species (*C. monilis* and *scheidleri*, Horion, 1950). This can be interpreted as indicating that the ancestral species differentiated into two distinct species during the Ice Age in two separate refugia, one in southeastern and one in southwestern Europe. On the other hand, Coope's investigations are mainly concerned with the widespread eurytopic and eurypotent species inhabiting large, continuous periglacial areas of Eurasia. It is possible that ecologically specialized forest species underwent quicker differentiation during the Pleistocene in isolated glacial forest refuges. This is strongly suggested by the zoogeographical investigations of Schweiger (1966) on the splitting up of systematic units encountered in the discontinuous mountain areas of the Near East.

The question as to whether repeated retreat into refugial areas during the Ice Age, resultant isolation and subsequent redistribution to ice-free regions, was responsible for speciation or at least for subspeciation of carabids and of many other animal groups remains highly controversial[17]. Schweiger (1966) is in no doubt whatsoever that closely related species of a genus of carabids found in mountain forest regions of Anatolia first evolved in the Pleistocene. Coope (1970), however, denies that the small territories over which endemic mountain species are distributed represent their site of origin. As an example, he quotes the case of the staphylinid *Oxytelus gibbulus:* although nowadays confined to a small area in the western Caucasus, it has been found in deposits from the last Ice Age in England, so that "it is more likely that the western Caucasus represents the last stand of the species rather than its place of birth." In North America the carabids *Pterostichus similis* and *P. parasimilis* are so

[17] According to Herre (1951) disjunction caused by the Ice Age did not lead to speciation in birds.

much alike that they can be termed sibling species. According to Coope (1970) Ball originally assumed (1963) that a large part of the process of speciation in the subgenus *Cryobius*, to which these species belong, did not take place until the Wisconsin glacial period. Following Matthews' (1968, cited from Coope, 1970) discovery of fossil material of the subgenus in Alaska, including both of the above-mentioned species, it appears that what had previously been held to be very recent species were, in fact, already differentiated more than 90,000 years ago, in the early Wisconsin period.

The Icelandic population of *Carabus problematicus* has only recently become the subject of more thorough investigation. Lindroth (1968) compared it with the four most closely neighbouring races occurring in Scandinavia, on the Faroe Islands and in Great Britain. The newly established race *C. problematicus islandicus* has a more primitive elytral sculpture than the others, the bands on the elytra being less interrupted than in the other races (the successive breakdown of the elytral sculpture is an evolutionary principle valid for other carabid species, too). Fossil material found in Scotland dating from between the Early and Main Würm Glacial period, about 57,000 years ago, also shows the interruption in elytral sculpture observable in recent members of the race procedens, so that subspeciation of *C. p. islandicus* must have taken place much earlier.

D. The Evolution of Carabids on Oceanic Islands

Isolated oceanic islands that have at no time been connected with the mainland provide valuable fields for phylogenetic research.

In the course of a study on the carabids of the Atlantic island of St. Helena, Basilewsky (1972) came to interesting conclusions concerning the endemic genus *Aplothorax*. The genus occupies a transitional position between the Calosomini and the Carabini and shows evidence of a neogenesis of derived characteristics. Since Calosomini were already fully developed in the middle Tertiary, Basilewsky assumes that the Aplothoracini originated in the Eocene or even earlier. They were presumably more widespread in the Southern Hemisphere at that time and probably soon reached the Upper Miocene volcanic island of St. Helena. Here they were preserved in isolation, whereas they became extinct in the rest of the world. Nothing is known with certainty about the evolution of derived characteristics on St. Helena, but it is possible that they accompanied the original population. *A. burcelli* is the only species of *Aplothorax* in existence today on the island.

On Aldabra, a recent atoll in the Indian Ocean, Basilewsky (1970) recorded 18 carabid species, none of which was endemic. All of them, however, are familiar from Madagascar and some have been found repeatedly on the African mainland.

It is fortunate that something is known about the carabids of the Galapagos, which has been the classical region for studying adaptive radiations on oceanic groups of islands since the time of Darwin. Again, it is to Basilewsky (1968) that we owe that most thorough investigation. He has reviewed the *Calosoma*

Table 54. The species and subspecies of the genus *Calosoma* (subgenus *Castrida*) on the Galapagos islands. (According Basilewsky, 1968)

Castrida granatense Géhin
- ssp. *granatense* Géhin: on nearly all islands of the archipelago—Espanola (Hood), San Cristobal (Chatham), Santa Fé (Barrington), Santa Cruz (Indefatigable), South Seymour, Pinzon (Duncan), San Salvador (James), Isabela (Albemarle): only in dry forests below 400 m, Pinta (Abingdon), Genovesa (Tower), Culpepper
- ssp. *darwinia* van Dyke: only in southeast of Isabela above 400 m
- ssp. *floreana* Basilewsky: only on Floreana = Sta. Maria (Charles)

Castrida leleuporum Basilewsky
only on Sta. Cruz (Indefatigable) in the subalpine grassy moors (prairies) above 500 m

Castrida galapageium Hope
only two specimens have been caught in the central highlands of San Salvador (James), Darwin's type specimens

Castrida linelli Mutchler
only three specimens are known from San Cristobal (Chatham). At high altitudes

species of the archipelago and brought order into their confused systematics (see Table 54). The *Calosoma* of the Galapagos belong to the subgenus *Castrida* that also occurs on the mainland of South America. The ssp. *granatense* of *C. granatense* inhabits the lower zones (dry forests below 400 m) of nearly all of the Galapagos islands. Only two further forms have differentiated out of this subspecies, the one, *floreana*, on the somewhat isolated southernmost island of the archipelago, and the other, *darwinia*, in the wet forest zone of the higher levels on Isabela. At similar altitudes on three other islands *C. galapageium*, *leleuporum* and *linelli* are encountered, all of them possessing the status of independent species.

The genus *Pterostichus* seems to be rather more differentiated on the Galapagos islands. As far as is known (van Dyke, 1953), it has differentiated into eight species, each confined to one particular island. All of them belong to the subgenus *Poecilus* and are held to be close relations of *Pterostichus peruviana* Dej. of the South American mainland (Table 55).

Table 55. The species and subspecies of the genus *Pterostichus* (subgenus *Poecilus*) on the Galapagos islands. (According to van Dyke, 1953)

P. insularius Boheman	Isabela (Albemarle)
P. calathoides Waterhouse	San Cristobal (Chatham)
P. waterhousei van Dyke	Floreana (Charles)
P. duncani van Dyke	Pinzon (Duncan)
P. williamsi van Dyke	Santa Cruz (Indefatigable)
P. galapagoensis Waterhouse	San Salvador (James)
P. blairi van Dyke	San Salvador (James)
P. mutchleri van Dyke	Isabela (Albemarle)

Most authors are of the opinion that the Galapagos group was never connected with the mainland. It is of volcanic origin and dates back to the Tertiary (older than the Pliocene, Lack, 1947)[18]. In all probability, therefore, the carabids on the Galapagos have had ample time in which to evolve, although compared with certain other animal groups, however, their radiation is not very impressive. The tortoise species *Testudo elephantopus*, for example, has differentiated to 12 races (all, however, within the framework of *one* species), whilst the Darwin finches form a subfamily (Geospizinae) endemic to the Galapagos and consisting of 14 species. Six of these have differentiated to between two and eight subspecies each (Lack, 1947). Whether the differences in radiation are due to the fact that the Darwin finches reached the Galapagos earlier than the carabids or whether the speed of evolution of the latter was so much slower cannot be resolved. It should be placed on record that the various forms of *Calosoma* found on the Galapagos are clearly differentiated as to their ecological behaviour.

E. Possible Clues as to the Evolution of Carabids from Studies on Behaviour and Parasitism

An analysis of the *defensive behaviour of carabids by means of their defence glands* has led to the chemical elucidation of large numbers of defence substances (Schildknecht *et al.*, 1968; see p. 103f.). The above authors have correlated the distribution of classes of defence substances over the subfamilies and tribes of carabids with their systematics and evolution.

According to their scheme, carabids possessing isovaleric acid and isobutyric acid are considered to be primitive since such substances are only weakly acid and slightly toxic. The two acids were found in members of each subfamily investigated, i.e. in the Omophroninae, the Carabinae (Notiophilini) and Harpalinae (in the very tribes that are held to be primitive, i.e. the Elaphrini and Loricerini, as well as in the Broscini and some Bembidiini). Since in all probability isovaleric acid also occurs in the Hygrobiidae, a relic family of the ancient Adephaga, the occurrence of this weakly toxic substance in representatives of three carabid subfamilies is taken as an indication that such substances are an old heritage from common ancestors.

Formic acid, a much more poisonous substance, occurs almost exclusively in association with paraffins, which function as carriers in facilitating the entry of nonlipophilic formic acid into the insect cuticle. The "formic acid group" comprises the Harpalini, Licinini, Lebiini, Odocanthini, Dryptini and Zuphiini, as well as some Pterostichini (subdivision Sphodrina), i.e. exclusively Harpalinae. Since these tribes are not particularly closely connected, the concept of the "formic acid group" as a natural unit, as postulated by Schildknecht *et al.* (1968) is, to me, incomprehensible.

[18] The age of the Galapagos Islands may have been overestimated in the past. According to calcium-argon dating, the oldest lava so far known is only 1.2 million years old. The age of the archipeligo is estimated at a few, and at the most 10 million years (Kramer, personal communication).

In most of the Pterostichini investigated, and in the Amarini, paraffins fulfil a similar carrier role for methacrylic or tiglinic acid. The absence of paraffins in *Abax* and *Molops*, in contrast to the other Pterostichini, confirms the assumption already made on the basis of morphological and zoogeographical observations, that the two genera are of recent origin and closely related to one another. Quinones as defence substances probably developed independently at various points in evolution since they are found in two widely separated tribes, the Scaritini and the Chlaeniini.

Although providing confirmation for the system, the biochemistry of the defensive secretions has so far revealed no new aspects. On the other hand, "a sharp dividing line cuts through the middle of the genus *Chlaenius*: *Chlaenius vestitus* produces quinone and three other species of *Chlaenius* produce m-cresol. *Callistus*, too, produces quinone, but *Panagaeus*, in contrast, cresol" (this genus belongs to the closely related tribe Panagaeini) (Schildknecht *et al.*, 1968). The authors draw the conclusion that the skeleto-morphological system either gives a distorted picture of natural relationships or the genus *Chlaenius* represents the point, in itself almost unchanged, from which chemical development took two different directions. A clearly recognizable "parting of the ways" in the chemical, and thus probably also in the natural system, can be seen in *Agonum (Platynus) dorsale*, in which salicylic acid methyl ester, lacking in *Agonum (Platynus) assimile*, is found in addition to formic acid and paraffins. Apart from unsaturated fatty acids, salicylaldehyde, lacking in *Carabus* and *Cychrus*, also occurs in *Calosoma*.

A good correlation exists in carabids between increasing toxicity of defensive secretions and greater structural complexity of the defence glands. The Brachinini (see p. 104) are undoubtedly highly evolved in both respects.

A comparison of the usually *host-specific parasitic mites of the Podapolipidae family with various carabid taxa* (Regenfuss, 1968; see p. 83ff.) revealed a good correlation between the systematics of the parasites and their hosts.

The systematics of the mites involved suggest that the allocation of *Poecilus* species to the genus *Pterostichus* is unwarranted, since beetles of the subgenera *Omaseus*, *Melanius* and *Bothriopterus* harbour very closely related *Eutarsopolipus* species, the latter being morphologically quite different from *Eutarsopolipus* species parasitizing hosts of the genus *Poecilus*.

It is impossible to decide here whether this justifies ranking *Poecilus* as a genus, or whether it is merely a particularly well differentiated subgenus within the genus *Pterostichus*, which would not be contested even by those systematicists who prefer to leave *Poecilus* among the genus *Pterostichus*. This is a good illustration of the fact that investigations of the kind described in this section, although complementing a type of systematics based upon morphological features, are still insufficient basis for amending the system or for dating individual carabid groups. Nevertheless, the striking fact emerges that, within the Pterostichini (seven genera investigated), four different classes of defence substances are encountered as compared with only two in seven Harpalini genera investigated. This can be interpreted as confirming the theory that the Pterostichini are phylogenetically an ancient tribe (see p. 301). In the light of what is to be said later in this book (see Chap. 10.E.) it is worth mentioning that the defence secretions of closely related and morphologically similar species within one and the same genus can be very different *(Chlaenius)*.

F. Concluding Remarks Concerning the Evolution of the Carabids

The foregoing review, of necessity incomplete, shows that many questions still remain to be answered in this field of research. The only unequivocal evidence is that provided by fossils, and such finds are a matter of luck. On the other hand, a systematic search for evidence of this kind, and its examination, has only just begun. But even supposing that much more attention is devoted to the field, it remains questionable whether, without entering too far into the realms of speculation, a clear picture of the evolution and history of distribution of the carabids and their systematic subunits will ever be drawn in other than very broad outlines.

G. Digression: Studies on Genetics and Population Genetics of Carabids

Caryotype Analysis. Weber (1966a) analysed the caryotype of 22 *Carabus* species from eight subgenera and reviewed the findings of earlier authors on these and other species. Weber makes some corrections to previous results and states that all species so far analyzed possess $2n=28$ chromosomes and are heterogametous in the males. It follows "that mechanisms of chromosome reorganization (centric fusions, dissociations, translocations, pericentric inversions) that might lead to chromosome polymorphism were not involved in the race and species formation of *Carabus*". It is interesting that the chromosome number $2n=28$ also occurs in a Japanese species of Carabinae, *Campalita chinense* (Kudoh *et al.*, 1970).

In each of a number of *Carabus* species Weber (1968b) found one large pair of autosomes (A-chromosomes) with a heterochromatic arm. The heterochromatin is not devoid of genes, but its share (determined from the ratio of the length of the heterochromatic to the euchromatic arm) varies considerably within the species *Carabus auronitens* (Mossakowski and Weber, 1972). Morphological features (metric characteristics, number of bristles on various parts of the body) are correlated with the heterochromatin content. An identical heterochromatin content can, however, influence a character in different populations in opposite ways, but whether genetic or ecological factors are responsible for this is unknown. The chromatin content is of no consequence in the formation of races: *C. a. auronitens* (from Westphalia, West Germany), *C. a. festivus* (from France) and *C. a. punctatoauratus* (from the Pyrenées) do not differ in their average heterochromatin content.

A completely different situation from that seen in *Carabus* was found by Nettmann (1976) and to some extent by Wilken (1973) before him, in the genera *Pterostichus* and *Agonum*. Chromosome numbers between 2n=27 and 46 were found in male *Pterostichus*. Large differences exist even between species of a sub-genus, as for example in *Poecilus*, with 31 and 44. Intraspecific polymorphism could be observed in *Pterostichus nigrita:* some populations had 2n=40 and some 2n=46 chromosomes. Such polymorphism might be connected with the fact that the species is extremely eurytopic. In some populations of *Pterostichus cupreus* the males had 2n=43 chromosomes and in others 2n=44. It can be concluded that XY and XO types of sex determination are involved.

In the genus *Agonum* the chromosome numbers of males lie between 2n=26 and 44: in contrast to *Pterostichus* the largest differences are found between species of different sub-genera. Despite such a large range of variation in chromosome numbers 2n=37 is by far the most frequent number in males of both genera.

Crossing Experiments. Highly successful crossing experiments were carried out by Puisségur (1964). He obtained bastards from seven different crossings with five species of the subgenus *Chrysocarabus*. He also successfully crossed bastards of *Chrysocarabus auratus* × *C. splendens* with *C. auronitens* and obtained hybrids of *C. hispanus* with two races of *C. rutilans*, all from the subgenus *Chrysotribax*. Members of both subgenera could be bastardized as follows: *Chrysotribax hispanus* with *Chrysocarabus splendens, lineatus, auronitens* and *punctatoauratus*, and *Chrysotribax rutilans* with *Chrysocarabus splendens*. Experimental crossings between these species and *Carabus intricatus* from the more remote subgenus *Chaetocarabus* were unsuccessful.

In species bastards the males were without exception sterile in F 1, whereas some of the females were fertile.

His results led Puisségur to a partial revision of the systematic grouping and raised doubts as to the validity of some forms so far treated as distinct species.

Using the hybrid index method, Freitag (1965) in Canada was able to demonstrate a zone of introgressive hybridization between *Cicindela oregona* and *C. duodecimguttata*, for the origin of which isolation of populations of the parent forms on either side of the Rockies during the Pleistocene is assumed. When the two forms reunited in the post-glacial era generic isolation was still incomplete, and accounts for the hybrid belt.

Statistical Analysis of the Combinations of Characteristics in Various Populations. A multivariate analysis of the combinations of characteristics (42) in usually small populations of *Carabus arcensis* on heaths and heather moors confirms a good biometrical separation (Mossakowski, 1971), which can only partially be put down to environmental influences. Since these populations are small and are geographically isolated from one another the differentiation is attributed mainly to gene drift. In northern Germany the disruption of the area of distribution of the species was due to human interference and thus of recent origin, which means that genetic changes have taken place within a very short period of time.

Chapter 10

Concerning the Reasons Underlying Species Profusion Manifest by the Carabids

How are we to explain the great number of carabid species interpreted at the beginning of this book as being evidence of phylogenetic success?

A. Powers of Dispersal and Speciation

The high degree of running and walking activity shown by most carabids and the fact that many species are capable of flight, make it a simple matter for isolated individuals or even entire populations to reach new habitats where they are then subject to fresh conditions of selection. Given effective geographical isolation, speciation can readily ensue.

As Mayr (1963) put it (p. 621): "The number of successful shifts into new adaptive zones will be directly proportional to the total number of new species that come into existence".

The ability of whole populations to migrate is a significant factor in the preservation of the species: According to den Boer (1972, see p. 71ff.) it spreads the risk and levels out fluctuations within populations, thus acting as a stabilizing factor.

Geographical isolation of carabids is well possible. Although, as already mentioned, ground beetles are very mobile, many species have either entirely lost their ability to fly or they exhibit wing dimorphism, which means that a part of a population may possess stunted alae and be incapable of flight. As a result, relatively small and isolated populations of carabids are frequently encountered (see e.g. Thiele, 1971b), in which rapid evolution would appear to be feasible. Mayr (1963) recognized the possibility of a "burst of speciation in flightless beetles". On the other hand, unlike the fully winged drosophilids, no carabid species is of cosmopolitan distribution.

It may well be that the interplay of a well-developed ability of a population to migrate with a temporary tendency to stay put, is the answer to the question as to why the Earth is not populated by a smaller number of widely distributed eurytopic and eurypotent carabids. This is obviously connected with the high degree of flexibility of physiology and behaviour involved in the evolution of the carabids, to be discussed in more detail later. Wing dimorphism may thus be *one* important principle responsible for the development of such an enormous wealth of carabid species in the course of evolution. Macropterous (long winged) individuals effect the rapid distribution of members of a population over large areas. Some of these animals land in habitats where they are faced with new conditions of selection. *They differentiate genetically with respect to their ecological requirements and in their behaviour and adapt to the new environment.* Once this has been achieved brachyptery becomes of selective advantage if the localities are small and stable, since the insects that fly away usually perish. Isolation is enhanced and speciation accelerated in this way. Alterations in the environment that tend to decrease stability favour again the macropterous individuals, and the hypothetical process described above is once more set into action. Lindroth, in a personal communication, comments: "It should also be pointed out that brachyptery within any one species can arise polyphyletically from various gene mutations (cf. *Drosophila*), or from 'recurrent' mutation."

This hypothesis is very well illustrated in the subgenus *Antilliscaris* of the genus *Scarites* in Puerto Rico (Hlavac, 1969). Four species of the subgenus are endemic to the montane rain forests of Puerto Rico: *A. darlingtoni, danforthi, mutchleri* and *megacephalus*. In contrast to other digging and flying Scarites species (see p. 10), members of the subgenus *Antilliscaris* do not burrow, as can be deduced from their body structure. Their antennae are longer than those of the diggers, and modifications in their mandibles indicate that they are not used for tunneling in the ground. One individual of *S. (A.) mutchleri* had no underground system. The transition to an epigaeic habit—in this case as a result of lack of competitors—seems to be of economical advantage in the evolution of the subgenus, since it requires less energy than a subterranean way of life. Particularly striking is the fusion of the elytra and the extreme reduction of the entire flight apparatus. Hlavac arrives at the following preliminary evolutionary diagnosis, "From an assumed lowland, winged and burrowing ancestor, *Antilliscaris* has entered the montane rain forests of Puerto Rico, become flightless, undergone an ecological shift from burrowing to ground crawling, and speciated." His considerations fit in well with the generalization of Mac Arthur and Wilson (1967): "The steps taken by a species after colonizing a depauperate habitat island are: initial adaptation, ecological shift, loss of dispersal power, speciation, and adaptive radiation."

Experiments and observations in the field have shown that neither interspecific competition nor in fact any other biotic factors play a primary role in establishing the ecological pattern of distribution of carabids, this rather depending upon adaptation to abiotic factors.

What degree of importance attaches to competition between carabids in the processes involved in their evolution? According to Darlington (1971) its role is a large one.

"Although *competition is difficult to demonstrate in particular cases* (present author's italics), the general evidences of it in the animal world are overwhelming. The strongest evidence comes from the general level and balance of faunas in all parts of the world ... For example, every habitable part of the world has a carabid fauna roughly proportional to area and climate, and the Carabidae in each part show a reasonable range of size and include representatives of all the principal ecological groups for which habitats are available. No substantial part of the world is overfull of Carabidae, and no part has a notable deficiency of them. This balance cannot be due to chance. Something must hold the size and composition of carabid faunas everywhere within certain limits in spite of continual multiplication and dispersals of successive phylogenetic groups. Only competition can do this...".

A statement of this nature is not easily reconcilable with what has already been said in Chapter 3. It is not necessary, however, to identify the factor "competition" so completely with the evolutionary factor of "selection". The many physiological adaptations of carabids to their habitats indicate a large number of potential possibilities for physiological mutation, thus supporting the idea put forward above that it is the abiotic factors that are of paramount importance in selection and species formation in this particular insect family. This only means that no absolutely dominant role can be attributed to interspecific competition of carabid against carabid. It should be emphasized, moreover, that this is a statement which should only be taken in this context as applying to carabids. As emphasized already by Darwin (1859), in the course of geological history there must have been many cases of a more highly evolved taxon replacing a less differentiated and more primitive group, and where, in all probability, competition between the two was involved. Nevertheless, we are well advised to bear in mind that our assumptions as to the nature of such competition are usually of a speculative nature. We have no exact idea as to how placentalia superceded the marsupials. Darwin wrote in 1859: "This ought to convince us of our ignorance on the mutual relations of all organic beings: a conviction as necessary as it is difficult to acquire".

In my opinion, competition seems to play a greater role in the long-term replacement of larger taxonomic groups by more advanced groups and a lesser role in speciation within a family. This embraces the possibility that competition might provide the mechanism by which carabids could, in the future, be replaced by another animal group capable of completely occupying its place in nature. Let us now take a look at the potential rivals of carabids for their ecological niche.

B. The Ecological Niche of the Carabids

It appears that carabids the world over are almost uncontested in their ecological niche. It is a niche occupied by epigaeic, mainly predatory arthropods that hunt their prey on the ground and consume considerable quantities of carrion and plant matter. This characterization holds for the adults only, the larvae being more endogaeic.

Within this wide niche[19], differentiations have taken place and probably still are taking place, making possible the occupation of yet other sub-divisions by carabids. The view that the number of niches open to carabids is predetermined and unalterable is certainly incorrect. This would imply that for newly evolved species of carabids a niche would only be available if the species already occupying it were to be ousted in competition. There is no evidence that carabids, in the course of their evolution, have crowded out any other, older, arthropods

[19] As suggested by Simpson, this could be termed the "adaptive zone," see Mayr (1963).

from the niche that they themselves now occupy. It is far more likely that the adaptive zone of the carabids has developed during the course of their evolution and that its continued development is closely linked with the evolution of other organisms that serve as prey or food plants, for example, for the carabids. We can define a niche as being the field of influence of a species within its ecosystem. New niches arise when an ecosystem increases in complexity. An analogous situation can be seen in the explosive increase in professions which has accompanied the rising complexity of human civilization. Despite the innumerable new branches that have been opened up, the already existing professions have in no way been supplanted. But the same example shows that competition does in fact exist and that certain trades have been ousted from their ecological niche by competitors armed with machines and computors. Competition, however, does not necessarily play a universal and paramount role in evolution.[20]

C. Potential Competitors for the Carabid Niche

I. Insects

The *staphylinids* are the only Coleoptera whose niche partially overlaps that of the carabids. Their build and highly endogaeic manner of locomotion renders them competitors of the larvae rather than of the adult carabids. Nevertheless, the nutritional habits of the staphylinids, about which we unfortunately know very little, differ considerably from those of the carabids in that the former are predominantly saprophagous and some even parasitic. Some authors are of the opinion that ants compete with carabids for their niche (see p. 100), particularly in the tropics. Again, this can only be a question of partial overlapping of niches. Not only do ants penetrate more deeply into the ground than carabids, they also swarm further up into the vegetational layer. Thus their behaviour differs substantially from that of the carabids and in some cases they even occupy different nutritional niches. Potential competitors of carabids, therefore, as far as nutrition is concerned, are mainly to be sought in other classes of arthropods.

II. Spiders—the Lycosidae Family

To a large extent the habitat of the carabids is shared by the predatory epigaeic walking lycosids. Schaefer (1972) has dealt with the question of competition

[20] In connection with the view held by various authors that no further speciation of carabids took place in the late Tertiary and Pleistocene (at least not in temperate latitudes), the question arises as to whether, with the deterioration in climate that has since taken place, no new niches have been created suitable for occupation by new species of carabids. This would apply to other groups of organisms as well and has in fact been postulated for birds, for example by Herre (1951).

between these two groups of animals in great detail. In a region consisting of salt meadows and coastal dunes he observed that the activity maxima of carabids and lycosids occurred at different seasons (Fig. 151), but he does not interpret this as resulting from competition. He states: "Lycosids and carabids apparently seldom have any direct connection with one another". "Experiments with *Amara lunicollis* and *Pardosa pullata* in the laboratory suggest that encounters between the two lead neither to a heightened tendency to flee nor to a preying reaction". Schaefer summarized his observations in a most vivid manner: "The long-legged lycosid species 'stalk' along above the carabids, which move close to the ground, in such a way that the two 'biological types' in no way seriously disturb each other".

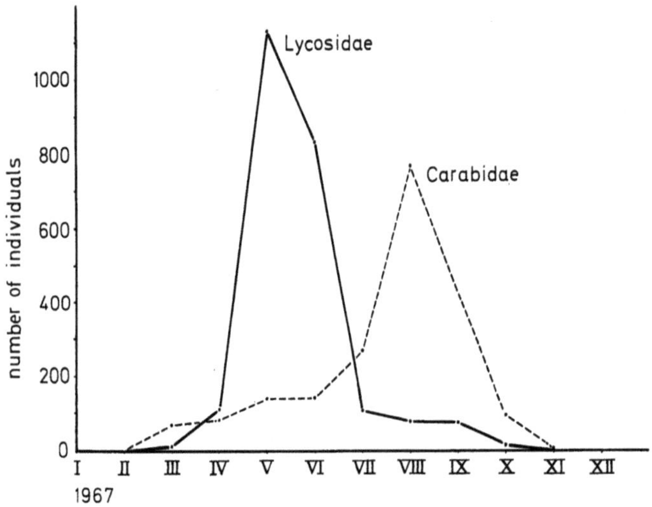

Fig. 151. Occurrence of lycosid spiders and carabids in a west German coastal region in the 12 months of one year. Number of individuals = number of the animals caught in 50 pitfall traps/month. From Schaefer, 1972

III. Chilopods

On account of their build these, too, are also at least semiendogaeic in type and therefore offer little competition to the epigaeic carabids.

IV. Opiliones

The numbers of harvest spiders are much lower than the above-mentioned groups. Even more than of the lycosids it can be said that although opilones are, like the carabids, more or less predators, they represent a completely different "biological type".

Schaefer (1972) sums up his investigations on this topic as follows: "Since lycosids and carabids are on the same trophic level they can, to some extent, be regarded as complementary groups ..., but competitive phenomena are apparently of no decisive importance in determining their pattern of distribution".

It therefore seems that carabids are in fact not exposed to competition for the occupation of their ecological niche, and thus there is no strong selective pressure of morphological changes for them to evolve out of this wide niche into a narrow and more specialized one. In the course of evolution no great advantage for carabids has attached to such a process. For this reason it is all the more important that, with an increasing diversity of species, the carabids are able continually to take possession of new niches ("subdivisions") within their adaptive zone (due to their ability to change their behaviour). It is not so much pressure of competitors that takes them to new niches but rather their ability to adapt to new habitats due to the genetic flexibility underlying their behaviour.

D. Differentiation of the Physiology and Behaviour of the Carabids

Two further aspects seem to me to be of particular importance in the development of the great number of species of carabids: (1) They have conservatively held on to a simple, little-specialized basic body plan. (2) At the same time they have remained very flexible as regards behaviour and environmental requirements. Extreme specializations in behaviour, however, are rare in carabids and would also limit the possibility of the development of a great number of species.

On the whole, the body plan of carabids varies little within the entire family and is seldom highly specialized. Wherever such specializations occur, their bearers form groups that are relatively poor in species. The Cychrini tribe, structurally adapted to a specific manner of catching its prey, comprises only 89 species. As compared with other Carabidae the Cicindelinae appear to be highly specialized as regards build, markedly predatory way of life, daytime activity and a need for high temperatures. Whereas their description occupies 302 pages in the Coleopterorum Catalogus (Junk and Schenkling, 1926–33) that of the Carabinae covers 551 pages. Even this is modest when compared with the 1640 pages devoted to the Harpalinae. At least with respect to their mode of nutrition (extraintestinal digestion mostly) the Carabinae are more specialized than the Harpalinae, which have retained far more possibilities for phylogenesis.

Apparently the simple basic body plan of the carabids and of beetles in general is especially robust and suited to existence in widely differing environments.

So far, this type of body plan has not been superseded in the carabid niche by another more successful type.[21] The diversity of types of behaviour which

[21] Lindroth, in a personal communication, says: "In some respects the carabids seem to be replaced by Tenebrionidae in warm, dry regions".

are compatible with this body plan has made it continually possible for them to widen their niche and to occupy many ramifications by means of a wide variety of behavioural adaptations in morphologically similar and closely related species.

Published investigations show that, although carabids exhibit great variability in behaviour, extreme specialization is rarely encountered. A study of preferences in microclimate has revealed a uniform preference for cold, moisture and darkness, or for warmth, dryness and light in only a few species occupying very extreme habitats (p. 277). As a rule eurypotency exists with respect to at least one factor. Where the processes of development and annual rhythmicity are governed by photoperiodism it has been shown that temperature can provide a substitute, although a less potent one, for the influence of the length of day (*Pterostichus nigrita*, see p. 253). It is as if, in a manner of speaking, a physiological master key fits a particular lock. A similar interpretation can be applied to the fact that short-day–long-day animals can also mature without a change in photoperiod if exposed to constant conditions for very long periods (see Chap. 6.B.III.). Day- or night-active species are so flexible in their daily rhythm that, especially in the breeding season, additional activity is possible in the phase otherwise used for rest (Chap. 6.A.). In recent years it has become increasingly clear that the types of annual rhythm exhibited by carabids are not rigidly distinguishable from one another but that, particularly in the case of all larval overwinterers, an additional overwintering is possible for parts of the population in the adult stage (p. 247). No further doubt exists to-day, but that the majority of carabid species are highly polyphagous and that a true nutritional specialization is rather rare (Chap. 3.F.IX.). Accordingly, key stimuli connected with catching of prey are fairly simple. Complicated instinctive behaviour connected with brood care is sporadically encountered and can, as we have seen already, limit the choice of habitat (see p. 272f.).

1. Differences in the Daily Time of Activity

The frequent occurrence of day- and night-active species within one and the same family is not necessarily to be expected. Entire orders and classes of animals are much more consistent in this respect than the Carabidae. For example, all Odonata (dragonflies and damsel flies) are active during the daytime, as well as butterflies (Rhopalocera) and the majority of Hymenoptera and Diptera. Members of other families of insects, in contrast, are nearly all active at night, an example being provided by the hawk moths (Sphingidae). An analysis of the pattern of daily activity of the soil arthropods in a forest gave the following results:

Day-active only	Night-active only	Day- and night-active
Diptera	Diplopoda	Carabidae
Hymenoptera (chiefly Formicidae)	Isopoda	Collembola
Araneae		

It appears that in any one habitat, completely different groups of animals are, as a rule, active at night and during the daytime. But within the carabid family itself, however, a definite "change of shift" can be observed, one set of species being active during daylight and another in the dark. It seems that the ecological niche inhabited by the carabids is exploited by two distinct groups of species, each active at a different time of day, one or the other group dominating according to the type of habitat. The occurrence of day- and night-active species cuts right through systematic units and demonstrates the polyphyletic origin of the patterns in diurnal activity of the Carabidae.

2. *Differences in Activity Season*

Carabids are usually univoltine. The annual periodicity of the activity of the one generation can be very different from species to species. In temperate climates groups of species reproducing throughout the autumn exist side by side with groups reproducing throughout the springtime. Nuances within the groups account for the fact that active adult carabids are to be found over the larger part of the year and larvae at all times. Throughout the whole of the annual cycle, therefore, the carabid niche is occupied by species of variously differentiated physiological types. Larval overwintering seems to be an adaptation to regions with mild winters, ensuring activity and accumulation of energy over the larger part of the year (see p. 249).

It is by no means usual to meet such varying types of annual rhythm in one and the same family. Whole orders of insects exhibit uniform patterns: all Ephemeroptera and almost all Odonata[22] for example, overwinter as larvae.

3. *Physiological Adaptations to Abiotic Factors*

Another characteristic not generally possessed by Arthropoda is the broad spectrum of adaptation to abiotic factors shown by the Carabidae, which explains its occurrence in such a wide variety of habitats. It extends from cold- to warm-stenothermy, from extreme preference for moisture to an equally extreme preference for dryness, and from dark- to light-preference. In these respects, whole classes and orders of animals exhibit a much more uniform type of behaviour than the carabids. Isopods (nevertheless an entire class) are predominantly moisture-loving animals, as are the members of the vertebrate class of Amphibia. In moist ecosystems these groups exhibit a sharp maximum in occurrence.

In many distinctly warmth-loving groups of animals, for example the Buprestidae family of beetles, all species prefer a more or less high temperature. They occur solely in warmer ecosystems of the warmer climatic zones of the Earth.

The broad span of adaptability to climate that typifies the carabids is probably a particularly important factor in differentiation into many species.

[22] Exception: the species of the genus *Sympegma* that overwinter as adults.

E. Physiological Differentiation in Carabids Compared With Other Animal Groups

As a family exhibiting considerable morphological homogeneity and comprising an astounding number of species (40,000 carabid species as compared with approx. 1000–2000 Drosophilidae), the ground beetles confront both ecologist and physiologist with a wide range of physiological adaptations that are of ecological significance as regards habitat affinity.

In this, the group as a whole provides confirmation of Krumbiegel's (1932) statements connected with his investigations on the physiological basis of race formation in the one species *Carabus nemoralis:* "If we survey the morphology of the races of *Carabus nemoralis* and compare it with the physiological findings, it can be said that the latter exhibit more variety: physiologically the races ... can be regarded as separate species although morphologically they all conform to the species characteristics of *C. nemoralis.* The morphological property reveals itself as the more conservative, lagging behind the physiological characteristic."

Krumbiegel demonstrated this very clearly using various populations of *C. nemoralis* from different geographical regions of Europe. Given a choice of black or white (see legend to Fig. 152) the choice of white rose from northeast towards the southwest in correlation with increasing daytime activity of the populations concerned.

A good example of an intraspecific physiological differentiation of populations of various geographical distribution, or, in other words, of physiological race

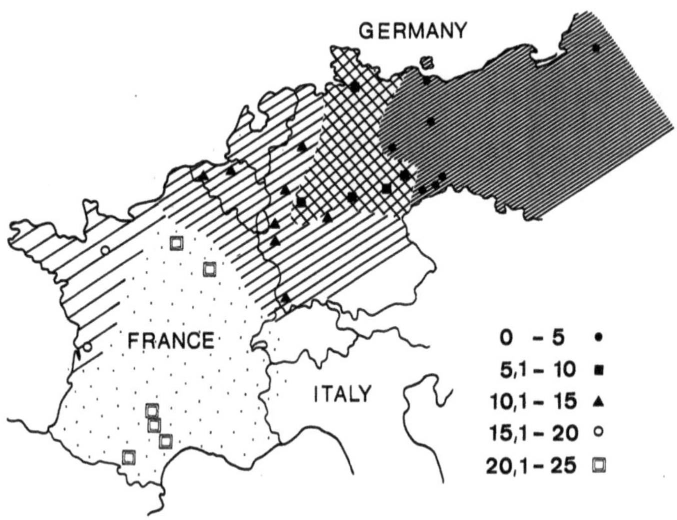

Fig. 152. Behaviour of *Carabus nemoralis* populations from central Europe in a choice apparatus in which they can choose between white and black boxes. *Below right:* percent choice of "white" by animals originating in different regions. The decrease in density of the *shading* from north east to south west indicates an increasing choice of white in the populations of Europe. Redrawn from Krumbiegel, 1932

formation, is offered by *Pterostichus nigrita* (Ferenz, 1975b). At low temperatures (+10°) the mortality of animals from Swedish Lapland (Messaure, 66° N) is much lower than in central European animals, although at other temperatures it is about equal (Table 56).

Table 56. Mortality in the pre-adult stages of *Pterostichus nigrita* from two different regions in % of initial population. (According to Ferenz, 1975b)

°C	Central Europe (Cologne: 50° N)	Lapland (Messaure: 66° N)
10	87	42
15	11	28
20	18	12
25	12	16
30	63	66

At all experimental temperatures listed in the table the animals from Lapland developed more quickly ($p<0.001$). The difference was greatest at 10° (duration of development reduced by 27%; by 9% at 20°; 23% at 30°). In absolute terms this means a reduction from 137 to 100 days at 10°, from 31.5 to 28.7 at 20° and from 25.7 to 19.8 days at 30°. At 25° and at 30° the growth rate of animals from Lapland is higher, and in spite of the shorter time required for development they achieve greater weights than the central European animals. These idiosyncrasies can be regarded as adaptations to the necessity of completing their development in the short subarctic summer. Central European males of *P. nigrita* mature in short day with 17h light phase or less. Females also take the first step towards egg-ripening (previtellogenesis) in short day (15h light phase or less). In northern Scandinavia the animals would be unable to mature under these conditions: "The warm summer is marked by extremely long light phases or continuous daylight. Shortening of the days is accompanied by a very rapid drop in temperature." And in spring, when it is warm enough for the beetles to develop, the days are already long. Accordingly, males from northern Scandinavia matured at all photoperiods of less than 22h light phase. Previtellogenesis in the females set in at a light phase of 19.5h or less. The four to five h shift toward long day in the critical photoperiod in both males and females of these populations is an adaptation to the subarctic light conditions (Ferenz, 1975b).

In 1964 Engelhardt wrote: "It is generally supposed today that the first step of differentiation of species is a physiological one," and his investigations on spiders offer supporting evidence for this view. One of the aims of the present author is the documentation of this principle for a large animal group comprising a wealth of species, since, in his opinion, the view cited above is *not* in fact "generally supposed". Engelhardt shows that four central European species of the genus *Trochosa* cannot be satisfactorily distinguished on the basis of the morphological characteristics that are usually important criteria in most spiders. It is interesting that the females can only be distinguished by means of their colour and that the copulatory apparatus of the males is so little differentiated that it can offer no explanation for the generative isolation of the species. On the other hand, the microclimatic requirements of the various species are well differentiated and narrowly adapted to the conditions prevailing in their habitats (Table 57).

Table 57. Ecological behaviour of spiders of the genus *Trochosa*. (According to Engelhardt, 1964)

Species	Ecological type (habitat affinity)	Experimental data		
		Loss of weight in % at 30° C and 75 % RH after 20 h.		Preferred temperature of females carrying cocoons in °C
		♀♀ (mean)	♂♂ (mean)	mean (min–max)
Trochosa spinipalpis	photobiontic, eurythermic, hygrobiontic	14.6	20.9	22.6 (20.9–24.3)
Trochosa terricola	hemiombrophilic, eurythermic, hemihygrophilic	9.9	15.8	23.4 (20.0–27.9)
Trochosa ruricola	photophilic, eurythermic, hemihygrophilic	11.5	18.2	24.8 (23.1–27.2)
Trochosa robusta	photobiontic, warm stenothermic, xerophilic	7.0	11.5	29.2 (24.0–33.0)

Engelhardt, in summing up, says that "the results of evaporation experiments are in excellent agreement with the biotopic requirements of the four species involved: Given identical conditions, the species that lives in the wettest biotope, *spinipalpis*, shows the highest transpiration, whereas *robusta*, that inhabits the driest habitats, has the lowest transpiration value. The other two species occupy a position between these two extremes, although the value for *ruricola* is slightly higher than that for *terricola*, exactly as would be expected from the degree of moisture in the biotopes of their choice... As with transpiration, the four species also differ in their P. T. (preferred temperature) and again, just as for transpiration, the relative position of the P. T. corresponds exactly with the conditions, in this case, thermal ones, in the biotope of their preference. *Spinipalpis* undoubtedly occupies the relatively coldest biotope and *robusta* the warmest, whilst the habitats of *terricola* and *ruricola* are somewhere between the two and are relatively similar as far as thermal conditions are concerned".

Haacker (1968) investigated the significance of preference and resistance for the habitat binding of 14 species of diplopods in comparison with carabids. "In contrast to Thiele's (1964a) findings for carabids moisture is less important than temperature as a controlling factor in the diplopods investigated: the former influenced the biotope-binding of ten species, the latter that of 13... Within the limits of tolerance the reaction of a species to microclimate is not automatic but is determined by its ecological organization. As Thiele points out, this is just as much a result of evolution as any other characteristic involving form or behaviour. All of this is very well demonstrated in cases where distribution follows one environmental factor but is inverse to another. The diplopods *Tachypodoiulus niger*, *Chromatoiulus proiectus*, *Cylindroiulus londinensis* and *C. latestriatus* are distributed in accordance with their temperature preferences but contrary to their preferences with regard to moisture... Among the diverging microclimatic requirements temperature appears to be dominant. In some carabids, however (Thiele, 1964a), the moisture requirements seem to dominate. Contradictions of this nature possibly result from the differences in type of life form: for purely surface-active animals such as the ground beetles that are only moderately resistant to drying-out the air humidity is of much greater importance than to the distinctly edaphic diplopods... A divergence in factors governing distribution means that some species do not live in what would be for them the 'best world of all'. The selection and evolution of better adapted ecotypes, although impossible to observe or demonstrate experimentally, has to be assumed, since only in this way can a maximum harmony between habitat and species such as is seen, for example, in the stenotopic forest type, *Orthochordeuma germanicum*, be accounted for".

The importance of ecological behaviour in habitat binding can also be demonstrated in a group of flying insects of very uniform body structure, i.e. in species of the genus *Panorpa* (Sauer, 1970). Using the procedure followed by Lauterbach (1964) for carabids, Sauer calculated index values for the experimentally observed behaviour of these species with respect to temperature, moisture and light, and from these values worked out an index of preference reaching, theoretically, in the case in question, from three (for cold, moisture and dark preference) to 15 (for warmth, dryness and light preference). The very different habitats occupied by the four species of *Panorpa* investigated, despite their great similarities in body structure, is clearly a reflection of their differing microclimatic requirements (Table 58).

An excellent example of differentiation in behaviour of populations living in different geographical regions but not morphologically distinguishable as

Table 58. Ecological behaviour of scorpion flies of the genus *Panorpa*. (According to Sauer, 1970)

Species	Behaviour towards Temperature	Humidity	Light	Preference Index ♂♂/♀♀	Habitat
Panorpa alpina	cold stenothermic	slightly moisture preferent	dark preferent	7.3/7.6	cool, moist, shady, dark
Panorpa communis	mesosteno-thermic	eury-hygric	eury-photic	9.5/9.4	temperate conditions
Panorpa germanica	warm stenothermic	eury-hygric	light preferent	10.5/10.5	dry, warm sunny, light
Panorpa cognata	warm stenothermic	dry preferent	light preferent	10.8/10.3	dry, warm sunny, light

sub-species, is provided by the midge, *Clunio marinus* Hal., which inhabits the tidal region on many European coasts. Since the animals are so short-lived (a few hours at most), concentrated emergence is of immense adaptive value if the two sexes are to find each other and reproduce. Furthermore most populations require a dry substrate for egg deposition. The suitable spots, however, are in the lower tidal zone and are covered with a growth of algae. Along coasts with a large tidal range they are only left dry at low water during spring tides. The animals have, therefore, to be provided with a special mechanism for programming their emergence to coincide with this low water level. Such mechanisms have been investigated in detail by Neumann and his co-workers.

Populations from southern and central European Atlantic coasts exhibited an endogenous circasemilunar rhythm in emergence, or, in other words, an endogenous oscillator governs readiness to emerge every fourteen days. A semilunar periodicity of this type is synchronized by the light of the full moon in such a way that emergence invariably occurs at full or new moon, when the places suitable for egg deposition are left dry by the spring tides. A circadian rhythm governed by the diurnal light changes also ensures that the animals actually emerge at low water which, at spring tides, always occurs at the same time of day. The phase relationship between LD (see Chap. 6.A.I.) and the circadian rhythm in emerging is fixed for each tribe according to the tidal conditions prevailing on that particular stretch of coast, and is governed by a small number of genes (Neumann, 1966, 1967, 1971).

Frequent cloud on the northern coasts of Europe makes the zeitgeber function of the moon unreliable. According to experimental investigations on populations from Helgoland (practically identical with southern populations in external appearance) emergence in parallel with the cycle of spring-neap tides is not governed by the light of the full moon. In animals from Helgoland and other north European tribes "however, the semilunar rhythms can be induced by an artificial tidal turbulence superimposed on a 24 h light cycle" (Neumann, 1976), whereby every 14 days, as in nature, the same phase relationship between tidal period (connected with turbulence period) and daylight period occurs. An Arctic population (Tromsö in northern Norway) showed no semilunar periodicity in emerging; the eclosion rhythm is merely parallel to the tidal rhythm and the conditions are suitable for the midges to emerge twice daily at intervals of about 12.4 h when the tide is on the ebb (Neumann and Honegger, 1969). Instead of an endogenous oscillator, an hour-glass principle is responsible for this tidal rhythm and is largely governed by temperature. A temporary experimental rise in temperature of about 1° C and more elicited immediate emergence and a second peak in emergence after 11–13 h (e.g. when the hour-glass had run out). When the sites are left dry at low tide their temperature rises and this can be regarded as the most important zeitgeber in triggering the instantaneous effect and in programming emergence for the next low tide (Pflüger, 1973). As would be expected, no trace of the rhythms observed in Atlantic populations could be found in a Baltic population (where there are no tides) from habitats that are perpetually submerged: emergence is merely confined to a particular time of day (Neumann, 1966). So far, all these populations are morphologically identical and they provide a particularly good example of a differentiation of ecologically relevant forms of behaviour preceding any detectable morphological subspeciation or even speciation.

Some species of cicadas belonging to the genus *Euscelis* (Fam. Delphacidae) differ considerably in both ecology and behaviour despite great morphological similarity. The west European *Euscelis ohausi* Wagn. lives on *Genista anglica* as a monophage, whereas the widespread species *Euscelis plebejus* Fall. is polyphagous, probably on Leguminosae mainly, but to some extent on grasses as well. The males of the two species have very similar mating songs but those of the females differ. Although females could be induced to utter their mating song by males of the other species, the males were never lured by the call of females other than of their own species, so that copulation leading to bastardization did not occur (Strübing, 1965). On the basis of its behaviour alone *Euscelis alsius* Ribaut from the western Mediterranean region could be shown to justify its recognition as a species (Strübing, 1970). On account of considerable morphological similarities it has sometimes been held to be only one of the seasonal variants of *E. plebejus*. *E. alsius*, however, avoids Leguminosae and perhaps lives as a monophage on grasses of the genus *Lolium*. But most important of all, the mating songs of both sexes of both species show species-specific patterns. Males and females of the two species nevertheless "understand" one another sufficiently for bastardization to occur under laboratory conditions, although no cases have been reported from the field.

At this point another family of insects, the Drosophilidae, deserves a more thorough consideration for purposes of comparison. So far, they have mainly been known for the important role they have played in another branch of biology, namely genetics. But in a recent comprehensive study, "Behavioural and ecological genetics," Parsons (1973) reviewed the ways in which these three disciplines converge in *Drosophila*. The aims of Parsons' study resemble in some ways those of the present author for the Carabidae: "... an attempt is being made to understand all the processes determining the distribution and abundance of species in the wild"! A need has arisen for a better interpretation of laboratory investigations and this can only be achieved by more thorough knowledge of the ecological behaviour of *Drosophila* species in the field. Parsons states: "... for example, fitnesses of karyotypes in *D. pseudoobscura* are often temperature-dependent. The classical models (namely of population genetics: author) do not take these complications into account: at the present stage many models exist at the theoretical level, *but biological reality is turning out to be more complex* (present author's italics). Although information on the behavioural and ecological genetics of *Drosophila* has been appearing for many years, *mainly as a by-product of other investigations* (present author's italics), the emphasis in the past few years has changed more to a direct investigation of the behavioural and ecological genetics of *Drosophila*. Even so, behavioural and ecological genetics have been developing on the basis of studies in other organisms, perhaps at stages in advance of studies in *Drosophila*". (Rodents, plants, butterflies and snails are listed as examples.) "Unfortunately, in many cases, knowledge on the formal genetics of many of these species is restricted as compared with *Drosophila*, but they possess the advantage of readily identifiable phenotypes in the wild as assessed by visual differences. This is an advantage that *Drosophila* lacks, since, in the field, species of the genus are remarkably uniform in colour and form...".

In this respect carabids differ from drosophilids. Investigations in the two families of insects had completely different points of departure. In the case of drosophilids the original question concerned their genetics and could be investigated in the laboratory whereas the extensive, quantitative investigations on the ecological distribution of the carabids were carried out in the field. Most of the studies on behaviour and ecology of the drosophilids have involved only a few species of the genus *Drosophila*. The world fauna, however, according to Parsons, includes 1000–2000 species of Drosophilidae. A remarkable fact is that 475 species occur on Hawaii, and about 400 of them are endemic. Apparently the eurytopic, cosmopolitan and, as regards behaviour, "polymorphous" species of drosophilids have been the subject of investigations so far. A broader and more precise knowledge of the ecological behaviour of a wide spectrum of species is still lacking. On the other hand, exact information is available concerning the chromosome make-up of tribes and species of different ecological behaviour.

Parsons described the situation as follows: "Therefore, although most of the work so far published has been on species easy to breed in the laboratory, these are just the species about which we have little information in the wild... With time ... those species which are easy to breed in the laboratory will be studied progressively more from *the point of view of their behaviour and ecology in the wild*" (present author's italics). To this end Parsons suggests the following programme: "There are two main approaches to the behavioural and ecological genetics of *Drosophila* which will be followed: first, the study of flies in natural environments in order to correlate species and genotypes within species, and their behaviour, with the environment, and second, the study of populations in artificial laboratory environments in order to assess what may be factors of behavioural and ecological importance in nature." Both paths have been followed more or less independently of one another and the second has brought the deeper insight into the problem.

The mating behaviour of drosophilids (and its dependence upon environmental factors) as well as its genetical basis are well known. Many experiments have been carried out on competition under various environmental conditions. In drosophilids, aspects of behaviour such as photo- and geotaxis have been studied more from the point of view of the speed with which positive and negative "selection lines" can be stabilized than in relation to the ecological environment of the species.

The examples cited with respect to daily rhythms reveal the paucity of knowledge on this topic. Overwintering and annual rhythms in reproduction have apparently scarcely been touched upon, and even their ecological distribution in the wild has only seldom been the subject of thorough investigation. Most is known, relatively speaking, of the ecology of certain endemic species in Hawaii which are of limited distribution and specialized requirements, "the main controlling factors being probably wind intensity, food sources, and acceptable ovipositional sites" (Parsons, 1973)—again, primarily microclimatic factors.

Remarkably enough, although the investigations on drosophilids and carabids had completely different points of origin they have led to convergent views as to the connections between ecological behaviour and evolution! In conclusion Parsons states: "The likely importance of behavioural phenomena in evolutionary change may help to explain the high level of variation of behavioural traits compared with morphological traits. Probably permanent morphological changes follow behavioural changes, and, as we have seen, there is an association between behavioural and morphological divergence".

Wickler (1970) lists numerous examples of birds and mammals where, within one species or within a group of related species, differences in social or reproductive behaviour occur as adaptations to greater or smaller, constant or varying availability of food during the course of the year. Wickler emphasizes "the way in which morphological, physiological and ethological traits are interwoven in ecological adaptation". A few cases taken from these animal groups will serve to illustrate how species of utmost morphological similarity fit their various habitats by means of differences in behaviour.

Two species of the genus *Emberiza* were studied by Wallgren (1954). Of these, the yellowhammer *Emberiza citrinella* is a nonmigratory species distributed over most of Eurasia, even remaining in Siberia during the winter. The breeding region of the closely related ortolan *(Emberiza hortulana)* extends further to the south but not so far to the north. The metabolic balance of both species is at its best and lowest within a particular range of temperature, above and below which the intensity of respiration increases. For the yellowhammer this is from 25 to 33° C, and for the ortolan 32 to 38° C. The ortolan can only compensate down to −15° C by increasing its oxygen consumption, and at temperatures below this the regulatory system breaks down. The yellowhammer, on the other hand, can compensate down to −40° C by a rise in respiratory intensity. An upper temperature region of 33 to 34° C, however, constitutes a danger, whereas the upper danger level for the ortolan is 38 to 39° C. These data show that temperature requirements determine not only the breeding area but also migratory behaviour. In migrating, the cold-sensitive ortolan seeks winter quarters with similar temperature conditions to those prevailing in summer in its breeding area. A morphological expression of their differing physiological behaviour is to some extent found in the plumage of the two species.

Ludescher (1973) reported a study on the sibling species *Parus palustris*, the marsh-tit, and *Parus montanus*, the willow-tit. They frequently share the same nesting territory and, as the author points out, provide an example of "closely related sibling species that can coexist even without a pronounced econutritional differentiation". The pecking instinct has developed quite differently in the two species. Both sexes of the willow-tit peck and they nest almost exclusively in holes that they have pecked out for themselves, whereas only the female marsh-tit exhibits the pecking trait and its nest-building is confined to extending already existent holes in trees. The fact that the willow-tit is the more eurytopic of the two species is probably connected with this difference in behaviour. The marsh-tit is not found at altitudes greater than 1200–1400 m above sea level; it inhabits deciduous and mixed forests, whereas the willow-tit shows no preference for any particular kind of forest and is found up to the tree limit in the Alps.

Wecker (1963) has studied the habitat affinities of two subspecies of prairie deer mouse, one of which, *Peromyscus maniculatus bairdi*, inhabits fields, whereas *P. m. gracilis* lives in forests. In experiments the animals were faced with a choice between two types of enclosure, one offering field vegetation and the other a forest type. *P. m. bairdi*, recently captured in the field, unhesitatingly chose the field vegetation, whereas a laboratory population consisting of the 12th to 20th generations of progeny from initially wild *P. m. bairdi* exhibited no preference for one or the other type of vegetation. The first laboratory generation from wild field specimens, however, chose the field enclosure. It seems that the preference for fields shown by *P. m. bairdi* is inherited but can be lost after a larger number of generations of laboratory breeding.

An interesting experiment throws some light on the nature of the habitat binding in this species. If animals from the laboratory population were reared in the field enclosure they continued to choose it for the rest of their lives, so that their original preference can apparently be reactivated. Laboratory animals of the same species reared in a forest enclosure, in contrast, remained indifferent, from which it can be concluded that early exposure to a foreign habitat does not alter the inherited affinity for the field type of habitat.

Although the factors responsible for the affinity to either field or forest are unknown, it appears that the two sub-species are bound to their respective habitats by inherited differences in behaviour rather than in morphology.

Behavioural adaptations in desert kangaroo rats have been reported by Kenagy (1973). *Dipodomys merriami* is known to be a seed specialist but *D. microps* is incapable of existing on a diet of this nature. Its ability to concentrate its urine is remarkably lower than that of other species of *Dipodomys*. For its nutrition it relies entirely upon the leaves of *Atriplex confertifolia* and in this connection has developed a unique pattern of behaviour. It nibbles off the outer salt-containing epidermis and eats only the inner leaf tissues which contains merely 3 % of the electrolytes found in the epidermis so that a water-containing diet is at its disposal throughout the year. Hand in hand with this adaptation in behaviour goes a chisel-like modification of the incisor teeth. In this way, *D. microps* has taken over an ecological niche that is inaccessible to other members of the genus and can probably occupy habitats otherwise unsuitable for its fellow species.

Apparently, therefore, the information concerning the significance of modes of behaviour in habitat binding and evolution gained from observations on ground beetles is of general applicability. The carabid family offers numerous opportunities both qualitative and quantitative for demonstrating the above-mentioned relationships.

The double role played by modes of behaviour in animal evolution has been pointed out by Mayr (1970).

1. Inborn behaviour is just as much exposed to the influence of evolutionary factors as are physical characteristics and undergoes the same kind of changes in the course of evolution. Wickler (1970) regards behaviour "so to say, as the most malleable organ of the organism ... The 'plasticity' of behaviour makes it especially useful as an adaptive feature and is one reason why behaviour

often acts as 'pacemaker' in evolution". This implies "that, in the process of adaptive evolution, alterations in behaviour precede changes in body structure...".

2. Considerable emphasis is placed by Mayr (1970) on the role of individual modifications in behaviour as causal factors in evolution. He writes: "The particular significance of behaviour as an evolutionary factor lies in the possibility offered to the individual of escaping certain environmental influences and of confronting others as they arise". We mentioned above that this holds true to a particularly high degree for carabids. Thanks to their powers of activity they are readily able to move on to new habitats that present them with changed conditions of selection, initiating new evolutionary processes and the occupation of new ecological niches.

In this connection Mayr considers it to be a "somewhat unexpected" discovery that "the aspects of behaviour that have been especially thoroughly studied by ethologists, i.e. inter- and intraspecific relationships... are just those that play a relatively insignificant role in macroevolution... Intraspecific behaviour consists mainly in an exchange of signals, and serves the purpose of communication or of misleading an enemy, or assists escape. *Only rarely does it play a role in the discovery of a new ecological niche or adaptive zone* (present author's italics). There is good reason for the ethologist's reluctance to concern himself with the ecological aspects of behaviour: they are extremely difficult to analyse. On the one hand, they often vary considerably even within one and the same population, and on the other, they always involve a strong, non-hereditary component. This is true of choice of food, choice of biotope and also of certain types of movement. The significance of types of behaviour in the larger evolutionary events becomes obvious when one considers the transition from aquatic to terrestrial life, from tree-life to one on the ground, or from an earthbound life to one involving flight".

The adaptive role of these types of behaviour in cladogenesis and speciation of any one animal group has seldom been the subject of such extensive investigation as in the carabids, and the studies on this group provide the zoologist with the means of bridging the gaps between ecology, behaviour and evolution.

Chapter 11

Summary

1. With something like 40,000 species already described, the Carabidae comprise more species than any other insect family.

2. The basic body structure varies little from species to species. The greatest modifications are found in adaptation to a particular form of obtaining food (specialized predators, carrion consumers), to an arboreal existence or to a burrowing mode of life. Ground-dwelling species of forest and field habitats, on the other hand, exhibit scarcely any morphological differentiations.

3. Many species—as shown by quantitative studies—are strictly bound to a certain type of habitat, often to a particular plant society. Such affinities occasioned the search for the underlying physiological or behavioural properties governing them.

4. Biotic factors such as competition, effects of predators, parasites and food supplies probably exert less influence on population dynamics than abiotic factors, although they influence the frequency of the species. Biotic factors do not, as a rule, limit the occurrence of a species to a particular habitat. An exception to this rule is seen where very high densities of ants in a habitat largely exclude all carabids.

As predatory organisms, the energy turnover of carabids in the ecosystem equals that of two other important animal groups of the soil surface, i. e. chilopods and spiders. Most carabids, however, are not exclusively predators but consume carrion and are also herbivorous, at least at certain times of year.

5. Experiments in the field and laboratory have revealed that carabids possess a regulatory effect on harmful insects, especially if the population density of the former is artificially raised. Even at natural densities they destroy some harmful insects (e. g. potato beetle) to an extent that makes it necessary to consider their preservation when applying insecticides.

The introduction of *Calosoma sycophanta* into North America had a regulatory effect on the destructive moths that had been brought from the Old World. The preservation of a varied type of scenery (e. g. hedges and small woods breaking up the fields) ensures a reservoir of many species even in industrialized regions. Damage to crop plants by carabids is rare and in very few cases is it of economic significance, e. g. damage to strawberries, to conifer seeds in nurseries and, due to the genus *Zabrus*, to grain crops.

As far as is known carabids do not function as conveyors of disease. Carabids are in a variety of ways damaged, reduced and destroyed by man and his economy measures. Cultivation measures employed on agricultural land cause relatively little harm, but insecticides destroy many species although the different agents have widely varying effects. Herbicides exert, above all, an indirect influence by changing the vegetational cover. Emissions from industry and traffic (near roads) change the composition of the carabid fauna considerably. The fauna of highly populated areas becomes more monotonous. Due to their differentiated requirements, ability to react rapidly and to the fact that their environmental requirements are well known, carabids are becoming important bioindicators of human interference in the environment.

6. Microclimatic factors have a decisive influence upon the ecological distribution of the species. Most carabids are relatively warmth-loving and warmth-requiring. Forest animals are almost all very hygrophilic and show little resistance to dryness whereas species of the open country are usually xerophilic or euryhygric. Forest species almost exclusively prefer the dark and are positively scototactic. Many species of the open country are euryphotic. Some shore species exhibit sun-compass orientation. Soil factors have little influence on the distribution of the species. The frequently observed affinity of some species for chalk is based upon microclimatic requirements. The question of an affinity for a certain pH value of the substrate remains unsettled, although in some cases an affinity for sodium chloride has been demonstrated. However, the powers of osmoregulation even of such "salt species" are small.

Only a few species voluntarily choose water for catching their prey. Resistance to submersion is higher in cold than warm water, and survival times are longer in fresh water than in sea water. Larvae survive submersion longer than adults. Although overwintering in ice can occur this is not possible under water. Ability to swim in fresh water and in slightly salty brackish water is so considerable that drifting is held to play a substantial role in dispersal.

7. Compared with many other insect families or even entire orders, carabids exhibit very different types as regards daily and annual rhythmicity. Some species are definitely day-active, others distinctly nocturnal. In a single habitat, therefore, a change of shift can be observed among the species in the course of the 24 h. Under constant conditions carabids show a circadian rhythm of activity whose most important zeitgeber is light-dark-change. Apart from some exceptions carabids are univoltine (species in which several generations mature in one year are rare). Species in which development takes several years are mainly to be found in subarctic, subantarctic and mountainous regions. Overwintering occurs either in a larval stage, and is thus dependent upon the influence of low temperature, or in the adult stage. In this case it is usually connected with a dormancy of the gonads, for which the carabids have developed very different regulatory types, usually governed by photoperiod. In some species thermic dormancy of the larvae and photoperiodic dormancy of the adults in the course of an individual life may succeed each other obligatorily.

8. The correlation between microclimatic requirements, daily and annual rhythmicity in the different types is so strict that these modes of behaviour can be assumed to have developed in dependence upon one another in the course of evolution. Forest dwellers are mainly hygrophilic, and avoid direct sunshine as a result of their preference for darkness and night-time activity. Since they thus live under cool conditions but require high temperatures for reproduction, the latter does not take place until late summer and autumn when the nights are warmer. This holds for temperate latitudes. Field dwellers

are often resistant to dryness, indifferent to light and are day-active, and can thus already reproduce in the spring.

9. Carabids are capable of such feats of movement on the ground that their distribution over wide areas appears to be possible without flight. However, a considerable number of species is indeed macropterous or dimorphic. Macropterous individuals settle new living space by flight. Brachyptery appears to be dominant. In stable microhabitats wing reduction is of selective advantage since individuals flying away would mainly land in unsuitable habitats and perish. Conversely, in unstable habitats distribution by flight is an essential adaptation.

10. The evolution of carabids probably began in the early Middle Ages of the Earth's history. Rapid evolution followed in the Tertiary. Further speciation in the Pleistocene is disputed but may have been possible in the forest fauna. Attempts to reconstruct the evolution and dispersal of carabids have so far proved unsatisfactory. Ecological considerations suggest that the most likely important centres of differentiation and distribution of carabids were in the tropics of the Old World, bearing in mind that in the epochs in question the tropics extended further to the north than today.

11. The great wealth of species encountered in the carabid family is closely connected with the manifold modes of behaviour and physiological properties of its species. Wing dimorphism may have been an important principle in the evolution of this profusion of species, since macropterous populations could reach new habitats and experience new conditions of selection. The plasticity of their physiology and behaviour facilitated adaptation to the new selective conditions, and brachyptery then became of selective advantage since the animals were adapted to the new habitats as long as these remained stable. Changes in the habitat again favoured macroptery, migration and renewed adaptations to a different type of habitat. Changes in physiology and behaviour thus—as can be demonstrated in other animal groups—precede morphological alterations. They are the pace-makers of evolution and in the case of the carabids have rendered possible the development of large diversity of species, the physiological requirements and behaviour of the species showing greater differentiation than their bodily structure.

References

Adeli, E.: Zur Kenntnis der Insektenfauna des Naturschutzgebietes bei der Sababurg im Reinhardswald. Z. Angew. Entomol. **53**, 345–410 (1963–64)

Adis, J.: Bodenfallenfänge in einem Buchenwald und ihr Aussagewert. Diplomarbeit, Göttingen (1974)

Adis, J., Kramer, E.: Formaldehyd-Lösung attrahiert *Carabus problematicus* (Coleoptera, Carabidae). Entomol. Germanica **2**, 121–125 (1975)

Allen, A. A.: The habit of aggregation in *Agonum dorsale* Pont. (Coleoptera, Carabidae). Entomol. Mon. Mag. **210**, 142 (1957)

Andersen, J.: The effect of inundation and choice of hibernation sites of Coleoptera living on river banks. Norsk Entomol. Tidsskr. **15**, 115–133 (1968)

Anderson, J. M.: Food and feeding of *Notiophilus biguttatus* F. (Coleoptera, Carabidae). Rev. Ecol. Biol. Sol. **9**, 177–184 (1972)

Aneshansley, D., Eisner, T., Widom, J., Widom, B.: Biochemistry at 100° C: Explosive secretory discharge of bombardier beetles *(Brachinus)*. Science **165**, 61–63 (1969)

Ankel, W.: Ein *Carabus* als Blütenbesucher. Z. Wiss. Insekt. Biol. **12**, 213 (1916)

Arnoldi, K. V., Ghilarov, M. S.: Die Wirbellosen im Boden und in der Streu als Indikatoren der Besonderheiten der Boden- und Pflanzendecke der Waldsteppenzone. Pedobiologia **2**, 183–222 (1963)

Asahina, E., Ohyama, Y.: Cold resistance in insects wintering in decayed wood. Low Temp. Sci. **27**, 143–152 (1969)

Aschoff, J.: Exogenous and endogenous components in circadian rhythms. Cold Spring Harb. Symp. Quant. Biol. **25**, 11–28 (1960)

Aschoff, J., Wever, R.: Resynchronisation der Tagesperiodik von Vögeln nach Phasensprung des Zeitgebers. Z. Vergleich. Physiol. **46**, 321–335 (1963)

Bachofen-Echt, A.: Der Bernstein und seine Einschlüsse. Vienna: Springer, 1949

Ball, G. E.: The distribution of the species of the subgenus *Cryobius* (Coleoptera, Carabidae, Pterostichus) with special reference to the Bering land bridge and Pleistocene refugia. In: Pacific Basin Biogeography. Gressit, J. L. (ed.). Hawaii: Bishop Museum Press, 1963, pp. 133–152

Bargagli, P.: Cenni biologici su due specie di *Percus*. Bull. Soc. Entomol. It. **6**, 27–30 (1874)

Basedow, T.: Der Einfluß epigäischer Raubarthropoden auf die Abundanz phytophager Insekten in der Agrarlandschaft. Pedobiologia **13**, 410–422 (1973)

Basedow, T., Borg, Å., de Clercq, R., Nijveldt, W., Scherney, F.: Untersuchungen über das Vorkommen der Laufkäfer (Coleoptera, Carabidae) auf europäischen Getreidefeldern. Entomophaga **21**, 59–72 (1976a)

Basedow, T., Borg, Å., Scherney, F.: Auswirkungen von Insektizidbehandlungen auf die epigäischen Raubarthropoden in Getreidefeldern, insbesondere die Laufkäfer (Coleoptera, Carabidae). Entomol. Exptl. Appl. **19**, 37–51 (1976b)

Basford, N. L., Butler, J. E., Leone, C. A., Rohlf, F. J.: Immunologic comparisons of selected Coleoptera with analyses of relationships using numerical taxonomic methods. Syst. Zool. **17**, 388–406 (1968)

Basilewsky, P.: Les Calosomes des îles Galapagos (Coleoptera, Carabidae). Miss. zool. belge aux îles Galapagos **1**, 179–207 (1968)

Basilewsky, P.: Les Coléoptères Carabidae de l'Ile d'Aldabra (Océan Indien). Bull. Ann. Soc. Roy. Entomol. Belg. **106**, 211–222 (1970)

Basilewsky, P.: La faune terrestre de l'île de Sainte-Hélène. II. Insectes. 9. Coleoptera. 1. Fam. Carabidae. Ann. Mus. R. Afr. Cent. **192**, 11–84 (1972)

Basilewsky, P.: Insectes Coléoptères: Carabidae, Scaritinae. In: Faune de Madagascar **37**. Paris: Orstom – CNRS 1973. 322 pp

Bathon, H.: Über das Formensehen bei der Verbergeorientierung der Laufkäfer (Coleoptera, Carabidae). Untersuchungen an *Carabus problematicus* Thoms., *Agonum assimile* Payk., *Abax ater* Vill. und *Bembidion litorale* Oliv. Z. Tierpsychol. **32,** 337–352 (1973)
Bathon, H.: Woran erkennen Laufkäfer einen Unterschlupf? Ber. Offenb. Ver. Naturkunde **78,** 34–40 (1974)
Bauer, T.: Zur Stridulation von Laufkäfern der Gattung *Elaphrus* Fabr. (Carabidae). Forma et Functio **6,** 177–190 (1973)
Bauer, T.: Ethologische, autökologische und ökophysiologische Untersuchungen an *Elaphrus cupreus* Dft. und *Elaphrus riparius* L. (Coleoptera, Carabidae). Zum Lebenformtyp des optisch jagenden Räubers unter den Laufkäfern. Oecologia (Berl.) **14,** 139–196 (1974)
Bauer, T.: Stridulation bei *Carabus irregularis* Fabr. (Coleoptera, Carabidae). Zool. Anz. **194,** 1–5 (1975a)
Bauer, T.: Zur Biologie und Autökologie von *Notiophilus biguttatus* F. und *Bembidion foraminosum* Strm. (Coleoptera, Carabidae) als Bewohner ökologisch extremer Standorte. Zum Lebensformtyp des visuell jagenden Räubers unter den Laufkäfern (II). Zool. Anz. **194,** 305–318 (1975b)
Bauer, T.: Experimente zur Frage der biologischen Bedeutung des Stridulationsverhaltens von Käfern. Z. Tierpsychol. **42,** 57–65 (1976)
Bauer, T., Völlenkle, W.: Hochfrequente Filmaufnahmen als Hilfsmittel bei der Analyse von Angriffs- und Fluchtverhalten in einer Räuber-Beute-Beziehung unter Bodentieren (Collembolenfang visuell jagender Carabiden). Wiss. Film (Vienna) **17,** 4–11 (1976)
Baust, J. G.: Temperature-induced neural adaptations. Motoneuron discharge in the Alaskan beetle *Pterostichus brevicornis* (Carabidae). Comp. Biochem. Physiol. **41A**, 205–213 (1972)
Baust, J. G., Miller, K.: Variations in glycerol content and its influence on cold hardiness in the Alaskan carabid *Pterostichus brevicornis*. J. Insect. Physiol. **16,** 979–990 (1970)
Becker, J.: Art und Ursachen der Habitatbindung von Bodenarthropoden (Carabidae—Coleoptera, Diplopoda, Isopoda) xerothermer Standorte in der Eifel. Beitr. Landespflege Rhld.-Pfalz Beih. **4,** 89–140 (1975)
Bergold, G.: Die Ausbildung der Stigmen bei Coleopteren verschiedener Biotope. Z. Morphol. Oekol. Tiere **29,** 511–526 (1935)
Beyer, R.: Über Beziehungen zwischen Standort, Witterung und Aktivität der Fauna der Laubstreu in einem nordwestsächsischen Waldgebiet. Pedobiologia **4,** 192–209 (1964)
Beyer, R.: Zur Fauna der Laubstreu einiger Waldstandorte im Naturschutzgebiet „Prinzenschneise“ bei Weimar. Arch. Naturschutz u. Landschaftsforsch. **12,** 203–229 (1972)
Bílý, S.: Larvae of the genus *Amara* from Central Europe. Studie Csl. Akad. Věd. **13,** 1–74 (1975)
Bílý, S., Pavlíček, J.: A comparison of the soil coleopterous fauna in three types of meadow in Bohemia. Acta Entomol. Bohemoslov. **67,** 287–303 (1970)
Blokhov, V. P., Mukhin, V. F.: An instance of the isolation of a *Salmonella typhi* culture from dead specimens of *Platysma vulgare*. (orig. Russian) Zhurn. mikrobiolog. epidemiolog. immunobiolog. **5,** 118–119 (1961)
Blunck, H., Mühlmann, H.: Coleoptera, Käfer: Carabidae, Laufkäfer. In: Handbuch der Pflanzenkrankheiten, 5th ed. 5, 2, Sorauer, P. (ed.). Berlin: Parey, 1954, pp. 3–12
Boer, P. J. den: Activiteitsperioden van loopkevers in Meijendel. Entomol. Ber. Amst. **18,** 80–89 (1958)
Boer, P. J. den: Vleugeldimorfisme bij loopkevers als indicator bij zoögeografisch onderzoek. Vakblad voor biologen **6,** 1–10 (1962)
Boer, P. J. den: Lebeort (Habitat)-Bindung einiger Wald-Carabidenarten in Drente (Holland) in Zusammenhang mit Waldtypus, Boden und Strukturelementen des Waldes. Paper read at the Coenological Colloquium, Zagreb, 9–14 Sept. 1963
Boer, P. J. den: Verbreitung von Carabiden und ihr Zusammenhang mit Vegetation und Boden. In: Biosoziologie. Tüxen, R. (ed.). Den Haag: Junk 1965, pp. 172–183
Boer, P. J. den: Spreading of risk and stabilization of animal numbers. Acta Biotheor. **18,** 165–194 (1968a)
Boer, P. J. den: Zoölogisch onderzoek op het Biologisch Station Wijster, 1959–1967. Misc. Papers Landbouwhogesch. Wageningen **2,** 161–181 (1968b)
Boer, P. J. den: On the significance of dispersal power for populations of carabid beetles (Coleoptera, Carabidae). Oecologia (Berl.) **4,** 1–28 (1970)

Boer, P. J. den: Stabilization of animal numbers and the heterogeneity of the environment: The problem of the persistence of sparse populations, 77–97. In: Dynamics of Populations. Proc. Advan. Study Inst. Dynamics Numbers Popul. den Boer, P. J., Gradwell, G. R. (eds.). Wageningen: Centre Agr. Publ. Docum. 1971a

Boer, P. J. den: On the dispersal power of carabid beetles and its possible significance. In: Dispersal and Dispersal Power of Carabid Beetles. Misc. Papers Landbouwhogesch. Wageningen **8,** 119–138 (1971b)

Boldori, L.: Appunti biologici su *Pterostichus multipunctatus.* Studi Trent. Sci. Nat. **14,** 222–223 (1933)

Bombosch, S.: Untersuchungen über die Auswertbarkeit von Fallenfängen. Z. Angew. Zool. **49,** 149–160 (1962)

Boness, M.: Die Fauna der Wiesen unter besonderer Berücksichtigung der Mahd (Ein Beitrag zur Agrarökologie). Z. Morphol. Oekol. Tiere **42,** 225–277 (1953)

Boness, M.: Biocoenotische Untersuchungen über die Tierwelt von Klee- und Luzernefeldern (Ein Beitrag zur Agrarökologie). Z. Morphol. Oekol. Tiere **47,** 309–373 (1958)

Bońkowska, T.: The effect of shelterbelts on the distribution of Carabidae. Ekol. Pol. **18,** 559–569 (1970)

Boyer-Lefèvre, N. H.: Influence de la dessiccation sur trois coléoptères de la tribu des Sphodrini vivant dans des milieux différents. Bull. Soc. Hist. Nat. Toulouse **107,** 595–605 (1971)

Brandmayr, P., Brandmayr, T.: Sulle cure parentali e su altri aspetti della biologia di *Carterus (Sabienus) calydonius* Rossi, con alcune considerazioni sui fenomeni di cura della prole sino ad oggi riscontrati in carabidi (Coleoptera, Carabidae). Redia **55,** 143–175 (1974)

Brandmayr, Z. T., Brandmayr, P.: Biologia di *Ophonus puncticeps* Steph. Cenni sulla fitofagia delle larve e loro etologia (Coleoptera, Carabidae). Ann. Fac. Sci. Agr. Univ. Torino **9,** 421–430 (1975)

Brandt, H.: Der Goldlaufkäfer als Getreideschädling. Anz. Schädl. Kunde **22,** 87–89 (1949)

Brehm, E., Hempel, G.: Untersuchungen tagesperiodischer Aktivitätsschwankungen bei Käfern. Naturwissenschaften **39,** 265–266 (1952)

Briggs, J. B.: A comparison of pitfall trapping and soil sampling in assessing populations of two species of ground beetles (Coleoptera, Carabidae). Rep. East Malling Res. Stn. for 1960, 108–112 (1961)

Briggs, J. B.: Biology of some ground beetles (Coleoptera, Carabidae) injurious to strawberries. Bull. Entomol. Res. **56,** 79–93 (1965)

Briggs, J. B., Tew, R. P.: Insecticides for the control of strawberry seed beetle, *Harpalus rufipes* (DEG.). Rep. East Malling Res. Stn. for 1968, 149–154 (1969)

Bro Larsen, E.: Biologische Studien über die tunnelgrabenden Käfer auf Skallingen. Vidensk. Medd. Dansk Naturh. Foren. **100,** 1–231 (1936)

Broen, B. von: Vergleichende Untersuchungen über die Laufkäferbesiedlung einiger norddeutscher Waldbestände und angrenzender Kahlschlagflächen. Deut. Entomol. Z. NF **12,** 67–82 (1965)

Bückmann, D.: Zur Leistung des Schweresinnes bei Insekten. Naturwissenschaften **42,** 78–79 (1955)

Burakowski, B.: Biology, ecology and distribution of *Amara pseudocommunis* Burak. (Coleoptera, Carabidae). Ann. Zool., Warsaw **24,** 485–526 (1967)

Burgess, A. F.: *Calosoma sycophanta:* its life history, behavior, and successful colonization in New England. Bull. Bur. Entomol. U. S. Dept. Agr. **101,** 1–94 (1911)

Burmeister, F.: Biologie, Ökologie und Verbreitung der europäischen Käfer auf systematischer Grundlage. 1: Adephaga, Caraboidea. Krefeld: Goecke and Evers, 1939

Clausen, C. P.: Entomophagous Insects. New York: Hafner, 1962

Cloudsley-Thompson, J. L.: Studies in diurnal rhythms. VI. Bioclimatic observations in Tunisia and their significance in relation to the physiology of the fauna, especially woodlice, centipedes, scorpions and beetles. Ann. Mag. Nat. Hist. **9,** 305–329 (1956)

Conradi-Larsen, E. M., Sømme, L.: Anaerobiosis in the overwintering beetle *Pelophila borealis* (Coleoptera, Carabidae). Nature (London) **245,** 388–390 (1973a)

Conradi-Larsen, E. M., Sømme, L.: The overwintering of *Pelophila borealis* Payk. (Coleoptera, Carabidae) II. Aerobic and anaerobic metabolism. Norsk. Entomol. Tidsskr. **20,** 325–332 (1973b)

Coope, G. R.: Interpretations of quarternary insect fossils. Ann. Rev. Entomol. **15,** 97–120 (1970)

Critchley, B. R.: A laboratory study of the effects of some soil-applied organophosphorus pesticides on Carabidae (Coleoptera). Bull. Entomol. Res. **62,** 229–242 (1972)

Critchley, B. R.: Parasitism of the larvae of some Carabidae (Coleoptera). J. Entomol. (A) **48,** 37–42 (1973)
Crombie, A.: Further experiments on insect competition. Proc. Roy. Soc. Ser. B **133,** 76–109 (1946)
Crombie, A.: Interspecific competition. J. Animal Ecol. **16,** 44–73 (1947)
Danilevski, A. S.: Photoperiodism and Seasonal Development of Insects. Edinburgh and London: Oliver and Boyd, 1965
Darlington, P. J.: Carabidae of mountains and islands: data on the evolution of isolated faunas, and on atrophy of wings. Ecol. Monographs **13,** 37–61 (1943)
Darlington, P. J.: Biogeography of the Southern End of the World. Cambridge, Mass.: Harvard Univ. Press, 1965
Darlington, P. J.: Carabidae on tropical islands, especially the West Indies. Biotropica **2,** 7–15 (1970)
Darlington, P. J.: The carabid beetles of New Guinea. Part IV. General considerations, analysis and history of fauna, taxonomic supplement. Bull. Mus. Comp. Zool. **142,** 129–337 (1971)
Darwin, C.: The Origin of Species by Means of Natural Selection or the Preservation of Favoured Races in the Struggle for Life. London: John Murray, 1859
Darwin, C.: A Naturalist's Voyage. London 1860
Davies, L.: Two *Amblystogenium* species (Coleoptera, Carabidae) co-existing on the Subantarctic Possession Island, Crozet Islands (Indian Ocean). Entomol. Scand. **3,** 275–286 (1972)
Davies, M.: The contents of the crops of some British carabid beetles. Entomol. Mon. Mag. **89,** 18–23 (1953)
Davies, M.: A contribution to the ecology of species of *Notiophilus* and allied genera (Coleoptera, Carabidae). Entomol. Mon. Mag. **95,** 25–28 (1959)
Davies, M.: A woodland-floor food-chain in the laboratory. Entomol. Mon. Mag. **103,** 187–189 (1967)
Dawson, N.: A comparative study of the ecology of eight species of fenland Carabidae (Coleoptera). J. Animal Ecol. **34,** 299–314 (1965)
Deleurance, S.: La neurosécrétion chez les Coléoptères cavernicoles. Imago. C. R. Acad. Sci. Paris **264,** 392–394 (1967)
Deleurance, S., Deleurance, E. P.: Reproduction et cycle évolutif larvaire des *Aphaenops* (*A. cerberus* Dieck, *A. crypticola* Lindner), insectes Coléoptères cavernicoles. C. R. Acad. Sci. Paris **258,** 4369–4370 (1964a)
Deleurance, S., Deleurance, E. P.: L'absence de cycle saisonnier de reproduction chez les insectes Coléoptères troglobies (Bathysciines et Trechines). C. R. Acad. Sci. Paris **258,** 5995–5997 (1964b)
Delkeskamp, K.: Biologische Studien über *Carabus nemoralis* Müll. Z. Morphol. Oekol. Tiere **19,** 1–58 (1930)
Delkeskamp, K.: Die flugunfähige Gattung *Plagiopisthen* (Erotyl. Col.). Rec. Zool. Bot. Afr. **25,** 305 (1934)
Dick, J., Johnson, N. E.: Carabid beetles damage Douglas-fir seed. J. Econ. Entomol. **51,** 542–544 (1958)
Dicker, G. H. L.: *Agonum dorsale* Pont. (Coleoptera, Carabidae): an unusual egg-laying habit and some biological notes. Entomol. Mon. Mag. **87,** 33–34 (1951)
Dijk, T. S. van: The significance of the diversity in age composition of *Calathus melanocephalus* L. (Coleoptera, Carabidae) in space and time at Schiermonnikoog. Oecologia (Berl.) **10,** 111–136 (1972)
Dinther, J. B. M. van: Residual effect of a number of insecticides on adults of the carabid *Pseudophonus rufipes* (Dej.). Entomophaga **8,** 43–48 (1963)
Dinther, J. B. M. van: Laboratory experiments on the consumption capacities of some Carabidae. Meded. Rijksfac. Landbouwwetensch. Gent **31,** 730–739 (1966)
Dinther, J. B. M. van, Mensink, F. T.: Egg consumption by *Bembidion ustulatum* and *Bembidion lampros* (fam. Carabidae) in laboratory prey density experiments with house fly eggs. Meded. LandbHoogesch. OpzoekStns Gent **30,** 1542–1554 (1965)
Dinther, J. B. M. van, Mensink, F. T.: Use of radioactive phosphorus in studying egg predation by carabids in cauliflower fields. Meded. Fak. Landbouwwetensch. Gent **36,** 283–293 (1971)
Doskočil, J., Hůrka, K.: Entomofauna der Wiese (Verband Arrhenatherion elatioris) und ihre Entwicklung. Rozpr. Csl. Akad. Věd. **7,** 1–99 (1962)

Drift, J. van der: Analysis of the animal community in a beech forest floor. Tijdschr. Entomol. **94,** 1–168 (1951)
Drift, J. van der: Field studies on the surface fauna of forests. Bijdr. Dierkde. **29,** 79–103 (1959)
Drift, J. van der: A comparative study of the soil fauna in forests and cultivated land on sandy soils in Suriname. Studies on the Fauna of Suriname and other Guyanas **6,** 1–42 (1963)
Dubrovskaya, N. A.: Field carabid beetles (Coleoptera, Carabidae) of Byelorussia. Entomol. Rev. **49,** 476–483 (1970)
Dürkop, H.: Die Tierwelt der Anwurfzone der Kieler Förde. Schriften Naturw. Ver. Schleswig-Holstein **20** (1934)
Dusaussoy, G.: Observations sur le comportement de *Calosoma sycophanta* L. en élevage. Rev. Pathol. Vég. Entomol. Agr. France **42,** 53–65 (1963)
Dyke, E. C. van: The Coleoptera of the Galapagos Islands. Occ. Pap. Calif. Acad. Sci. San Francisco, 1953
Eidmann, H.: Zur Theorie der Bevölkerungsbewegung der Insekten. Anz. Schädlingskunde **13,** 47–52 (1937)
Eisner, T.: Chemical ecology: on arthropods and how they live as chemists. Verhandl. Deut. Zool. Ges. Helgold. **1971,** 123–137 (1972)
Elton, C.: Competition and the structure of ecological communities. J. Animal Ecol. **15,** 54–68 (1946)
Emmerich, H., Thiele, H. U.: Wirkung von Farnesylmethyläther auf die Ovarienreifung von *Pterostichus nigrita* F. Naturwissenschaften **56,** 641 (1969)
Engelhardt, W.: Die mitteleuropäischen Arten der Gattung *Trochosa* C. L. Koch, 1848 (Araneae, Lycosidae). Morphologie, Chemotaxonomie, Biologie, Autökologie. Z. Morph. Oekol. Tiere **54,** 219–392 (1964)
Erwin, T. L.: Bombardier beetles (Coleoptera, Carabidae) of North America: Part II. Biology and behavior of *Brachinus pallidus* Erwin in California. Coleopts Bull. **21,** 41–55 (1967)
Erwin, T. L.: Carabid beetles, mountain tops, and trees. Proc. Entomol. Soc. Wash. **75,** 127 (1973)
Evans, M. E. G.: The feeding method of *Cicindela hybrida* L. (Coleoptera, Cicindelidae). Proc. Roy. Entomol. Soc. Lond. (A) **40,** 61–66 (1965)
Evans, P. D., Ruscoe, C. N. E., Treherne, J. E.: Observations on the biology and submergence behaviour of some littoral beetles. J. Mar. Biol. Ass. U. K. **51,** 375–386 (1971)
Faasch, H.: Beobachtungen zur Biologie und zum Verhalten von *Cicindela hybrida* L. und *Cicindela campestris* L. und experimentelle Analyse ihres Beutefangverhaltens. Zool. Jb. Syst. **95,** 477–522 (1968)
Ferenz, H. J.: Steuerung der Larval- und Imaginalentwicklung von *Pterostichus nigrita* F. (Coleoptera, Carabidae) durch Umweltfaktoren und Hormone. Dissertation, Cologne (1973)
Ferenz, H. J.: Photoperiodic and hormonal control of reproduction in male beetles, *Pterostichus nigrita.* J. Insect Physiol. **21,** 331–341 (1975a)
Ferenz, H. J.: Anpassungen von *Pterostichus nigrita* F. (Coleoptera, Carabidae) an subarktische Bedingungen. Oecologia (Berl.) **19,** 49–57 (1975b)
Ferenz, H. J.: Two-step photoperiodic and hormonal control of reproduction in the female beetle, *Pterostichus nigrita.* J. Insect. Physiol. **23,** 671–676 (1977)
Frank, J. H.: The insect predators of the pupal stage of the winter moth, *Operophtera brumata* (L.) (Lepidoptera, Hydriomenidae). J. Animal Ecol. **36,** 375–389 (1967a)
Frank, J. H.: The effect of pupal predators on a population of winter moth, *Operophtera brumata* (L.) (Hydriomenidae). J. Animal Ecol. **36,** 611–621 (1967b)
Frank, J. H.: Carabidae (Coleoptera) of an arable field in central Alberta. Quaest. Entomol. **7,** 237–252 (1971a)
Frank, J. H.: Carabidae (Coleoptera) as predators of the redbacked cutworm (Lepidoptera, Noctuidae) in central Alberta. Can. Entomologist **103,** 1039–1044 (1971b)
Franz, H.: Morphologische und phylogenetische Studien an *Carabus* L. und nächstverwandten Gattungen. Z. Wiss. Zool. **135,** 163–213 (1929)
Franz, H., Gunhold, P., Pschorn-Walcher, H.: Die Kleintiergemeinschaften der Auwaldböden der Umgebung von Linz und benachbarter Flußgebiete. Naturkundl. Jb. Stadt Linz 1–63 (1959)
Freitag, R.: A revision of the North American species of the *Cicindela maritima* group with a study of hybridization between *Cicindela duodecimguttata* and *oregona.* Quaest. Entomol. **1,** 87–170 (1965)

Freitag, R., Hastings, L., Mercer, W. R., Smith, A.: Ground beetle populations near a kraft mill. Can. Entomologist **105**, 299–310 (1973)

Freitag, R., Lee, S. K.: Sound producing structures in adult *Cicindela tranquebarica* (Coleoptera, Cicindelidae) including a list of tiger beetles and ground beetles with flight wing files. Can. Entomologist **104**, 851–857 (1972)

Freitag, R., Ozburn, G. W., Leech, R. E.: The effects of Sumithion and Phosphamidon on populations of five carabid beetles and the spider *Trochosa terricola* in northwestern Ontario and including a list of collected species of carabid beetles and spiders. Can. Entomologist **101**, 1328–1333 (1969)

Freitag, R., Poulter, F.: The effects of the insecticides Sumithion and Phosphamidon on populations of five species of carabid beetles and two species of lycosid spiders in northwestern Ontario. Can. Entomologist **102**, 1307–1311 (1970)

Freude, H.: Carabidenstudien. NachrBl. Bayer. Entomol. **19**, 25–28 (1970)

Fuchs, G.: Die ökologische Bedeutung der Wallhecken in der Agrarlandschaft Nordwestdeutschlands, am Beispiel der Käfer. Pedobiologia **9**, 432–458 (1969)

Ganagarajah, M.: The neuro-endocrine complex of adult *Nebria brevicollis* F. and its relation to reproduction. J. Insect Physiol. **11**, 1377–1387 (1965)

Gause, G. F.: The struggle for existence. Baltimore 1934

Geiler, H.: Zur Ökologie und Phänologie der auf mitteldeutschen Feldern lebenden Carabiden. Wiss. Z. Karl-Marx-Universität Leipzig **6**, 35–61 (1956/57)

Geiler, H.: Die Coleopteren des Luzerne-Epigaions von Nordwestsachsen. Faun. Abh. **2**, 19–36 (1967)

Gersdorf, E.: Ökologisch-faunistische Untersuchungen über die Carabiden der mecklenburgischen Landschaft. Zool. Jb. Syst. **70**, 17–86 (1937)

Gersdorf, E.: Die Carabidenfauna einer Moorweide und der umgebenden Hecke. Z. Angew. Entomol. **52**, 475–489 (1965)

Gersdorf, E.: Käfer (Coleoptera) aus dem Jungtertiär Norddeutschlands. Geol. Jb. **87**, 295–332 (1969)

Gersdorf, E.: Weitere Käfer (Coleoptera) aus dem Jungtertiär Norddeutschlands. Geol. Jb. **88**, 629–670 (1971)

Gese, K.: Der Einfluß einiger Insektizide bei verschiedenen Anwendungsverfahren auf die Laufkäferfauna (Coleoptera, Carabidae) von Rübenfeldern. Dissertation, Gießen (1974)

Ghilarov, M. S.: Die Veränderung der Steppenbodenfauna unter dem Einfluß der künstlichen Bewaldung. Beitr. Entomol. **11**, 256–269 (1961)

Ghilarov, M. S., Arnoldi, K. V.: Steppe elements in the soil arthropod fauna of north-west Caucasus Mountains. Memorie Soc. Entomol. Ital. **47**, 103–112 (1969)

Giers, E.: Die Habitatgrenzen der Carabiden (Coleoptera, Insecta) im Melico-Fagetum des Teutoburger Waldes. Abhandl. Landesmus. Naturkunde Münster **35**, 1–36 (1973)

Gilbert, O.: The natural histories of four species of *Calathus* (Coleoptera, Carabidae) living on sand dunes in Anglesey, North Wales. Oikos **7**, 22–47 (1956)

Gilbert, O.: Notes on the breeding seasons of some Illinois carabid beetles. Pan-Pacif. Entomologist **33**, 53–58 (1957)

Gilbert, O.: The life history patterns of *Nebria degenerata* Schaufuss and *N. brevicollis* Fabricius (Coleoptera, Carabidae). J. Soc. Brit. Entomol. **6**, 11–14 (1958)

Gorny, M.: Dynamics of the soil insect communities in two biotopes of an agricultural landscape. Ekol. Pol. (A) **16**, 705–727 (1968a)

Gorny, M.: Faunal and zoocenological analysis of the soil insect communities in the ecosystem of shelterbelt and field. Ekol. Pol. (A) **16**, 297–324 (1968b)

Gorny, M.: Studies on the Carabidae (Coleoptera) of fieldhedges and the crops they bound. Polskie Pismo Entomol. **41**, 387–415 (1971)

Goulet, H.: Biology and relationship of *Pterostichus adstrictus* Eschscholtz and *Pterostichus pennsylvanicus* Leconte (Coleoptera, Carabidae). Quaest. Entomol. **10**, 3–34 (1974)

Green, J.: The mouthparts of *Eurynebria complanata* (L.) and *Bembidion laterale* (Sam.) (Coleoptera, Carabidae). Entomol. Mon. Mag. **92**, 110–113 (1956)

Greene, A.: Biology of five species of Cychrini (Coleoptera: Carabidae) in the steppe region of Southeastern Washington (USA). Melanderia **19**, 1–43 (1975)

Greenslade, P. J. M.: Daily rhythms of locomotor activity in some Carabidae (Coleoptera). Entomol. Exptl. Appl. **6,** 171–180 (1963a)

Greenslade, P. J. M.: Further notes on aggregation in Carabidae (Coleoptera), with especial reference to *Nebria brevicollis* (F.). Entomol. Mon. Mag. **99,** 109–114 (1963b)

Greenslade, P. J. M.: On the ecology of some British carabid beetles with special reference to life histories. Trans. Soc. Brit. Entomol. **16,** 149–179 (1965)

Grüm, L.: Seasonal changes of activity of the Carabidae. Ekol. Pol. (A) **7,** 255–268 (1959)

Grüm, L.: Diurnal activity rhythm of starved Carabidae. Bull. Pol. Acad. Sci. (Cl. II) **14,** 405–411 (1966)

Grüm, L.: Egg production of some Carabidae species. Bull. Pol. Acad. Sci. **21,** 261–268 (1973)

Grüm, L.: Mortality patterns in carabid populations. Ekol. Pol. **23,** 649–665 (1975a)

Grüm, L.: An attempt to estimate production of a few *Carabus* L. species (Coleoptera, Carabidae). Ekol. Pol. **23,** 673–680 (1975b)

Grüm, L.: Biomass production of carabid-beetles in a few forest habitats. Ekol. Pol. **24,** 37–56 (1976)

Haacker, U.: Deskriptive, experimentelle und vergleichende Untersuchungen zur Autökologie rhein-mainischer Diplopoden. Oecologia (Berl.) **1,** 87–129 (1968)

Haeck, J.: The immigration and settlement of carabids in the new Ijsselmeer-polders. In: Misc. Papers Landbouwhogesch. Wageningen **8,** 33–53 (1971)

Hamilton, C. C.: Studies on the morphology, taxonomy, and ecology of the larvae of holarctic tiger beetles. Proc. U. S. Nat. Mus. Washington **65,** 1925

Hasselmann, E. M.: Über die relative spektrale Empfindlichkeit von Käfer- und Schmetterlingsaugen bei verschiedenen Helligkeiten. Zool. Jb. Physiol. **69,** 537–576 (1962)

Hauchecorne, F.: Studien über die wirtschaftliche Bedeutung des Maulwurfs *(Talpa europaea).* Z. Morph. Oekol. Tiere **9,** 439–571 (1927)

Heerdt, P. F. van: The temperature and humidity preferences of certain Coleoptera. Proc. K. Ned. Akad. Wet. (Sect. C) **53,** 347–360 (1950)

Heerdt, P. F. van, Blokhuis, B., Haaften, C. van: The reproductive cycle and age composition of a population of *Pterostichus oblongopunctatus* (Fabricius) in the Netherlands (Coleoptera: Carabidae). Tijdschr. Entomol. **119,** 1–13 (1976)

Heerdt, P. F. van, Isings, J., Nijenhuis, L. E.: Temperature and humidity preferences of various Coleoptera from the duneland area of Terschelling I. Proc. K. Ned. Akad. Wet. (Sect. C) **59,** 668–676 (1956)

Heerdt, P. F. van, Isings, J., Nijenhuis, L. E.: Temperature and humidity preferences of various Coleoptera from the duneland area of Terschelling II. Proc. K. Ned. Akad. Wet. (Sect. C) **60,** 99–106 (1957)

Heerdt, P. F. van, Mörzer-Bruyns, M. F.: A biocenological investigation in the yellow dune region of Terschelling. Tijdschr. Entomol. **103,** 225–275 (1960)

Hempel, G., Hempel, I.: Über die tägliche Verteilung der Laufaktivität von Käfern des Hohen Nordens. Naturwissenschaften **42,** 77–78 (1955)

Hempel, W., Hiebsch, H., Schiemenz, H.: Zum Einfluß der Weidewirtschaft auf die Arthropoden-Fauna im Mittelgebirge. Faun. Abh. Mus. Tier. Dresden **3,** 235–281 (1971)

Hennig, W.: Die Stammesgeschichte der Insekten. Frankfurt: Waldemar Kramer, 1969

Herre, W.: Tierwelt und Eiszeit. Biologia Gen. **19,** 469–489 (1951)

Herter, K.: Über den Temperatursinn einiger Insekten. Z. Vergl. Physiol. **1,** 221–288 (1924)

Herter, K.: Der Temperatursinn der Insekten. Berlin: Duncker und Humblot, 1953

Herting, B.: Biologie der westpaläarktischen Raupenfliegen Dipt., Tachinidae. Monogr. Angew. Entomol. **16,** 1–188 (1960)

Heydemann, B.: Agrarökologische Problematik. Dissertation, Kiel (1953)

Heydemann, B.: Carabiden der Kulturfelder als ökologische Indikatoren. Ber. 7. Wanderversamm. Deut. Entomol. 172–185 (1955)

Heydemann, B.: Über die Bedeutung der "Formalinfallen" für die zoologische Landesforschung. Faun. Mitt. Norddeutschland **6,** 19–24 (1956)

Heydemann, B.: Die Biotopstruktur als Raumwiderstand und Raumfülle für die Tierwelt. Verh. Deut. Zool. Ges. Hamburg **1956,** 332–347 (1957)

Heydemann, B.: Erfassungsmethoden für die Biozönosen der Kulturbiotope. In: Lebensgemeinschaften der Landtiere, Balogh, J. (ed.). Budapest: Verl. Ungar. Akad. Wiss., 1958, pp. 453–537

Heydemann, B.: Der Einfluß des Deichbaus an der Nordsee auf Larven und Imagines von Carabiden und Staphyliniden. Ber. 9. Wanderversamm. Deut. Entomol. 237–274 (1962a)
Heydemann, B.: Die biozönotische Entwicklung vom Vorland zum Koog. Vergleichend-ökologische Untersuchungen an der Nordseeküste. II. Teil: Käfer (Coleoptera). Abh. Math.-Naturw. Kl. Akad. Wiss. Mainz **11,** 765–964 (1962b)
Heydemann, B.: Deiche der Nordseeküste als besonderer Lebensraum (Ökologische Untersuchungen über die Arthropoden-Besiedlung). Die Küste **11,** 90–130 (1963)
Heydemann, B.: Die Carabiden der Kulturbiotope von Binnenland und Nordseeküste – ein ökologischer Vergleich (Coleoptera, Carabidae). Zool. Anz. **172,** 49–86 (1964)
Heydemann, B.: Die Biologische Grenze Land–Meer im Bereich der Salzwiesen. Wiesbaden: Steiner, 1967a
Heydemann, B.: Der Überflug von Insekten über Nord- und Ostsee nach Untersuchungen auf Feuerschiffen. Deut. Entom. Z. **14,** 185–215 (1967b)
Heydemann, B.: Über die epigäische Aktivität terrestrischer Arthropoden der Küstenregionen im Tagesrhythmus, 249–263. In: Progress in Soil Biology. Graff, O., Satchell, J. E. (eds.). Braunschweig: Vieweg, 1967
Heydemann, B.: Das Freiland- und Laborexperiment zur Ökologie der Grenze Land–Meer. Verh. Deut. Zool. Ges. Heidelberg **1967,** 256–309 (1968)
Hiebsch, H.: Faunistisch-ökologische Untersuchungen in Steinrücken, Windschutzhecken und den angrenzenden Wiesen und Feldflächen. Tag.-Ber. Deut. Akad. Landw. Wiss. Berl. **60,** 25–35 (1964)
Hlavac, T. F.: A review of the species of *Scarites (Antilliscaris)* (Coleoptera, Carabidae), with notes on their morphology and evolution. Psyche **76,** 1–17 (1969)
Hölters, W.: Der Einfluß der Photoperiode auf die Gonadenreifung und Laufaktivität von *Pterostichus angustatus* Dft. (Coleoptera, Carabidae). Diplomarbeit, Cologne (1974)
Hoffmann, H. J.: Neuro-endocrine control of diapause und oöcyte maturation in the beetle, *Pterostichus nigrita.* J. Insect Physiol. **16,** 629–642 (1969)
Holste, G.: *Calosoma sycophanta* L. Seine Lebensgeschichte und -Gewohnheiten und seine erfolgreiche Ansiedlung in Neuengland. Eine Besprechung nebst einigen Bemerkungen über *Calosoma inquisitor* L. Z. Angew. Entomol. **2,** 413–421 (1915)
Honczarenko, J.: The effects of high doses of nitrogenous fertilizers on the Insecta of meadow soils. Pedobiologia **16,** 58–62 (1975)
Horion, A.: Faunistik der deutschen Käfer. 1: Adephaga – Caraboidea. Krefeld: Goecke and Evers, 1941
Horion, A.: Diskontinuierliche Ost-West-Verbreitung Mitteleuropäischer Käfer. Trans. 8th Intern. Congr. Entomol. (1950)
Horion, A.: Brutpflege bei montanen und alpinen Carabiden. Nachr. Bl. Bayer. Entomol. **5** (1956)
Horion, A.: Die halobionten und halophilen Carabiden der deutschen Fauna. Wiss. Z. Univ. Halle, Math.-Nat. **8,** 549–556 (1959)
Horstmann, K.: Untersuchungen über den Nahrungserwerb der Waldameisen (*Formica polyctena* Foerster) im Eichenwald. III. Jahresbilanz. Oecologia (Berl.) **15,** 187–204 (1974)
Hossfeld, R.: Synökologischer Vergleich der Fauna von Winter- und Sommerrapsfeldern. Ein Beitrag zur Agrarökologie. Z. Angew. Entomol. **52,** 209–254 (1963)
Humphrey, B. J., Dahm, P. A.: Chlorinated hydrocarbon insecticide residues in Carabidae and the toxicity of dieldrin to *Pterostichus chalcites* (Coleoptera, Carabidae). Environ. Entomol. **5,** 729–734 (1976)
Hůrka, K.: Fortpflanzung und Entwicklung der mitteleuropäischen *Carabus*- und *Procerus*-Arten. Studie čsl. Akad. Věd. **9,** 1–78 (1973)
Hurlbert, S.: The nonconcept of species diversity: a critique and alternative parameters. Ecology **52,** 577–586 (1971)
Jarmer, G.: Ein Vergleich der Carabidenfauna an eutrophen und dystrophen Gewässern in der Umgebung der Station Grietherbusch am Niederrhein. Staatsexamensarbeit, Cologne (1973)
Jeannel, R.: Coléoptères Carabiques 1. Faune de France **39,** 1–571. Paris (1941)
Jeannel, R.: Sur deux larves de Carabiques. Revue Fr. Entomol. **15,** 74–78 (1948)
Johnson, N. E., Cameron, R. S.: Phytophagous ground beetles. Ann. Entomol. Soc. Am. **62,** 909–914 (1969)
Jung, W.: Ernährungsversuche an *Carabus*-Arten. Entomol. Bl. **36,** 117–124 (1940)

Junk, W., Schenkling, S.: Coleopterorum Catalogus. Carabidae 1: W. Horn: Carabidae, Cicindelinae. Berlin: Junk, 1926. E. Csiki: Carabidae, Carabinae. Berlin: Junk, 1927. Carabidae 2: E. Csiki: Mormolycinae, Harpalinae I. Berlin: Junk, 1928–1931. Carabidae 3: E. Csiki: Harpalinae II. Berlin: Junk, 1932–1933

Kabacik, D.: Beobachtungen über die Quantitätsveränderungen der Laufkäfer (Carabidae) in verschiedenen Feldkulturen. Ekol. Pol. (A) **10**, 307–323 (1962)

Kabacik-Wasylik, D.: Ökologische Analyse der Laufkäfer (Carabidae) einiger Agrarkulturen. Ekol. Pol. (A) **18**, 137–209 (1970)

Kabacik-Wasylik, D.: Research into the number, biomass and energy flow of Carabidae (Coleoptera) communities in rye and potato fields. Pol. Ecol. Stud. **1**, 111–121 (1975)

Kaczmarek, W.: An analysis of interspecific competition in communities of the soil macrofauna of some habitats in the Kampinos National Park. Ekol. Pol. (A), **21**, 422–483 (1963)

Kaczmarek, W.: Elements of organization in the energy flow of forest ecosystems (preliminary notes), 663–678. In: Secondary Productivity of Terrestrial Ecosystems 2. Petrusewicz, K. (ed.). Warsaw/Krakow 1967

Karvonen, V. J.: Beobachtungen über die Insektenfauna in der Gegend von Vaaseni am mittleren Lauf des Syväri (Swir). Ann. Entomol. Fenn. **11** (1945)

Kasischke, P.: Freiland- und Laborversuche zur Aktivitätsrhythmik der tagaktiven Carabiden eines Waldbiotops (mit besonderer Berücksichtigung der Gattung *Notiophilus*). Staatsexamensarbeit, Cologne (1975)

Kaufmann, O.: Einige Bemerkungen über den Einfluß von Temperaturschwankungen auf die Entwicklungsdauer und Steuerung bei Insekten und seine graphische Darstellung durch Kettenlinie und Hyperbel. Z. Morph. Oekol. Tiere **25**, 353–361 (1932)

Kempf, W.: Zur Biologie von *Broscus cephalotes* L. (Carabidae). Zool. Anz. **155**, 30–33 (1954)

Kenagy, G. J.: Adaptations for leaf eating in the Great Basin kangaroo rat, *Dipodomys microps*. Oecologia (Berl.) **12**, 383–412 (1973)

Kern, P.: Über die Fortpflanzung und Eibildung bei einigen Caraben. Zool. Anz. **40**, 345–351 (1912)

Kerstens, G.: Coleopterologisches vom Lichtfang. Entomol. Bl. **57**, 119–138 (1961)

Kirchner, H.: Untersuchungen zur Ökologie feldbewohnender Carabiden. Dissertation, Cologne (1960)

Kirchner, H.: Tageszeitliche Aktivitätsperiodik bei Carabiden. Z. Vergl. Physiol. **48**, 385–399 (1964)

Kirk, V. M.: Ground beetles in cropland in South Dakota. Ann. Entomol. Soc. Am. **64**, 238–241 (1971a)

Kirk, V. M.: Biological studies of a ground beetle, *Pterostichus lucublandus*. Ann. Entomol. Soc. Am. **64**, 540–544 (1971b)

Kirk, V. M.: Seed-caching by larvae of two ground beetles, *Harpalus pennsylvanicus* and *H. erraticus*. Ann. Entomol. Soc. Am. **65**, 1426–1428 (1972)

Kirk, V. M.: Biology of *Pterostichus chalcites*, a ground beetle of cropland. Ann. Entomol. Soc. Am. **68**, 855–858 (1975)

Kirk, V. M., Dupraz, B. J.: Discharge by a female ground beetle, *Pterostichus lucublandus* (Coleoptera, Carabidae), used as a defense against males. Ann. Entomol. Soc. Am. **65**, 513 (1972)

Klein, A.: Studien zur Kenntnis der Insekten bestimmter Standorte des Bruchberges (Oberharz). Z. Angew. Entomol. **56**, 148–238 (1965)

Klein-Krautheim, F.: Zur Ökologie des Kartoffelkäfers, seine natürlichen Feinde und ihre Schädigung durch moderne Insektizide. Mitt. Biol. ZentrAnst. Ld- u. Forstw. Dahlem **75**, 37–41 (1953)

Kless, J.: Tiergeographische Elemente in der Käfer- und Wanzenfauna des Wutachgebietes und ihre ökologischen Ansprüche. Z. Morphol. Oekol. Tiere **49**, 541–628 (1961)

Klots, A. B., Klots, E. B.: Knaurs Tierreich in Farben: Insekten. Munich/Zürich: Droemer/Knaur, 1959

Klug, H.: Histo-physiologische Untersuchungen über die Aktivitätsperiodik bei Carabiden. Wiss. Z. Humboldt-Univ. Berl. Math.-Nat. Reihe **8**, 405–434 (1958/59)

Kmitowa, K., Kabacik-Wasylik, D.: An attempt at determining the pathogenicity of two species of entomopathogenic fungi in relation to Carabidae. Ekol. Pol. (A) **19**, 727–733 (1971)

Knopf, H. E.: Vergleichende ökologische Untersuchungen an Coleopteren aus Bodenoberflächenfängen in Waldstandorten auf verschiedenem Grundgestein. Z. Angew. Entomol. **49**, 353–362 (1962)

Kolb, A.: Nahrung und Nahrungsaufnahme bei Fledermäusen. Z. Säugetierkunde **23**, 84–95 (1958)
Kolbe, W.: Der Einfluß der Waldameise auf die Verbreitung von Käfern in der Bodenstreu eines Eichen-Birken-Waldes. Natur u. Heimat **3**, 120–124 (1968a)
Kolbe, W.: Vergleich der bodenbewohnenden Coleopteren aus zwei Eichen-Birken-Wäldern. Entomol. Z. **78**, 140–144 (1968b)
Kolbe, W.: Käfer im Wirkungsbereich der Roten Waldameise. Entomol. Z. **79**, 269–280 (1969)
Kolbe, W.: Vergleichende Coleopterenfänge in zwei Siegerländer Laubwäldern. Natur u. Heimat **30**, 22–25 (1970)
Kolbe, W.: Aktivitätsverteilung bodenbewohnender Coleopteren in einem Laubwald und drei von diesem eingeschlossenen Wertmehrungshorsten mit exotischen Coniferen. Decheniana **125**, 155–164 (1972)
Komárek, J.: Mutterpflege bei *Molops piceus* Panz. Acta Soc. Entomol. Čslv. **51**, 130–134 (1954)
Kovačević, J., Danon, M.: Mageninhalte der Vögel. Larus **4–5**, 185–217; **11**, 111–129 (1952 and 1959)
Kreckwitz, H.: Nahrungsaufnahme und Gewichtszunahme des Carabiden *Pterostichus nigrita* F. in Abhängigkeit von Temperatur und Photoperiode. Staatsexamensarbeit, Cologne (1974)
Krehan, I.: Die Steuerung von Jahresrhythmik und Diapause bei Larval- und Imagoüberwinterern der Gattung *Pterostichus* (Coleoptera, Carabidae). Oecologia (Berl.) **6**, 58–105 (1970)
Krieg, A.: Grundlagen der Insektenpathologie. Darmstadt: Steinkopff, 1961
Krogerus, H.: Ökologische Untersuchungen über Uferinsekten. Acta Zool. Fenn. **53**, 1–157 (1948)
Krogerus, R.: Über die Ökologie und Verbreitung der Arthropoden der Triebsandgebiete an den Küsten Finnlands. Acta Zool. Fenn. **12**, 1–308 (1932)
Krogerus, R.: Ökologische Studien über nordische Moorarthropoden. Commentat. Biol. **21**, 1–238 (1960)
Krumbiegel, I.: Untersuchungen über physiologische Rassenbildung (Ein Beitrag zum Problem der Artbildung und der geographischen Variation). Zool. Jb. Syst. **63**, 183–280 (1932)
Krumbiegel, I.: Morphologische Untersuchungen über Rassenbildung. Zool. Jb. Syst. **68**, 105–178 (1936)
Krumbiegel, I.: Die Rudimentation. Stuttgart: Fischer, 1960
Kudoh, K., Abe, A., Kondoh, I., Satoh, N., Saitoh, K.: Some cytological aspects of three species of beetles. Kontyu **38**, 232–238 (1970)
Kühnelt, W.: Wege zu einer Analyse der ökologischen Valenz. Verh. Deut. Zool. Ges. Tübingen **1954**, 292–299 (1955)
Kullmann, E., Nawabi, S.: Versuche zur Trägerfunktion aasfressender Käfer (Silphidae, Carabidae) bei der Trichinellosis. Z. Parasitenkunde **35**, 234–240 (1971)
Kulman, H. M.: Comparative ecology of North American Carabidae with special reference to biological control. Entomophaga **7**, 61–70 (1974)
Lack, D.: Darwin's Finches. Cambridge: University Press, 1947
Lampe, K. H.: Die Fortpflanzungsbiologie und Ökologie des Carabiden *Abax ovalis* Dft. und der Einfluß der Umweltfaktoren Bodentemperatur, Bodenfeuchtigkeit und Photoperiode auf die Entwicklung in Anpassung an die Jahreszeit. Zool. Jb. Syst. **102**, 128–170 (1975)
Lamprecht, G., Weber, F.: Nachweis der Selbst-Erregung der circadianen Periodizität bei Carabiden (Coleoptera, Ins.). Experientia **26**, 149–151 (1970)
Lamprecht, G., Weber, F.: Die Synchronisation der Laufperiodik von *Carabus*-Arten (Ins., Coleoptera) durch Zeitgeber unterschiedlicher Frequenz. Z. Vergl. Physiol. **72**, 226–259 (1971)
Lamprecht, G., Weber, F.: Die Circadian-Rhythmik von drei unterschiedlich weit an ein Leben unter Höhlenbedingungen adaptierten *Laemostenus*-Arten (Coleoptera, Carabidae). Ann. Spéléol. **30**, 471–482 (1975)
Larochelle, A.: Notes on the food of Cychrini (Coleoptera, Carabidae). Gt. Lakes Entomol. **5**, 81–83 (1972a)
Larochelle, A.: Collecting hibernating ground beetles in stumps (Coleoptera, Carabidae). Col. Bull. **26**, 30 (1972b)
Larochelle, A.: Stridulation and flight in some *Omophron* (Coleoptera, Carabidae). Proc. Entomol. Soc. Wash. **74**, 320 (1972c)
Larochelle, A.: The food of Cicindelidae of the world. Cicindela **6**, 21–43 (1974a)
Larochelle, A.: A world list of prey of *Chlaenius* (Coleoptera, Carabidae). Gt. Lakes Entomol. **7**, 137–142 (1974b)

Larochelle, A.: The American toad as champion carabid beetle collector. Pan-Pacif. Entomol. **50,** 203–204 (1974c)
Larochelle, A.: Carabid beetles (Coleoptera, Carabidae) as prey of North American frogs. Gt. Lakes Entomol. **7,** 147–148 (1974d)
Larochelle, A.: A list of mammals as predators of Carabidae. Carabologia **3,** 95–98 (1975a)
Larochelle, A.: A list of amphibians and reptiles as predators of North American Carabidae. Carabologia **3,** 99–103 (1975b)
Larson, D. J.: A revision of the genera *Philophuga* Motschoulsky and *Tecnophilus* Chaudoir with notes on the North American Callidina (Coleoptera, Carabidae). Quaest. Entomol. **5,** 15–84 (1969)
Larsson, S. G.: Entwicklungstypen und Entwicklungszeiten der dänischen Carabiden. Entomol. Meddr. **20,** 277–560 (1939)
Larsson, S. G., Gigja, G.: Coleoptera 1. Synopsis. In: The Zoology of Iceland. 3, Part 46a. Copenhagen, Reykjavik: E. Munksgaard, 1959
Lauterbach, A. W.: Verbreitungs- und aktivitätsbestimmende Faktoren bei Carabiden in sauerländischen Wäldern. Abhandl. Landesmus. Naturkunde Münster **26,** 1–100 (1964)
Lavigne, R. J.: Cicindelids as prey of robber flies (Dipt., Asilidae). Cicindela **4,** 1–7 (1972)
Lecordier, C., Pollet, A.: Les Carabiques (Coleoptera) d'une lisière forêt-falerie/savane à Lamto (Côte-d'Ivoire). Ann. Univ. Abidjan **4,** 251–286 (1971)
Lehmann, H.: Ökologische Untersuchungen über die Carabidenfauna des Rheinufers in der Umgebung von Köln. Dissertation, Cologne (1962)
Lehmann, H.: Ökologische Untersuchungen über die Carabidenfauna des Rheinufers in der Umgebung von Köln. Z. Morphol. Oekol. Tiere **55,** 597–630 (1965)
Lengerken, H. von: *Carabus auratus* und seine Larve. Arch. Naturgesch. (A) **87,** 31–113 (1921)
Lengerken, H. von: Coleoptera, Käfer. In: Schulze, P.: Biologie der Tiere Deutschlands. 12, Part 40. Berlin, 1924
Leśniak, A.: Investigations of the composition and structure of associations of carabid beetles (Carabidae, Coleoptera) in dependence on the intensity of occurrence of some primary injurious insects. Pr. Inst. badaw. Leśn. **407,** 4–44 (1972)
Leyk, G.: Untersuchungen zur Steuerung der Laufaktivität von *Pterostichus nigrita* F. (Coleoptera, Carabidae) durch exogene und endogene Faktoren. Staatsexamensarbeit, Cologne (1975)
Lindroth, C. H.: *Oodes gracilis* Villa. Eine thermophile Carabide Schwedens. Notul. Entomol. **22,** 109–157 (1943)
Lindroth, C. H.: Inheritance of wing dimorphism in *Pterostichus anthracinus* Ill. Hereditas **32,** 27–40 (1946)
Lindroth, C. H.: Notes on the ecology of Laboulbeniaceae infesting carabid beetles. Svensk Bot. Tidskr. **42,** 34–41 (1948)
Lindroth, C. H.: Die Fennoskandischen Carabidae. Kungl. Vetensk. Vitterh. Samh. Handl. (Ser. B 4) 1, Spezieller Teil, 1–709 (1945); 3, Allgemeiner Teil, 1–911 (1949)
Lindroth, C. H.: Die Larve von *Lebia chlorocephala* Hoffm. (Coleoptera, Carabidae). Opusc. Entomol. **19,** 29–33 (1954)
Lindroth, C. H.: The faunal connections between Europe and North America. Stockholm, New York: Almquist and Wiksell/Wiley and Sons, 1957
Lindroth, C. H.: Coleopteren – hauptsächlich Carabiden – aus dem Diluvium von Hösbach. Opusc. Entomol. **25,** 112–128 (1960)
Lindroth, C. H.: The ground-beetles (Carabidae, excl. Cicindelinae) of Canada and Alaska. 1–6. Opusc. Entomol. (Suppl.) **20, 24, 29, 33, 34, 35** (1961–69)
Lindroth, C. H.: The fauna history of Newfoundland. Illustrated by carabid beetles. Opusc. Entomol. (Suppl.) **23,** 1–112 (1963)
Lindroth, C. H.: The Icelandic form of *Carabus problematicus* Hbst. (Coleoptera, Carabidae). A statistic treatment of subspecies. Opusc. Entomol. **33,** 157–182 (1968)
Lindroth, C. H.: The theory of glacial refugia in Scandinavia. Comments on present opinions. Notul. Entomol. **49,** 178–192 (1969)
Lindroth, C. H.: Survival of animals and plants on ice-free refugia during the Pleistocene glaciations. Endeavour **29,** 129–134 (1970a)
Lindroth, C. H.: Surtsey, Island – Untersuchungen über terrestrische Biota. Schr. Naturw. Ver. Schlesw.-Holst. Sonderband, 11–19 (1970b)

Lindroth, C. H.: Biological investigations on the new volcanic island Surtsey, Iceland, 65–69. In: Dispersal and Dispersal Power of Carabid Beetles. den Boer, P. J. (ed.). Misc. Papers Landbouwhogesch. Wageningen **8** (1971 a)
Lindroth, C. H.: Disappearance as a protective factor. A supposed case of Bates'ian mimicry among beetles (Coleoptera, Carabidae and Chrysomelidae). Entomol. Scand. **2**, 41–48 (1971 b)
Lindroth, C. H.: On the elytral microsculpture of carabid beetles (Coleoptera, Carabidae). Entomol. Scand. **5**, 251–264 (1974 a)
Lindroth, C. H.: Coleoptera. Family Carabidae. In: Handbooks for the Identification of British Insects. London: Roy. Entomol. Soc., 1974 b
Löser, S.: Brutfürsorge und Brutpflege bei Laufkäfern der Gattung *Abax*. Verh. Deut. Zool. Ges. Würzburg **1969**, 322–326 (1970)
Löser, S.: Art und Ursachen der Verbreitung einiger Carabidenarten (Coleoptera) im Grenzraum Ebene–Mittelgebirge. Zool. Jb. Syst. **99**, 213–262 (1972)
Lohmeyer, W., Rabeler, W.: Aufbau und Gliederung der mesophilen Laubmischwälder im mittleren und oberen Wesergebiet und ihre Tiergesellschaften, pp. 238–257. In: Biosoziologie. Tüxen, R. (ed.). The Hague: Junk, 1965
Louda, J.: Über den Einfluß vom Bebauen und von der chemischen Zusammensetzung des Bodens auf das Erscheinen von Carabidae. Sborník PF v Hradci Králové **5**, 219–236 (1968)
Louda, J.: Beitrag zur Verbreitung der Carabiden auf bewirtschafteten Feldern im Überschwemmungsgebiet des Flusses Želivka in Mittelböhmen. Zprávy čsl. spol. entomol. ČSAV **7**, 3–6 (1971)
Lücke, E.: Die epigäische Fauna auf Zuckerrübenfeldern unterschiedlicher Bodenverhältnisse im Göttinger Raum. Z. Angew. Zool. **47**, 43–90 (1960)
Ludescher, F. B.: Sumpfmeise (*Parus p. palustris* L.) und Weidenmeise (*P. montanus salicarius* Br.) als sympatrische Zwillingsarten. J. Ornithol. **114**, 3–56 (1973)
Ludwig, W.: Zur Theorie der Konkurrenz. Die Annidation (Einnischung) als fünfter Evolutionsfaktor. In: Neue Erg. Probl. Zool. = Zool. Anz. Erg.bd. zu **145**, 516–537 (1950)
Luff, M. L.: Some effects of Formalin on the numbers of Coleoptera caught in pitfall traps. Entomol. Mon. Mag. **104**, 115–116 (1968)
Luff, M. L.: The annual activity pattern and life cycle of *Pterostichus madidus* (F.) (Coleoptera, Carabidae). Entomol. Scand. **4**, 259–273 (1973)
Luff, M. L.: Adult and larval feeding habits of *Pterostichus madidus* (F.) (Coleoptera, Carabidae). J. Nat. Hist. **8**, 403–409 (1974)
Lumaret, J. P.: Cycle biologique et comportement de ponte de *Percus (Pseudopercus) navaricus* (Coleoptera, Carabique). Entomologiste **27**, 49–52 (1971)
MacArthur, R. H., Wilson, E. O.: The Theory of Island Biogeography. Princeton: University Press, 1967
Mandl, K.: Wiederherstellung des Familienstatus der Cicindelidae (Coleoptera). Beitr. Entomol. **21**, 507–508 (1971)
Manga, N.: Population metabolism of *Nebria brevicollis* (F.) (Coleoptera, Carabidae). Oecologia (Berl.) **10**, 223–242 (1972)
Manley, G. V.: A seed-cacheing carabid, *Synuchus impunctatus* Say (Coleoptera, Carabidae). Ann. Entomol. Soc. Am. **64**, 1474–1475 (1971)
Mansfeld, K.: Zur Ernährung des Rotrückenwürgers (*Lanius collurio* L.) besonders hinsichtlich der Nestlingsnahrung, der Vertilgung von Nutz- und Schadinsekten und seines Einflusses auf den Singvogelbestand. Beitr. Vogelkunde **6**, 271–292 (1957–60)
Maurer, R.: Die Vielfalt der Käfer- und Spinnenfauna des Wiesenbodens im Einflußbereich von Verkehrsimmissionen. Oecologia (Berl.) **14**, 327–351 (1974)
Mayr, E.: Animal Species and Evolution. Cambridge (Mass.): Belknap Press, 1963
Mayr, E.: Evolution und Verhalten. Verh. Deut. Zool. Ges. Köln **1970**, 322–336 (1970)
Meijer, J.: Immigration of arthropods into the new Lauwerszee Polder, 53–65. In: Dispersal and Dispersal Power of Carabid Beetles. den Boer, P. J. (ed.). Misc. Papers Landbouwhogesch. Wageningen **8** (1971)
Meijer, J.: A comparative study of the immigration of carabids (Coleoptera, Carabidae) into a new polder. Oecologia (Berl.) **16**, 185–208 (1974)
Meijer, J.: Carabid (Coleoptera, Carabidae) migration studied with Laboulbeniales (Ascomycetes) as biological tags. Oecologia (Berl.) **19**, 99–103 (1975)
Miller, L. K.: Freezing tolerance in an adult insect. Science **166**, 105–106 (1969)

Mitchell, B.: Ecology of two carabid beetles, *Bembidion lampros* (Herbst) and *Trechus quadristriatus* (Schrank). II. Studies on populations of adults in the field, with special reference to the technique of pitfall trapping. J. Animal Ecol. **32,** 377–392 (1963)

Mletzko, G.: Beitrag zur Carabiden-Fauna des NSG Burgholz Halle/S. Hercynia (Leipzig) N.F. **7,** 92–110 (1970)

Mletzko, G.: Ökologische Valenzen von Carabidenpopulationen im Fraxino-Ulmetum (Tx 52, Oberst 53). Beitr. Entomol. **22,** 471–485 (1972)

Mobedi, I., Arfaa, F.: Probable role of ground beetles (Coleoptera, Carabidae) in the transmission of *Capillaria hepatica* (Nematoda). J. Parasitol. **57,** 1144–1145 (1971)

Mossakowski, D.: Ökologische Untersuchungen an epigäischen Coleopteren atlantischer Moor- und Heidestandorte. Z. Wiss. Zool. **181,** 233–316 (1970a)

Mossakowski, D.: Das Hochmoor-Ökoareal von *Agonum ericeti* (Panz.) (Coleoptera, Carabidae) und die Frage der Hochmoorbindung. Faun.-Ökol. Mitt. **3,** 378–392 (1970b)

Mossakowski, D.: Zur Variabilität isolierter Populationen von *Carabus arcensis* Hbst. (Coleoptera). Z. Zool. Syst. Evolutionsforsch. **9,** 81–106 (1971)

Mossakowski, D., Weber, F.: Korrelationen zwischen Heterochromatingehalt und morphologischen Merkmalen bei *Carabus auronitens* (Coleoptera). Z. Zool. Syst. Evolutionsforsch. **10,** 291–300 (1972)

Müller, G.: Faunistisch-ökologische Untersuchungen der Coleopterenfauna der küstennahen Kulturlandschaft bei Greifswald. Teil I: Die Carabidenfauna benachbarter Acker- und Weideflächen mit dazwischenliegendem Feldrain. Pedobiologia **8,** 313–339 (1968)

Müller, G.: Laboruntersuchungen zur Wirkung von Herbiziden auf Carabiden. Arch. Pflanzenschutz **7,** 351–364 (1971)

Müller, G.: Faunistisch-ökologische Untersuchungen der Coleopterenfauna der küstennahen Kulturlandschaft bei Greifswald. Die Wirkung der Herbizide UVON-Kombi (II) und Elbanil (III) auf die epigäische Fauna von Kulturflächen. Pedobiologia **12,** 169–211 (1972)

Müller, H. J.: Formen der Dormanz bei Insekten. Nova Acta Leopoldina **35,** 7–25 (1970)

Müller, P., Klomann, U., Nagel, P., Reis, H., Schäfer, A.: Indikatorwert unterschiedlicher biotischer Diversität im Verdichtungsraum von Saarbrücken. Verh. Ges. Ökol. Saarbrücken **1974,** 113–128 (1975).

Murdoch, W. W.: Aspects of the population dynamics of some marsh Carabidae. J. Animal Ecol. **35,** 127–156 (1966)

Murdoch, W. W.: Life history patterns of some British Carabidae (Coleoptera) and their ecological significance. Oikos **18,** 25–32 (1967)

Nettmann, H. K.: Karyotyp-Analysen bei Carabiden (Coleoptera). Mitt. dtsch. ent. Ges. **35,** 113–117 (1976)

Neudecker, C.: Lokomotorische Aktivität von *Carabus glabratus* Payk. und *Carabus violaceus* L. am Polarkreis. Oikos **22,** 128–130 (1971)

Neudecker, C.: Das Präferenzverhalten von *Agonum assimile* Payk. (Carabidae, Coleoptera) in Temperatur-, Feuchtigkeits- und Helligkeitsgradienten. Zool. Jb. Syst. **101,** 609–627 (1974)

Neudecker, C., Thiele, H. U.: Die jahreszeitliche Synchronisation der Gonadenreifung bei *Agonum assimile* Payk. (Coleoptera, Carabidae) durch Temperatur und Photoperiode. Oecologia (Berl.) **17,** 141–157 (1974)

Neumann, D.: Die lunare und tägliche Schlüpfperiodik der Mücke *Clunio*. Steuerung und Abstimmung auf die Gezeitenperiodik. Z. Vergl. Physiol. **53,** 1–61 (1966)

Neumann, D.: Genetic adaptation in emergence time of *Clunio* populations to different tidal conditions. Helgoländer Wiss. Meeresunters. **15,** 163–171 (1967)

Neumann, D.: The temporal programming of development in the intertidal chironomid *Clunio marinus* (Diptera, Chironomidae). Can. Entomologist **103,** 315–318 (1971)

Neumann, D.: Adaptation of chironomids to intertidal environments. Ann. Rev. Entomol. **21,** 387–414 (1976)

Neumann, D., Honegger, H. W.: Adaptations of the intertidal midge *Clunio* to Arctic conditions. Oecologia (Berl.) **3,** 1–13 (1969)

Neumann, U.: Die Sukzession der Bodenfauna (Carabidae—Coleoptera, Diplopoda und Isopoda) in den forstlich rekultivierten Gebieten des Rheinischen Braunkohlenreviers. Pedobiologia **11,** 193–226 (1971)

Nickerl, O.: *Carabus auronitens* Fab. Ein Beitrag zur Kenntnis vom Lebensalter der Insekten. Stettin. Entomol. Z. **50,** 155–163 (1889)

Niemann, G.: Zum biotopmäßigen Vorkommen von Coleopteren. Teil I: Kiefern-Altbestände auf hügeligen (grundwasserfernen) und auf grundwasserbeeinflußten Standorten. Z. Angew. Entomol. **53,** 82–110 (1963–64)

Noonan, G. R.: The anisodactylines (Insecta, Coleoptera, Carabidae, Harpalini): classification, evolution, and zoogeography. Quaest. Entomol. **9,** 266–480 (1973)

Novák, B.: Zur Ethologie der Laufkäfer (Carabidae). Sb. Vys. Šk. Pedagog. V Olomouci Přírodní Vědy **7,** 93–96 (1959)

Novák, B.: Saisonmäßiges Vorkommen und Synökologie der Carabiden auf Zuckerrübenfeldern von Haná (Coleoptera, Carabidae). Acta Univ. Palackianae Olomucensis Fac. Rer. Nat. **13,** 101–251 (1964)

Novák, B.: Bindung der Imagines von manchen Feldcarabidenarten an die Lebensbedingungen in einem Gerstebestand (Coleoptera, Carabidae). Acta Univ. Palackianae Olomucensis Fac. Rer. Nat. **25,** 77–94 (1967)

Novák, B.: Bindungsgrad der Imagines einiger Feldcarabidenarten an die Lebensbedingungen in einem Winterweizenbestand (Coleoptera, Carabidae). Acta Univ. Palackianae Olomucensis Fac. Rer. Nat. **28,** 99–132 (1968)

Novák, B.: Bodenfallen mit großem Öffnungsdurchmesser zur Untersuchung der Bewegungsaktivität von Feldcarabiden (Coleoptera, Carabidae). Acta Univ. Palackianae Olomucensis Fac. Rer. Nat. **31,** 71–86 (1969)

Novák, B.: Diurnale Aktivität der Carabiden in einem Feldbiotop. (Coleoptera, Carabidae). Acta Univ. Palackianae Olomucensis Fac. Rer. Nat. **34,** 129–149 (1971 a)

Novák, B.: Verhaltensanalysen zur genaueren Auswertung der Bodenfallenfänge von Carabiden (Coleoptera, Carabidae). Acta Univ. Palackianae Olomucensis Fac. Rer. Nat. **34,** 119–128 (1971 b)

Novák, B.: Saisondynamik der tageszeitlichen Aktivität bei Carabiden in einem Feldbiotop (Coleoptera, Carabidae). Acta Univ. Palackianae Olomucensis Fac. Rer. Nat. **39,** 59–97 (1972)

Novák, B.: Jahreszeitliche Dynamik der diurnalen Aktivität bei Carabiden in einem Waldbiotop (Coleoptera, Carabidae). Acta Univ. Palackianae Olomucensis Fac. Rer. Nat. **43,** 251–280 (1973)

Novák, K., Skuhravý, V., Hrdý, I., Hůrka, K.: Versuch zur Bestimmung der Einwirkung einer Bestäubung der Waldränder mit HCH auf die Biocoenosis der Insekten. Zool. Entomol. Listy **2,** 3–16 (1953)

Novák, K., Skuhravý, V.: Der Einfluß von DDT in Aerosolform auf einige Insektenarten des Kartoffelfeldes. Zool. Listy **6,** 41–51 (1957)

Novák, K., Skuhravý, V., Zelený, J.: Der Einfluß von Systox auf einige Insektenarten des Zuckerrübenfeldes. Anz. Schädl. Kunde **35,** 17–20 (1962)

Novák, V.: Die natürlichen Feinde und Krankheiten des Gemeinen Nutzholzborkenkäfers *Trypodendron lineatum* Olv. Zool. Listy **9,** 309–322 (1960)

Obrtel, R.: Carabidae and Staphylinidae occurring on soil surface in Lucerne fields (Coleoptera). Acta Entomol. Bohemoslov. **65,** 5–20 (1968)

Obrtel, R.: Number of pitfall traps in relation to the structure of the catch of soil surface Coleoptera. Acta Entomol. Bohemoslov. **68,** 300–309 (1971 a)

Obrtel, R.: Soil surface coleoptera in a lowland forest. Acta Sci. Nat. Brno. **5,** 1–47 (1971 b)

Obrtel, R.: Soil surface coleoptera in a reed swamp. Acta Sci. Nat. Brno. **6,** 1–35 (1972)

Obrtel, R.: Animal food of *Apodemus flavicollis* in a lowland forest. Zool. Listy **22,** 15–30 (1973)

Ohnesorge, B.: Beziehungen zwischen Regulationsmechanismus und Massenwechselablauf bei Insekten (Ein Beitrag zur Theorie der Populationsdynamik). Z. Angew. Zool. **50,** 427–483 (1963)

Ohyama, Y., Asahina, E.: Frost resistance in adult insects. J. Insect Physiol. **18,** 267–282 (1972)

Oppermann, J.: Die Nahrung des Maulwurfs (*Talpa europaea* L. 1758) in unterschiedlichen Lebensräumen. Pedobiologia **8,** 59–74 (1968)

Paarmann, W.: Vergleichende Untersuchungen über die Bindung zweier Carabidenarten (*Pterostichus angustatus* Dft. und *Pterostichus oblongopunctatus* F.) an ihre verschiedenen Lebensräume. Z. Wiss. Zool. **174,** 83–176 (1966)

Paarmann, W.: Prothetelie bei Carabiden der Gattung *Pterostichus*. Zool. Beitr. **13,** 121–135 (1967)

Paarmann, W.: Untersuchungen über die Jahresrhythmik von Laufkäfern (Coleoptera, Carabidae) in der Cyrenaika (Libyen, Nordafrika). Oecologia (Berl.) **5,** 325–333 (1970)

Paarmann, W.: Bedeutung der Larvenstadien für die Fortpflanzungsrhythmik der Laufkäfer *Broscus laevigatus* Dej. und *Orthomus atlanticus* Fairm. (Coleoptera, Carabidae) aus Nordafrika. Oecologia (Berl.) **13**, 81–92 (1973)

Paarmann, W.: Der Einfluß von Temperatur und Lichtwechsel auf die Gonadenreifung des Laufkäfers *Broscus laevigatus* Dej. (Coleoptera, Carabidae) aus Nordafrika. Oecologia (Berl.) **15**, 87–92 (1974)

Paarmann, W.: Freilanduntersuchungen in Marokko (Nordafrika) zur Jahresrhythmik von Carabiden (Coleoptera, Carabidae) und zum Mikroklima im Lebensraum der Käfer. Zool. Jb. Syst. **102**, 72–88 (1975)

Paarmann, W.: Die Bedeutung exogener Faktoren für die Gonadenreifung von *Orthomus barbarus atlanticus* (Coleoptera, Carabidae) aus Nordafrika. Entomol. Exptl. Appl. **19**, 23–36 (1976a)

Paarmann, W.: Jahreszeitliche Aktivität und Fortpflanzungsrhythmik von Laufkäfern (Coleoptera, Carabidae) im Kivugebiet (Ost-Zaire, Zentralafrika). Zool. Jb. Syst. **103**, 311–354 (1976b)

Paarmann, W.: The annual periodicity of the polyvoltine ground beetle *Pogonus chalceus* Marsh. (Col., Carabidae) and its control by environmental factors. Zool. Anz. **196**, 150–160 (1976c)

Palm, T., Lindroth, C. H.: Die Coleopterafauna am Klarälven. 1, Allgemeiner Teil. Ark. Zool. (A) **28**, 1–42 (1937a)

Palm, T., Lindroth, C. H.: Die Coleopterafauna am Klarälven. 2, Spezieller Teil. Entomol. Tidskr. **58**, 115–145 (1937b)

Palmén, E.: Die anemohydrochore Ausbreitung der Insekten als zoogeographischer Faktor. Ann. Zool. Soc. Zool. Bot. Fenn. Vanamo **10**, 1–262 (1944)

Palmén, E.: Felduntersuchungen und Experimente zur Kenntnis der Überwinterung einiger Uferarthropoden. Ann. Entomol. Fenn. (Suppl.) **14**, 169–179 (1949)

Palmén, E.: Effect of soil moisture upon "temperature preferendum" in *Dyschirius thoracicus* Rossi (Coleoptera, Carabidae). Ann. Entomol. Fenn. **20**, 1–13 (1954)

Palmén, E., Platonoff, S.: Zur Autökologie und Verbreitung der ostfennoskandischen Flußuferkäfer. Ann. Entomol. Fenn. **9**, 74–195 (1943)

Palmén, E., Suomalainen, H.: Experimentelle Untersuchungen über die Transpiration bei einigen Arthropoden, insbesondere Käfern. Ann. Zool. Soc. Zool. Bot. Fenn. Vanamo **11**, 1–52 (1945)

Papi, F.: Orientamento astronomico in alcuni Carabidi. Memorie Soc. tosc. Sci. nat. (Ser. B) **62**, 83–97 (1955)

Parsons, P. A.: Behavioural and Ecological Genetics: A Study in *Drosophila*. Oxford: Clarendon Press, 1973

Penney, M. M.: Studies on certain aspects of the ecology of *Nebria brevicollis* (F.) (Coleoptera, Carabidae). J. Animal Ecol. **35**, 505–512 (1966)

Penney, M. M.: Diapause and reproduction in *Nebria brevicollis* (F.) (Coleoptera, Carabidae). J. Animal Ecol. **38**, 219–233 (1969)

Perttunen, V.: The humidity preferences of various carabid species (Coleoptera, Carabidae) of wet and dry habitats. Ann. Entomol. Fenn. **17**, 72–84 (1951)

Petruška, F.: Carabiden als Bestandteil der Entomofauna der Rübenfelder in der Unicov-Ebene (Coleoptera, Carabidae). Acta Univ. Palackianae Olomucensis Fac. Rer. Nat. **25**, 121–243 (1967)

Petruška, F.: On the possibility of escape of the various components of the epigaeic fauna of the fields from the pitfall traps containing Formalin (Coleoptera). Acta Univ. Palackianae Olomucensis Fac. Rer. Nat. **31**, 99–124 (1969)

Petruška, F.: The influence of the agricultural plants on the development of the populations of Carabidae living in the fields (Coleoptera, Carabidae). Acta Univ. Palackianae Olomucensis Fac. Rer. Nat. **34**, 151–191 (1971)

Peus, F.: Die Tierwelt der Moore. In: Handbuch der Moorkunde. Berlin 1932, Vol. III

Pfeifer, S., Keil, W.: Versuche zur Steigerung der Siedlungsdichte höhlen- und freibrütender Vogelarten und ernährungsbiologische Untersuchungen an Nestlingen einiger Singvogelarten in einem Schadgebiet des Eichenwicklers (*Tortrix viridana* L.) im Osten von Frankfurt am Main. Biol. Abh. **15/16** (1958)

Pflüger, W.: Die Sanduhrsteuerung der gezeitensynchronen Schlüpfrhythmik der Mücke *Clunio marinus* im arktischen Mittsommer. Oecologia (Berl.) **11**, 113–150 (1973)

Pittendrigh, C. S.: Adaptation, natural selection, and behavior, 390–416. In: Behavior and Evolution. Roe, A., Simpson, G. G. (eds.). New Haven: Yale University Press, 1958

Pollard, E.: Hedges. II. The effect of removal of the bottom flora of a hawthorn hedgerow on the fauna of the hawthorn. J. Appl. Ecol. **5**, 109–123 (1968a)
Pollard, E.: Hedges. III. The effect of removal of the bottom flora of a hawthorn hedgerow on the Carabidae of the hedge bottom. J. Appl. Ecol. **5**, 125–139 (1968b)
Pollard, E.: Hedges. IV. A comparison between the Carabidae of a hedge and field site and those of a woodland glade. J. Appl. Ecol. **5**, 649–657 (1968c)
Prilop, H.: Untersuchungen über die Insektenfauna von Zuckerrübenfeldern in der Umgebung von Göttingen. Z. Angew. Zool. **44**, 447–509 (1957)
Puisségur, C.: Recherches sur la génétique des Carabes. Vie et Milieu (Suppl.) **18**, 1–288 (1964)
Puisségur, C.: La kystogenèse de grégarines hémocoeliennes chez les carabes (Coleoptera, Carabidae) des Pyrénées et des Corbières. Bull. Biol. Fr. Belg. **106**, 101–122 (1972)
Rabeler, W.: Die Tiergesellschaft der trockenen *Calluna*heiden in Nordwestdeutschland. Jbr. Naturh. Ges. Hannover **94/98**, 357–375 (1947)
Rabeler, W.: Die Tiergesellschaft eines nitrophilen Kriechrasens in Nordwestdeutschland. Mitt. Flor.-Soz. ArbGemein **4**, 166–171 (1953)
Rabeler, W.: Die Tiergesellschaft eines Eichen-Birkenwaldes im nordwestdeutschen Altmoränengebiet. Mitt. Flor.-Soz. ArbGemein **6/7**, 297–319 (1957)
Rabeler, W.: Die Tiergesellschaften von Laubwäldern (Querco-Fagetea) im oberen und mittleren Wesergebiet. Mitt. Flor.-Soz. ArbGemein **9**, 200–229 (1962)
Rabeler, W.: Zur Kenntnis der nordwestdeutschen Eichen-Birkenwaldfauna. Schriftenreihe Vegkunde **4**, 131–154 (1969a)
Rabeler, W.: Über die Käfer- und Spinnenfauna eines nordwestdeutschen Birkenbruchs. Vegetatio **18**, 387–392 (1969b)
Rapp, A.: Zur Biologie und Ethologie der Käfermilbe *Parasitus coleoptratorum* L. 1758. (Ein Beitrag zum Phoresie-Problem.) Zool. Jb. Syst. **86**, 303–366 (1959)
Reddingius, J., Boer, P. J. den: Simulation experiments illustrating stabilization of animal numbers by spreading of risk. Oecologia (Berl.) **5**, 240–284 (1970)
Regenfuss, H.: Untersuchungen zur Morphologie, Systematik und Ökologie der Podapolipidae (Acarina, Tarsonemini). (Unter besonderer Berücksichtigung der Parallelevolution der Gattungen *Eutarsopolipus* und *Dorsipes* mit ihren Wirten (Coleoptera, Carabidae)). Z. Wiss. Zool. **177**, 183–282 (1968)
Regenfuss, H.: Die Antennen-Putzeinrichtung der Adephaga (Coleoptera), parallele evolutive Vervollkommnung einer komplexen Struktur. Z. Zool. Syst. Evolutionsforsch. **13**, 278–299 (1975)
Reichenbach-Klinke, H. H.: Die Abhängigkeit der Darmgestalt bei der Raubkäferfamilie der Carabiden von phylogenetischen und ökologischen Faktoren. Dissertation, Berlin (1938)
Reise, K., Weidemann, G.: Dispersion of predatory forest floor arthropods. Pedobiologia **15**, 106–128 (1975)
Remmert, H.: Über tagesperiodische Änderungen des Licht- und Temperaturpräferendums bei Insekten (Untersuchungen an *Cicindela campestris* und *Gryllus domesticus*). Biol. Zbl. **79**, 577–584 (1960)
Remmert, H.: Der Schlüpfrhythmus der Insekten. Wiesbaden: Franz Steiner, 1962
Renkonen, O.: Statistisch-ökologische Untersuchungen über die terrestrische Käferwelt der finnischen Bruchmoore. Ann. Zool. Soc. Zool. Bot. Fenn. Vanamo **6**, 1–231 (1938/39)
Rensch, B.: Aktivitätsphasen von *Cicindela*-Arten in klimatisch stark unterschiedenen Gebieten. Zool. Anz. **158**, 33–38 (1957)
Rensing, L.: Biologische Rhythmen und Regulation. Grundbegriffe der modernen Biologie **10**, 1–265. Stuttgart: Fischer, 1973
Richoux, P.: Ecologie et Ethologie de la faune des fissures intertidales de la Région malouine. (1). Bull. Lab. Marit. Dinard **1**, 145–206 (1972)
Rivard, I.: Carabid beetles (Coleoptera, Carabidae) from agricultural lands near Belleville, Ontario. Can. Entomologist **96**, 517–520 (1964a)
Rivard, I.: Notes on parasitism of ground beetles (Coleoptera, Carabidae) in Ontario. Can. J. Zool. **42**, 919–920 (1964b)
Röber, H., Schmidt, G.: Untersuchungen über die räumliche und biotopmäßige Verteilung einheimischer Käfer (Carabidae, Silphidae, Geotrupidae, Necrophoridae). Natur u. Heimat **9**, 1–19 (1949)
Roer, H.: *Carabus purpurascens* (Coleoptera, Carabidae) als Beute des Mausohrs *(Myotis myotis)* (Mamm. Chiroptera). Entomol. Bl. **67**, 62–63 (1971)

Roth, A.: Vergleichende biozönotische Untersuchungen über Insekten an Laub- und Nadelfeldgehölzen in der Magdeburger Börde. Hercynia (Leipzig) (N. F.) **1**, 51–81 (1963)

Rudolph, R.: Ökethologische und funktionsmorphologische Untersuchungen an *Nebria complanata* L. (Coleoptera, Carabidae). Forma et functio **2**, 189–237 (1970)

Rüschkamp, F.: Zum erdgeschichtlichen Alter unserer Coleopterenfauna. Tijdschr. Entomol. **75**, 16–20 (1932)

Sabrovsky, C. W.: How many insects are there. Yb. Agr. U.S. Dept. Agr. **1952**, 1–7 (1952)

Sauer, K. P.: Zur Monotopbindung einheimischer Arten der Gattung *Panorpa* (Mecoptera) nach Untersuchungen im Freiland und im Laboratorium. Zool. Jb. Syst. **97**, 201–284 (1970)

Schaefer, M.: Ökologische Isolation und die Bedeutung des Konkurrenzfaktors am Beispiel des Verteilungsmusters der Lycosiden einer Küstenlandschaft. Oecologia (Berl.) **9**, 171–202 (1972)

Schaerffenberg, B.: Die Nahrung des Maulwurfs (*Talpa europaea* L.). Z. Angew. Entomol. **27**, 1–70 (1940)

Schaller, F.: *Notiophilus biguttatus* F. (Coleoptera) und *Japyx solifugus* Haliday (Diplur.) als spezielle Collembolenräuber. Zool. Jb. Syst. **78**, 294–296 (1950)

Scheloske, H. W.: Biologie, Ökologie und Systematik der Laboulbeniales (Ascomycetes). Unter besonderer Berücksichtigung des Parasit-Wirtverhältnisses. Parasit. Schriftenreihe **19**, 1–176 (1969)

Scherf, H., Drechsel, U.: Faunistisch bemerkenswerte Nachweise von Coleopteren in Hessen durch Lichtfang. Entomol. Z. **83**, 28–32 (1973)

Scherney, F.: Untersuchungen über Vorkommen und wirtschaftliche Bedeutung räuberisch lebender Käfer in Feldkulturen. Z. Pflanzenbau Pflanzenschutz **6**, 49–73 (1955)

Scherney, F.: Über die Wirkung verschiedener Insektizide auf Laufkäfer (Coleoptera, Carabidae). Pflanzenschutz **10**, 87–92 (1958)

Scherney, F.: Unsere Laufkäfer. Wittenberg: Ziemsen, 1959a

Scherney, F.: Der biologische Wirkungseffekt von Carabiden der Gattung *Carabus* auf Kartoffelkäferlarven. Verh. 4. Intern. Pflanzenschutz Kongr. Hamburg 1957, **1**, 1035–1038 (1959b)

Scherney, F.: Beiträge zur Biologie und ökonomischen Bedeutung räuberisch lebender Käferarten. (Untersuchungen über das Auftreten von Laufkäfern (Carabidae) in Feldkulturen (Teil II)). Z. Angew. Entomol. **47**, 231–255 (1960a)

Scherney, F.: Kartoffelkäferbekämpfung mit Laufkäfern (Gattung *Carabus*). Pflanzenschutz **12**, 34–35 (1960b)

Scherney, F.: Über die Zu- und Abwanderung von Laufkäfern (Carabidae) in Feldkulturen. Pflanzenschutz **12**, 169–171 (1960c)

Scherney, F.: Beiträge zur Biologie und ökonomischen Bedeutung räuberisch lebender Käferarten. (Beobachtungen und Versuche zur Überwinterung, Aktivität und Ernährungsweise der Laufkäfer (Carabidae) (Teil III)). Z. Angew. Entomol. **48**, 163–175 (1961)

Scherney, F.: Laufkäfer als natürliche Helfer der Schädlingsbekämpfung. Pflanzenschutzinformationen (Bayer. Landesanst. Pflanzenbau, Pflanzenschutz) **3** (1962a)

Scherney, F.: Untersuchungen über das Vorkommen für die biologische Schädlingsbekämpfung wichtiger Laufkäfer-Arten (Coleoptera, Carabidae) in Bayern. Bayer. Landw. Jb. **39**, 193–218 (1962b)

Schildknecht, H., Maschwitz, E., Maschwitz, U.: Die Explosionschemie der Bombardierkäfer (Coleoptera, Carabidae). III. Mitt.: Isolierung und Charakterisierung der Explosionskatalysatoren. Z. Naturforschung **23 B**, 1213–1218 (1968a)

Schildknecht, H., Maschwitz, U., Winkler, H.: Zur Evolution der Carabiden-Wehrdrüsensekrete. Über Arthropoden-Abwehrstoffe XXXII. Naturwissenschaften **55**, 112–117 (1968b)

Schiller, W., Weber, F.: Die Zeitstruktur der ökologischen Nische der Carabiden (Untersuchungen in Schatten- und Strahlungshabitaten des NSG "Heiliges Meer" bei Hopsten). Abh. Landesmus. Naturkunde Münster **37**, 1–34 (1975)

Schindler, U.: Forleulenbekämpfung 1956 im Südosten der Lüneburger Heide. Nbl. Deut. Pflanzenschutzdienst. **10**, 17–21 (1958)

Schjøtz-Christensen, B.: The beetle fauna of the Corynephoretum in the ground of the Mols Laboratory. Natura Jutlandica **6–7**, 1–20 (1957)

Schjøtz-Christensen, B.: Biology and population studies of Carabidae of the Corynephoretum. Natura Jutlandica **11**, 1–173 (1965)

Schjøtz-Christensen, B.: Biology of some ground beetles (*Harpalus* Latr.) of the Corynephoretum. Natura Jutlandica **12**, 225–229 (1966)

Schjøtz-Christensen, B.: Some notes on the biology and ecology of *Carabus hortensis* L. (Coleoptera, Carabidae). Natura Jutlandica **14**, 127–154 (1968)

Schmidt, G.: Physiologische Untersuchungen zur Transpiration und zum Wassergehalt der Gattung *Carabus* (Ins., Coleoptera). Zool. Jb. Physiol. **65**, 459–495 (1954–55)

Schmidt, G.: Der Stoffwechsel der Caraben (Ins., Coleoptera) und seine Beziehung zum Wasserhaushalt. Zool. Jb. Physiol. **66**, 273–294 (1955–56)

Schmidt, G.: Der Einfluß des Wasserhaushalts und Stoffwechsels auf die Vorzugstemperatur der Insekten. Biol. Zbl. **75**, 178–205 (1956a)

Schmidt, G.: Zum Problem der physiologischen Rassenbildung. Zool. Anz. **157**, 14–19 (1956b)

Schmidt, G.: Die Bedeutung des Wasserhaushaltes für das ökologische Verhalten der Caraben (Ins., Coleoptera). Z. Angew. Entomol. **40**, 390–399 (1957)

Schmidt, G., Renner, K., Gernert, W.: Ein Beitrag zur Coleopteren-Fauna des Bayerischen Waldes mit Untersuchungen über ihre räumliche Verteilung. Zool. Anz. **176**, 327–348 (1966)

Schremmer, F.: Beitrag zur Biologie von *Ditomus clypeatus* Rossi, eines körnersammelnden Carabiden. Z. Arb. Gem. öst. Entomol. **3**, 140–146 (1960)

Schütte, F.: Untersuchungen über die Populationsdynamik des Eichenwicklers (*Tortrix viridana* L.). Teil I und Teil II. Z. Angew. Entomol. **40**, 1–36; 285–331 (1957)

Schweiger, H.: Die Insektenfauna des Wiener Stadtgebietes als Beispiel einer kontinentalen Gross-Stadtfauna. Proc. 11th Intern. Congr. Entomol. Vienna 1960, **3**, 184–193 (1962)

Schweiger, H.: Die Bedeutung Kleinasiens als Evolutionszentrum. Deut. Entomol. Z. (N. F.) **13**, 473–495 (1966)

Schweiger, H.: Gebirgssysteme als Zentren der Artbildung. Deut. Entomol. Z. (N. F.) **16**, 159–174 (1967)

Schwerdtfeger, F.: Ökologie der Tiere. 1, Autökologie. Hamburg, Berlin: Parey, 1963

Schwerdtfeger, F.: Ökologie der Tiere. 2, Demökologie. Hamburg, Berlin: Parey, 1968

Sharova, I. K.: Morpho-ecological types of carabid larvae. (Orig. Russian.) Zool. Sh. **39**, 691–708 (1960)

Sharova, I. K.: Life forms of imago in Carabidae (Coleoptera). Zool. Sh. **53**, 692–709 (1974)

Shelford, V. E.: Life-histories and larval habits of the tiger beetles (Cicindelidae). J. Linn. Soc. London (Zool.) **30**, 157–184 (1908)

Silvestri, F.: Contribuzione alla connoscenza della metamorfosi e dei costumi della *Lebia scapularis* Fourc. con descrizione dell'apparato sericiparo della larva. Redia **2**, 68–84 (1904)

Skuhravý, V.: Fallenfang und Markierung zum Studium der Laufkäfer. Beitr. Entomol. **6**, 285–287 (1956)

Skuhravý, V.: Bewegungsareal einiger Carabidenarten. Acta Soc. Entomol. čsl. **53**, 171–179 (1957)

Skuhravý, V.: Die Fallenfangmethode. Acta Soc. Entomol. čsl. **54**, 27–40 (1957)

Skuhravý, V.: Einfluß landwirtschaftlicher Maßnahmen auf die Phänologie der Feldcarabiden. Fol. Zool. **7**, 325–338 (1958)

Skuhravý, V.: Die Nahrung der Feldcarabiden. Acta Soc. Entomol. čsl. **56**, 1–18 (1959)

Skuhravý, V.: Zur Anlockungsfähigkeit von Formalin für Carabiden in Bodenfallen. Beitr. Entomol. **20**, 371–374 (1970)

Skuhravý, V., Novák, K.: Entomofauna des Kartoffelfeldes und ihre Entwicklung. Rozpr. čsl. Akad. Věd. Řada MPV **67**, 1–49 (1957)

Smeenk, C.: Ökologische Vergleiche zwischen Waldkauz *Strix aluco* und Waldohreule *Asio otus*. Ardea **60**, 1–71 (1972)

Smit, H.: Onderzoek naar het voedsel von *Calathus erratus* Sahlb. en *Calathus ambiguus* Payk. aan de hand van hun magen inhouden. Entomol. Ber. Amst. **17**, 199–209 (1957)

Speight, M. R., Lawton, J. H.: The influence of weed-cover on the mortality imposed on artificial prey by predatory ground beetles in cereal fields (Col., Carabidae and Staphylinidae). Oecologia (Berl.) **23**, 211–223 (1976)

Stein, W.: Biozönologische Untersuchungen über den Einfluß verstärkter Vogelansiedlung auf die Insektenfauna eines Eichen-Hainbuchen-Waldes. I. Z. Angew. Entomol. **46**, 345–370 (1960); II. Z. Angew. Entomol. **47**, 196–230 (1960)

Stein, W.: Die Zusammensetzung der Carabidenfauna einer Wiese mit stark wechselnden Feuchtigkeitsverhältnissen. Z. Morphol. Oekol. Tiere **55**, 83–99 (1965)

Strenzke, K.: Die ökologische Umwelt. Erg. Biol. **27**, 79–97 (1964)

Strübing, H.: Das Lautverhalten von *Euscelis plebejus* Fall. und *Euscelis ohausi* Wagn. (Homoptera, Cicadina). Zool. Beitr. **11,** 289–341 (1965)
Strübing, H.: Zur Artberechtigung von *Euscelis alsius* Ribaut gegenüber *Euscelis plebejus* Fall. (Homoptera, Cicadina). Ein Beitrag zur Neuen Systematik. Zool. Beitr. **16,** 441–478 (1970)
Sturani, M.: Osservazioni e ricerche biologiche sul genere *Carabus* Linnaeus (Sensu Lato) (Coleoptera, Carabidae). Memorie Soc. Entomol. ital. **41,** 85–202 (1962)
Thiele, H. U.: Die Tiergesellschaften der Bodenstreu in den verschiedenen Waldtypen des Niederbergischen Landes. Z. Angew. Entomol. **39,** 316–357 (1956)
Thiele, H. U.: Experimentelle Untersuchungen über die Abhängigkeit bodenbewohnender Tierarten vom Kalkgehalt des Standorts. Z. Angew. Entomol. **44,** 1–21 (1959)
Thiele, H. U.: Zuchtversuche an Carabiden, ein Beitrag zu ihrer Ökologie. Zool. Anz. **167,** 431–442 (1961)
Thiele, H. U.: Zusammenhänge zwischen Jahreszeit der Larvalentwicklung und Biotopbindung bei waldbewohnenden Carabiden. Proc. 11th Intern. Congr. Entomol. Vienna 1960, **3,** 165–169 (1962)
Thiele, H. U.: Experimentelle Untersuchungen über die Ursachen der Biotopbindung bei Carabiden. Z. Morphol. Oekol. Tiere **53,** 387–452 (1964a)
Thiele, H. U.: Ökologische Untersuchungen an bodenbewohnenden Coleopteren einer Heckenlandschaft. Z. Morphol. Oekol. Tiere **53,** 537–586 (1964b)
Thiele, H. U.: Einflüsse der Photoperiode auf die Diapause von Carabiden. Z. Angew. Entomol. **58,** 143–149 (1966)
Thiele, H. U.: Ein Beitrag zur experimentellen Analyse von Euryökie und Stenökie bei Carabiden. Z. Morphol. Oekol. Tiere **58,** 355–372 (1967)
Thiele, H. U.: Formen der Diapausesteuerung bei Carabiden. Verh. Deut. Zool. Ges. Heidelberg **1967,** 358–364 (1968a)
Thiele, H. U.: Was bindet Laufkäfer an ihre Lebensräume? Naturw. Rdsch., Stuttg. **21,** 57–65 (1968b)
Thiele, H. U.: Zur Methode der Laboratoriumszucht von Carabiden. Decheniana **120,** 335–341 (1968c)
Thiele, H. U.: The control of larval hibernation and of adult aestivation in the carabid beetles *Nebria brevicollis* F. and *Patrobus atrorufus* Stroem. Oecologia (Berl.) **2,** 347–361 (1969a)
Thiele, H. U.: Zusammenhänge zwischen Tagesrhythmik, Jahresrhythmik und Habitatbindung bei Carabiden. Oecologia (Berl.) **3,** 227–229 (1969b)
Thiele, H. U.: Die Steuerung der Jahresrhythmik von Carabiden durch exogene und endogene Faktoren. Zool. Jb. Syst. **98,** 341–371 (1971a)
Thiele, H. U.: Wie isoliert sind Populationen von Waldcarabiden in Feldhecken? In: Dispersal and Dispersal Power of Carabid Beetles. den Boer, P. J. (ed.). Misc. Pap. Landbouwhogesch. Wageningen **8,** 105–111 (1971b)
Thiele, H. U.: Anpassungen an die unbelebte Natur. 1. Der Einfluß von Temperatur und Feuchtigkeit, 35–47. In: Unsere Umwelt als Lebensraum (Grzimeks Tierleben: Sonderband Ökologie). Illies, J., Klausewitz, W. (eds.). Zürich: Kindler (1973a)
Thiele, H. U.: Remarks about Mansingh's and Müller's classifications of dormancies in insects. Can. Entomologist **105,** 925–928 (1973b)
Thiele, H. U.: Physiologisch-ökologische Studien an Laufkäfern zur Kausalanalyse ihrer Habitatbindung. Verh. Ges. Ökol. Saarbrücken **1973,** 39–54 (1974)
Thiele, H. U.: Interactions between photoperiodism and temperature with respect to the control of dormancy in the adult stage of *Pterostichus oblongopunctatus* F. (Coleoptera, Carabidae). I. Experiments on gonad maturation under different climatic conditions in the laboratory. Oecologia (Berl.) **19,** 39–47 (1975)
Thiele, H. U.: Tageslängenmessung als Grundlage der Jahresrhythmik des Laufkäfers *Pterostichus nigrita* F. Verhandl. Deut. Zool. Ges. Hamburg **1976,** 218 (1976)
Thiele, H. U., Könen, H.: Interactions between photoperiodism and temperature with respect to the control of dormancy in the adult stage of *Pterostichus oblongopunctatus* F. (Coleoptera, Carabidae). II. The development of the reproduction potential during the winter months in the field. Oecologia (Berl.) **19,** 339–343 (1975)
Thiele, H. U., Kolbe, W.: Beziehungen zwischen bodenbewohnenden Käfern und Pflanzengesellschaften in Wäldern. Pedobiologia **1,** 157–173 (1962)

Thiele, H. U., Krehan, I.: Experimentelle Untersuchungen zur Larvaldiapause des Carabiden *Pterostichus vulgaris*. Entomol. Exptl. Appl. **12,** 67–73 (1969)

Thiele, H. U., Lehmann, H.: Analyse und Synthese im tierökologischen Experiment. Z. Morphol. Oekol. Tiere **58,** 373–380 (1967)

Thiele, H. U., Paarmann, W.: Versuche zur Schlüpfrhythmik bei Carabiden. Oecologia (Berl.) **2,** 7–18 (1968)

Thiele, H. U., Weber, F.: Tagesrhythmen der Aktivität bei Carabiden. Oecologia (Berl.) **1,** 315–355 (1968)

Thienemann, A.: Die Grundlagen der Biocoenotik und Monards faunistische Prinzipien. Festschrift Zschokke Basel **4,** 1–14 (1920)

Tietze, F.: Untersuchungen über die Beziehungen zwischen Flügelreduktion und Ausbildung des Metathorax bei Carabiden unter besonderer Berücksichtigung der Flugmuskulatur (Coleoptera, Carabidae). Beitr. Entomol. **13,** 87–163 (1963)

Tietze, F.: Ein Beitrag zur Laufkäferbesiedlung (Coleoptera, Carabidae) von Waldgesellschaften des Südharzes. Hercynia (Leipzig) N.F. **3,** 340–358 (1966a)

Tietze, F.: Zur Laufkäfer-Fauna der Rabeninsel bei Halle (Saale) (Coleoptera, Carabidae). Hercynia (Leipzig) N.F. **3,** 387–399 (1966b)

Tietze, F.: Zur Ökologie, Soziologie und Phänologie der Laufkäfer (Coleoptera, Carabidae) des Grünlandes im Süden der DDR. I. Teil. Die Carabiden der untersuchten Lebensorte. Hercynia (Leipzig) N.F. **10,** 3–76 (1973a)

Tietze, F.: Zur Ökologie, Soziologie und Phänologie der Laufkäfer (Coleoptera, Carabidae) des Grünlandes im Süden der DDR. II. Teil. Die diagnostisch wichtigen Carabidenarten des untersuchten Grünlandes und ihre Verbreitungsschwerpunkte. Hercynia (Leipzig) N.F. **10,** 111–126 (1973b)

Tietze, F.: Zur Ökologie, Soziologie und Phänologie der Laufkäfer (Coleoptera, Carabidae) des Grünlandes im Süden der DDR. III. Teil. Die diagnostisch wichtigen Artengruppen des untersuchten Grünlandes. Hercynia (Leipzig) N.F. **10,** 243–263 (1973c)

Tietze, F.: Zur Ökologie, Soziologie und Phänologie der Laufkäfer (Coleoptera, Carabidae) des Grünlandes im Süden der DDR. IV. Teil. Ökofaunistische und autökologische Aspekte der Besiedlung des Grünlandes durch Carabiden. Hercynia (Leipzig) N.F. **10,** 337–365 (1973d)

Tietze, F.: Zur Ökologie, Soziologie und Phänologie der Laufkäfer (Coleoptera, Carabidae) des Grünlandes im Süden der DDR. V. Teil (Schluß). Zur Phänologie der Carabiden des untersuchten Grünlandes. Hercynia (Leipzig) N.F. **11,** 47–68 (1974)

Tischler, W.: Biozönotische Untersuchungen an Wallhecken Schleswig-Holsteins. Zool. Jb. Syst. **77,** 283–400 (1948)

Tischler, W.: Grundzüge der terrestrischen Tierökologie. Braunschweig: Vieweg, 1949

Tischler, W.: Die Hecke als Lebensraum für Pflanzen und Tiere unter besonderer Berücksichtigung ihrer Schädlinge. Erdkunde (Arch. Wiss. Geogr.) **5,** 125–132 (1951)

Tischler, W.: Synökologie der Landtiere. Stuttgart: Fischer, 1955

Tischler, W.: Synökologische Untersuchungen an der Fauna der Felder und Feldgehölze (Ein Beitrag zur Ökologie der Kulturlandschaft). Z. Morphol. Oekol. Tiere **47,** 54–114 (1958)

Tischler, W.: Agrarökologie. Jena: Fischer, 1965

Tischler, W.: Untersuchungen über das Hypolithion einer Hausterrasse. Pedobiologia **6,** 13–26 (1966)

Tomlin, A. D.: Notes on the biology and rearing of two species of ground beetles, *Pterostichus melanarius* and *Harpalus pennsylvanicus* (Coleoptera, Carabidae). Can. Entomologist **107,** 67–74 (1975)

Topp, W.: Zur Ökologie der Müllhalden. Ann. Zool. Fenn. **8,** 194–222 (1971)

Topp, W.: Zur Besiedlung einer neu entstehenden Insel. Untersuchungen am "Hohen Knechtsand". Zool. Jb. Syst. **102,** 215–240 (1975)

Tostowaryk, W.: Coleopterous predators of the Swaine jackpine sawfly, *Neodiprion swainei* Middleton (Hymenoptera, Diprionidae). Can. J. Zool. **50,** 1139–1146 (1972)

Tretzel, E.: Technik und Bedeutung des Fallenfanges für ökologische Untersuchungen. Zool. Anz. **155,** 276–287 (1955a)

Tretzel, E.: Intragenerische Isolation und interspezifische Konkurrenz bei Spinnen. Z. Morphol. Oekol. Tiere **44,** 43–162 (1955b)

Tretzel, E.: Biologie, Ökologie und Brutpflege von *Coelotes terrestris* (Wider) (Araneae, Agelenidae). I: Biologie und Ökologie. Z. Morphol. Oekol. Tiere **49,** 658–745 (1961)

Uttendörfer, O.: Die Ernährung der deutschen Tagraubvögel und Eulen. Neudamm: Neumann, 1939

Vlijm, L.: Onderzoek aan Loopkevers op Schiermonnikoog 1960. Hektograph. Bericht, 1–19 (1960)

Vlijm, L., Hartsuijker, L., Richter, C. J. J.: Ecological studies on carabid beetles. Archs. néerl. Zool. **14**, 410–422 (1961)

Vlijm, L., Dijk, T. S. van, Wijmans, S. Y.: Ecological studies on carabid beetles. III. Winter mortality in adult *Calathus melanocephalus* (Linn.). Egg production and locomotory activity of the population which has hibernated. Oecologia (Berl.) **1**, 304–314 (1968)

Volterra, V.: Variations and fluctuations of the number of individuals in animal species living together. In: Chapman, R. N. Animal Ecology. New York, London: 1931, pp. 409–448

Volterra, V., d'Ancona, U.: Les associations biologiques au point de vue mathématique. Act. Sci. Industr. **241**, Paris 1935

Wallgren, H.: Energy metabolism of two species of the genus *Emberiza* as correlated with distribution and migration. Acta Zool. Fenn. **84**, 1–110 (1954)

Wautier, V.: Un phénomène social chez les Coléoptères: le grégarisme de *Brachinus* (Caraboidea, Brachinidae). Insectes Soc. **18**, 1–84 (1971)

Wautier, V., Viala, C.: La longévité imaginale des *Brachinus* (Coleoptera, Carabidae). Bull. Soc. Entomol. France **74**, 9–13 (1969)

Weber, F.: Feld- und Laboruntersuchungen zur Winteraktivität der Carabiden auf Kulturfeldern. Z. Morph. Oekol. Tiere **54**, 551–565 (1965a)

Weber, F.: Vergleichende Untersuchungen über das Verhalten von *Carabus*-Arten in Luftfeuchtigkeitsgefällen. Z. Morph. Oekol. Tiere **55**, 233–249 (1965b)

Weber, F.: Zur Tagaktivität von *Carabus*-Arten. Zool. Anz. **175**, 354–360 (1965c)

Weber, F.: Beitrag zur Karyotypanalyse der Laufkäfergattung *Carabus* L. (Coleoptera). Chromosoma **18**, 467–476 (1966a)

Weber, F.: Die tageszeitliche Aktivitätsverteilung des dunkelaktiven *Carabus problematicus* (Coleoptera, Ins.) nach Phasenumkehr des Zeitgebers. Zool. Beitr. **12**, 161–179 (1966b)

Weber, F.: Die tageszeitliche Aktivitätsverteilung von *Carabus cancellatus* Ill. und *nitens* L. unter künstlichen Belichtungsbedingungen. Zool. Anz. **177**, 367–379 (1966c)

Weber, F.: Zur tageszeitlichen Aktivitätsverteilung der *Carabus*-Arten. Zool. Jb. Physiol. **72**, 136–156 (1966d)

Weber, F.: Die Periodenlänge der circadianen Laufperiodizität bei drei *Carabus*-Arten (Coleoptera, Ins.). Naturwissenschaften **5**, 122 (1967)

Weber, F.: Circadian-Regel und Laufaktivität der Caraben (Ins., Coleoptera). Oecologia (Berl.) **1**, 155–170 (1968a)

Weber, F.: Die intraspezifische Variabilität des heterochromatischen Armes eines Chromosoms bei der Gattung *Carabus* L. (Coleoptera). Chromosoma **23**, 288–308 (1968b)

Weber, F.: Die circadiane Laufperiodik der *Carabus*-Arten bei konstanten Umweltbedingungen. Faun. Oekol. Mitt. **3**, 337–347 (1970)

Wecker, S. S.: The role of early experience in habitat selection by the prairie deer mouse, *Peromyscus maniculatus bairdi*. Ecol. Monographs **33**, 307–325 (1963)

Weidemann, G.: Ökologische und biometrische Untersuchungen an Proctotrupiden (Hymenoptera, Proctotrupidae S. Str.) der Nordseeküste und des Binnenlandes. Z. Morphol. Oekol. Tiere **55**, 425–514 (1965)

Weidemann, G.: Food and energy turnover of predatory arthropods of the soil surface. Ecol. Studies **2**, 110–118 (1971a)

Weidemann, G.: Zur Biologie von *Pterostichus metallicus* F. (Coleoptera, Carabidae). Faun. ökol. Mitt. **4**, 30–36 (1971b)

Weidemann, G.: Die Stellung epigäischer Raubarthropoden im Ökosystem Buchenwald. Verh. Deut. Zool. Ges. Helgold. **1971**, 106–116 (1972)

Weidner, H.: Carabidenlarven als Feinde der Termiten in Hamburg. Anz. Schädl. Kunde **30**, 109 (1957)

Wellenstein, G.: Beiträge zur Biologie der roten Waldameise (*Formica rufa* L.). Z. Angew. Entomol. **14**, 1–68 (1929)

Wellenstein, G.: Die Insektenjagd der Roten Waldameise (*Formica rufa* L.). Z. Angew. Entomol. **36**, 185–217 (1954)

Weseloh, R. M.: Relationships between different sampling procedures for the gypsy moth, *Porthetria dispar* (Lepidoptera, Lymantriidae) and its natural enemies. Can. Entomologist **106,** 225–231 (1974)

Wickler, W.: Soziales Verhalten als ökologische Anpassung. Verh. Deut. Zool. Ges. Köln **1970,** 291–304 (1970)

Wilken, U.: Karyotyp-Analysen bei Carabiden. Staatsexamensarbeit, Münster (1973)

Williams, C. B.: The logarithmic series and its application to biological problems. J. Ecol. **34,** (1947)

Williams, G.: Seasonal and diurnal activity of Carabidae, with particular reference to *Nebria, Notiophilus* and *Feronia.* J. Animal Ecol. **28,** 309–330 (1959)

Willis, H. L.: Species density of North American *Cicindela* (Coleoptera, Cicindelidae). Cicindela **4,** 29–43 (1972)

Wilms, B.: Untersuchungen zur Bodenkäferfauna in drei pflanzensoziologisch unterschiedenen Wäldern der Umgebung Münsters. Abhandl. Landesmus. Naturkunde Münster **23,** 1–15 (1961)

Witzke, G.: Beitrag zur Kenntnis der Biologie und Ökologie des Laufkäfers *Pterostichus (Platysma) niger* Schaller 1783 (Coleoptera, Carabidae). Z. Angew. Zool. **63,** 145–162 (1976)

Zeuner, F. E.: A Triassic insect fauna from the Molteno Beds in South Africa. Proc. 11th Intern. Congr. Entomol. Vienna **1,** 304–306 (1961)

Zimka, J.: The predacity of the field frog (*Rana arvalis* Nilsson) and food levels in communities of soil macro-fauna of forest habitats. Ekol. Pol. (A) **14,** 549–605 (1966)

Systematic Index of Cited Families, Subfamilies, Tribes, and Genera

The following system is based upon the classification of carabids by Horn and Csiki (1926–1933) in the "Coleopterorum Catalogus" of Junk and Schenkling. It is beyond the scope of this book to embark upon a discussion of the various systems of classification that have been employed for carabids. The uniformity of the family has, with only one exception, never been contested. Jeannel (1941), by raising subfamilies and tribes of carabids to the status of families, arrived at 26 families in place of our one large family. Apart from a few French authors who have adhered to Jeannel's *Faune de France* this view has won almost no supporters. In 1970 Freude convincingly pointed out the disadvantages of splitting up the family in this manner—without in any way belittling the enormous value of such a great systematicist as Jeannel. It should be mentioned, however, that some authors rank many tribes as subfamilies.

A subject of greater controversy is whether the tiger beetles are in fact a subfamily of the carabids or whether they should be ranked as a family on their own. Mandl (1971) called for a "reinstatement of the Cicindelinae to family status," supporting his arguments with skeletomorphological evidence (further literature on the topic in his article). Immunoelectrophoretic investigations on homogenized whole beetles again led Basford *et al.* (1968) to regard the Cicindelinae as a subfamily. That tiger beetles and ground beetles are closely related is universally accepted, and for this reason the inclusion of the former in a book of such an ecological nature as the present one seems to be completely justified. In the following therefore, five subfamilies of Carabidae will be distinguished (recently some systematicists have abandoned the subfamily Harpalinae, regarding its species as belonging to the Carabinae; see Lindroth 1974b).

Page numbers are given for references to tribes and genera only where individual species belonging to the groups are not mentioned on that page.

[23] The Pterostichini are often divided into a greater number of independent tribes.

Species Index

In brackets in the following are included such synonyma as are often used in the literature. The names given in the text are those mainly—often exclusively—cited in the bulk of the literature covering our field of investigation. In some cases, the names given here in brackets are perhaps valid on the basis of priority. But the author—being no systematicist—does not consider himself competent to abandon names which have been in uncontested use for many decades in all of the literature concerning our field, or where no doubt is possible about which species is meant.

Subject Index

Zoophysiology and Ecology

Coordinating Editor: D. S. FARNER
Editors: W. S. HOAR, B. HOELLDOBLER, H. LANGER, M. LINDAUER

Volume 1: P. J. BENTLEY
Endocrines and Osmoregulation
A Comparative Account of the Regulation of Water and Salt in Vertebrates
29 figures. XVI, 300 pages. 1971
ISBN 3-540-05273-9

Volume 2: L. IRVING
Arctic Life of Birds and Mammals
Including Man
59 figures. XI, 192 pages. 1972
ISBN 3-540-05801-X

Volume 3: A. E. NEEDHAM
The Significance of Zoochromes
54 figures. XX, 429 pages. 1974
ISBN 3-540-07081-8

Volume 4/5: A. C. NEVILLE
Biology of the Arthropod Cuticle
233 figures. XVI, 448 pages. 1975
ISBN 3-540-07081-8

Volume 6: K. SCHMIDT-KOENIG
Migration and Homing in Animals
64 figures, 2 tables. XII, 99 pages. 1975
ISBN 3-540-07433-3

Volume 7: E. CURIO
The Ethology of Predation
70 figures, 16 tables. X, 250 pages. 1976
ISBN 3-540-07720-0

Volume 8: W. LEUTHOLD
African Ungulates
A Comparative Review of their Ethology and Behavioral Ecology
55 figures, 7 tables. XIII, 307 pages. 1977
ISBN 3-540-07951-3

Volume 9: E. B. EDNEY
Water Balance in Land Arthropods
109 figures, 49 tables.
282 pages. 1977
ISBN 3-540-08084-8

Springer-Verlag
Berlin Heidelberg
New York

MIX
Papier aus verantwortungsvollen Quellen
Paper from responsible sources
FSC® C105338

If you have any concerns about our products,
you can contact us on
ProductSafety@springernature.com

In case Publisher is established outside the EU,
the EU authorized representative is:
Springer Nature Customer Service Center GmbH
Europaplatz 3, 69115 Heidelberg, Germany

Printed by Libri Plureos GmbH
in Hamburg, Germany